AF898048

The Sirex Woodwasp and its Fungal Symbiont

Bernard Slippers • Peter de Groot
Michael John Wingfield
Editors

The Sirex Woodwasp and its Fungal Symbiont

Research and Management of a Worldwide Invasive Pest

Springer

Editors
Bernard Slippers
Department of Genetics
Forestry and Agricultural
Biotechnology Institute
University of Pretoria
Cnr Lynnwood and University Roads
Hatfield, 0028 Pretoria
South Africa
Bernard.slippers@fabi.up.ac.za

Michael John Wingfield
Forestry and Agricultural
Biotechnology Institute
University of Pretoria
Cnr Lynnwood and University Roads
Hatfield, 0028 Pretoria
South Africa
Mike.wingfield@fabi.up.ac.za

Peter de Groot
Natural Resources Canada - Canadian
Forest Service
Great Lakes Forestry Centre
1219 Queen Street East
Sault Ste. Marie, Ontario
Canada P6A 2E5

ISBN 978-94-007-1959-0 e-ISBN 978-94-007-1960-6
DOI 10.1007/978-94-007-1960-6
Springer Dordrecht Heidelberg London New York

Library of Congress Control Number: 2011941198

© Springer Science+Business Media B.V. 2012
Chapters 10, 11 & 19 - © US Government 2012
No part of this work may be reproduced, stored in a retrieval system, or transmitted in any form or by any means, electronic, mechanical, photocopying, microfilming, recording or otherwise, without written permission from the Publisher, with the exception of any material supplied specifically for the purpose of being entered and executed on a computer system, for exclusive use by the purchaser of the work.

Printed on acid-free paper

Springer is part of Springer Science+Business Media (www.springer.com)

Dedication

Peter de Groot on top of Table Mountain during a visit to South Africa in May 2010 to work on editing of the book

The decision to assemble this book was that of our dear friend and colleague Dr. Peter de Groot. Peter who was passionate about this project and was determined to keep the other two editors continuously fired up to take it to its completion. Some years into the venture, Peter was diagnosed with cancer, but this did not quell his determination to complete the project. He fought his cancer courageously, convincing us that it would not conquer him. We were buoyed by his determination and he continued to work on revising and writing chapters, often under difficult circumstances. As a final burst of determination he visited us in South Africa in May 2010 in order to work on final elements of the project. At that time, we knew Peter was seriously ill, but he did not let this dull his enthusiasm. Sadly, the cancer caught up with Peter before the final touches had been put to this work. He would have loved to see it completed and we would have loved to celebrate this accomplishment with him. This was not to be, but we know that he is close and that he will be happy to see his work and that of many dear friends brought to conclusion.

This book is dedicated to the early pioneers of *Sirex noctilio* research, but especially to Peter de Groot (14 August 1954 to 22 October 2010), a giant of a man with a heart of gold and a dear friend and motivator to us all. May he rest in peace.

Bernard Slippers and Michael John Wingfield

Preface

The Sirex woodwasp, *Sirex noctilio*, first appeared outside its native range around 1900 in pine plantations of New Zealand. Since then this alien invasive pest, together with its fungal symbiont, *Amylostereum areolatum*, gradually spread, first to Australia (1950s) and later, in the 1980s to South American countries including Uruguay, Brazil, Argentina, and Chile. In the early 1990s the woodwasp appeared for the first time in South Africa. In 2004 *S. noctilio* was discovered in the USA and the following year in Canada. In every country where this woodwasp has become established, it has resulted in considerable economic and ecological damage, or it has the potential to do so. There is no doubt that the global spread of the Sirex woodwasp will continue. With this spread, there is an increased need for new research to address its behaviour, impact and management in novel environments.

Information on the Sirex woodwasp was scant until its appearance in the Southern Hemisphere, which led to the first period of intensive research on this pest. This included important fundamental studies on the biology of *Sirex* and its fungal symbiont. Substantial attention was given to control options, which included the discovery and deployment of various biological control agents, including wasp parasitoids and a parasitic nematode. These agents, together with silvicultural practices to reduce stress in plantations, have yielded impressive control in New Zealand and Australia. However, when biocontrol was first attempted in South America and South Africa, the initial results were variable and disappointing. Likewise, the situation in North America with its diverse and complex conifer ecosystems consisting of mostly native species with well-developed arthropod, plant and diseases communities, including natural guilds of insect parasitoids and competing insect communities, was very different to that in Southern Hemisphere plantations of even-aged single species. How much impact the woodwasp will have in North American forests is still an open question.

An International Sirex Symposium and Workshop was held in South Africa from 9 to 16 May 2007 in recognition of the continuing global threat from the woodwasp and the realization that a sustained research effort would be required to deal with the pest across its growing distribution range. Approximately 80 researchers and foresters from 18 different countries convened in Pretoria and later in Pietermaritzburg, to discuss research on the Sirex woodwasp and its control. The meeting was jointly organized by researchers at the Forestry and Agricultural Biotechnology Institute (FABI) at the University of Pretoria and the Institute for Commercial Forestry Research in Pietermaritzburg.

The Sirex meetings in Pretoria and Pietermaritzburg succeeded remarkably in bringing together a diverse mosaic of practical and academic experience from virtually all groups working in this domain around the world. The foremost researchers and foresters from Australia, Europe, Japan, Brazil, Chile and Argentina shared their extensive experience regarding Sirex woodwasp biology, its fungi, and especially its management. Furthermore, the meeting was attended by a large group of researchers and control managers from North America and where much work is currently being done to better understand and manage this complex problem. The shared experience from that meeting, as well as the research and management work undertaken subsequently to it, is the genesis of this book.

This book contains 20 chapters covering all aspects of Sirex woodwasp biology, ecology and management. It is organized thematically along these topics starting with four chapters on the insect's taxonomy, life history, host-plant relationships, and population dynamics. The woodwasp's fungal mutualist is discussed in the next three chapters addressing far ranging issues such as the nature of the symbiosis, and its distribution, diversity and biology in native and non-native areas. The following three chapters deal with insect and nematode parasites, two critical cornerstones of an integrated pest management system for this pest. The woodwasp and some of its parasites use chemicals emitted from the tree, the symbiotic fungus or the insect itself to find hosts or mates, which is covered in two chapters. These chemicals can be used in traps to provide an early warning system that the Sirex woodwasp has recently invaded the area. Early detection is critical to the success of all pest management programs. Seven chapters follow that show how each country or region has responded to the new invasions, in terms of detection and monitoring, quarantine, biological and silvicultural control, and public awareness. These chapters identify the current and emerging issues and challenges each region faces. It should be abundantly clear from reading these chapters that "silver bullets" do not exist for controlling the Sirex woodwasp. More importantly, that local adaptation of pest management strategies and tactics is needed to achieve this goal. In the final chapter, the editors have pulled together the major themes and issues of pest management and have identified many exciting opportunities for basic and applied research, and for international collaboration. Indeed, the global community need to face the increasing threats to forests from the Sirex woodwasp, and many other pests, together.

Some notes regarding common use of organism names are needed. There is uncertainty at present about the taxonomic position of the nematode biological control agent of *Sirex noctilio*, with both *Deladenus siricidicola* and *Beddingia siricidicola* being used in literature. The confusion results from the most recent taxonomic work suggesting that the latter genus is not valid, a view that is not shared by all researchers in the field. Until this uncertainty has been resolved, the original name should take precedence and is, therefore, used throughout the book. Furthermore, there is no official common name for *S. noctilio*. We have used the most common version, the Sirex woodwasp, throughout the book.

The authors are indebted to a number of organizations and individuals who have contributed to making this project possible. The South African forestry industry, through the Tree Protection Co-operative Programme, and University of Pretoria in South Africa have sponsored intensive Sirex research for many years, including the symposium that initiated this book project. We are also grateful to the authors who have unselfishly shared their expert knowledge and worked hard to produce a product of high quality.

Contents

Memories from the International Sirex Symposium

The images on this page and next page capture the spirit of the International Sirex Symposium held in South Africa from 10 to 16 May 2007 and that brought together approximately 80 researchers and foresters from 18 different countries. The meeting began with 2 days of presentations and intensive discussions at the University of Pretoria, including 'team building' fun with African drumming and dancing, where after the participants set off on a field trip.

The interactions in Pretoria established an incredibly congenial atmosphere prior to participants engaging in a 4 day field trip through forestry areas and the Kruger National Park, while concurrently continuing the intensive discussions. The week ended with a workshop to review Sirex management and research in South Africa and around the world. Those discussions were the genesis of this book.

Chapter 1
Sirex Systematics; Problems and Solutions

Henri Goulet

Abstract Earlier literature on the species of *Sirex* is confusing, also for North American species. Among the traditionally recognized species most are in need of some change in concepts. Fourteen species are known to occur in the Western Hemisphere and only one of these, *Sirex noctilio* F., is introduced. This review discusses characters that helped to clarify the species concepts in North America based independently on classical morphological and mitochondrial DNA (cytochrome c oxidase subunit 1 – CO1) data. Significant structural characters, especially the ovipositor pits in females, are discussed and illustrated based on this study of Western Hemisphere *Sirex* species.

1.1 Introduction

Key concepts on the systematics of the New World Siricidae have previously been published in Benson (1943; Benson 1962, 1963), Bradley (1913), Cresson (1880), Middlekauff (1960), Smith (1978, 1979, 1988), Smith and Schiff (2002) and Schiff et al. (2006). These concepts are currently under revision (Schiff et al. 2011) and will include numerous changes in the species numbers and concepts relative to previous works. This review contains a brief comparison of differences between the old and new concepts.

A number of examples are cited to illustrate some of the difficulties with previous concepts in the New World Siricidae. A first example is that of *Sirex juvencus juvencus* Linnaeus. Although this is a European species not found in North America, many

H. Goulet (✉)
Agriculture and Agri-Food Canada, K. W. Neatby Building, 960 Carling Avenue, Ottawa, ON, Canada K1A 0C6
e-mail: henri.goulet@agr.gc.ca

B. Slippers et al. (eds.), *The Sirex Woodwasp and its Fungal Symbiont: Research and Management of a Worldwide Invasive Pest*, DOI 10.1007/978-94-007-1960-6_1,
© Springer Science+Business Media B.V. 2012

papers applied this name to North American specimens (Benson 1963, Smith 1979; Smith and Schiff 2002; Schiff et al. 2006). Females of *S. juvencus* have a unique type of ovipositor not seen in any American specimens.

A second confusing species is *S. cyaneus* F., a North American species. The first problem with this taxon is that the type was unknown until lately (images of the ovipositor of the type specimen studied by the author, specimen in the Hunterian Museum in Glasgow, Scotland). For this reason, there has been lack of clarity regarding the identification of this species. The second problem is that five species in North America and one in Europe fall under this name. All have reddish brown femora and females are dark blue with black antennae.

Contrary to Benson (1945), the name *S. juvencus californicus* (Ashmead) is distinct from and not closely related to *S. juvencus*. Associated with *S. californicus* are three more species (*S. obesus* Bradley and two new ones). Moreover, females of *S. californicus* exist in two distinct color forms, dark and pale legged. This last color form is confused with *S. cyaneus* of previous authors.

Finally, classically distinct species, *S. edwardsii* Brullé and *S. nigricornis* F., are nothing more than discreet color forms of one species, *S. nigricornis* (based on our morphological and DNA results). Four species from previous studies remains the same (i.e., *S. areolatus* Cresson, *S. longicauda* Middlekauff, *S. behrensii* Cresson, and *S. noctilio* F.). Surprisingly, out of 14 species known from North America, meso-America and the Greater Antilles, four are new to science and only one, *Sirex noctilio*, is an accidental introduction.

1.2 Important Features for Solving Identification Problems

Three important features underpin the solution to resolving the systematics of the New World species: the discovery of the significance of the ovipositor structures, mitochondrial DNA, and large collections of reared specimens.

1.2.1 Ovipositor Morphology

Viitasaari (1984) published on the fauna of Siricidae of Finland and nearby countries. In his work, he consistently used a structure apparently not used frequently before, namely the pits on the ventral portion (lancet) of the ovipositor. This character was used for the first time by Kjellander (1945) to segregate females of *S. juvencus* from those of *S. noctilio*. Kjellander's discovery remained unknown as it was published in a Swedish journal that was difficult to locate. Viitasaari's paper was in Finish, but it was excellently illustrated. The character had great potential for resolving taxonomic problems of *Sirex* in the New World. In 1989, an English key was published by Viitasaari and Midtgaard confirming the interpretation of Viitasaari (1984).

It is curious that the ovipositor pits were not noticed by the numerous sawfly experts. After all, sawfly taxonomists have a special predilection for ovipositor details, especially those along the ventral edge. A likely reason is the lack of saw-like structures on the ventral edge of the lancet. Moreover, pits on the lateral surface of the lancet are difficult to observe unless one uses a diffused light source.

1.2.2 Mitochondrial DNA

The results of the mitochondrial DNA (cytochrome c oxidase subunit 1 – CO1) sequencing are promising to be very useful in distinguishing Siricid species. The establishment of *S. noctilio* around the east end of Lake Ontario (Hoebeke et al. 2005) in New York and Ontario was the trigger for our revision. The question was how one might distinguish a *S. noctilio* specimen from other North American species. For this purpose, one study developed a barcoding approach based on CO1 sequence data to distinguish *S. noctilio* from other native Siricid species (Wilson and Schiff 2010). This study is based on 22 native and non-native species of wood boring sawflies (Anaxyelidae, Siricidae and Xyphidriidae), in addition to *S. noctilio*. Ongoing work on the North American Siricids now include sequence data for 12 species of *Sirex*. The recognition of species by mitochondrial DNA and classical morphological studies are in complete agreement.

Not only did mitochondrial DNA data allow for the accurate identification of adults of both sexes, it is also useful for associating specimens with distinct color forms of the same species (e.g., the dark abdomen *S. edwardsii* and pale abdomen *S. nigricornis*). Furthermore, it segregates larvae accurately, which is important for screening intercepted specimens by quarantine systems.

1.2.3 Collections

The importance of having large and representative collections must be stressed. The Canadian National Collection in Ottawa, Ontario, holds over 2,000 specimens of *Sirex*, about 70% of which were reared specimens from identified host trees. As part of current studies, the author organized the collection (in 2007) by obvious species, then by provinces and states for the United States and Canada, and finally by field collected versus reared specimens. All original series were reassembled and reared specimens were organized by host trees from which they had emerged. This process included checking each specimen and dissecting many during the curatorial process. Most of the species now recognized were segregated during the curatorial process.

The Smithsonian Institute (Washington, DC) contains a large collection of Siricidae, with more species from the New World and Eurasia and more intercepted specimens at various ports of entry. Except for the Smithsonian collection, other collections were not so rich in number of field-collected and reared specimens of

Sirex. During five visits to the Smithsonian collection (2008–2010), the author discovered two more species, one from the Southwest United States and the nearby state of Chihuahua in Mexico and one from the Mexico City area. Recently, the author recognized another species from the Dominican Republic. Altogether, there are thus 14 species of *Sirex*, based on more than 4,000 specimens from the New World.

1.3 Diversity and Biogeographic Coverage

In the New World, the range of *Sirex* covers all coniferous regions of Canada, the United States, and as far south as Guatemala in Central America. The species of *Sirex* from the Dominican Republic was collected in the pine forest zone above 1,000 m. The fauna of Canada and the United States is quite satisfactorily sampled and obvious new species are not likely, but many species are likely to be discovered in Mexico and Guatemala wherever conifers grow in temperate and montane regions. The discovery of a species in the Greater Antilles suggests that other species may be found, perhaps in Cuba at higher elevations in places where pines and junipers grow.

Regarding Old World specimens, more time and resources will be needed to resolve systematic issues. The species in Europe are quite well defined, but eastward they become very confusing. In eastern Asia, *Sirex* is greatly in need of study. This work must be done with large collections of specimens covering the Palaearctic region systematically using an approach similar to that used in the New World and Viitasaari (1984).

1.4 Morphological Characters Used to Recognize Species

The aim of the following section is to stress practical aspects of the morphological characters that can be readily assessed by researchers using only a dissection microscope. The ovipositor is the key character in distinguishing most species. In the New world, each species has an ovipositor with special attributes. Building on this foundation, other characters are useful in recognizing species. Once females are segregated and identified, males can be associated.

1.4.1 Ovipositor Characters

The structures associated with egg laying are illustrated in Fig. 1.1c. The ovipositor consists of a dorsally fused section, the lance, and ventrally of two isolated and parallel sections, the lancets. The lancets slide back and forth independently to move the egg and to penetrate the wood (Figs. 1.1c, and 1.2a–c). The important characters

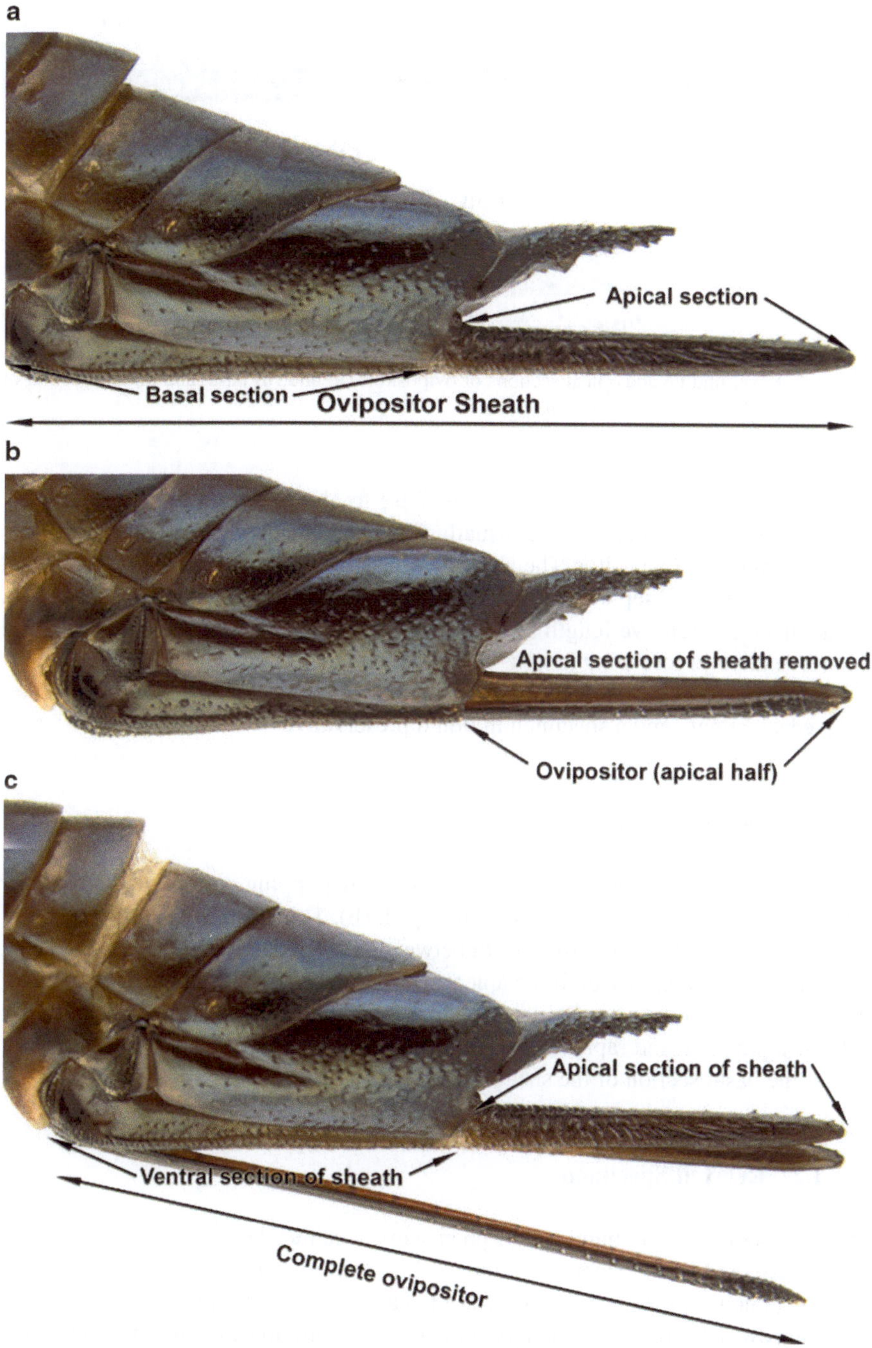

Fig. 1.1 Female external reproductive structures of *Sirex nitidus* (Harris); (**a**) ovipositor hidden by sheath, (**b**) apical half of ovipositor visible after breaking off apical section of sheath, (**c**) sheath and complete ovipositor visible after dissection

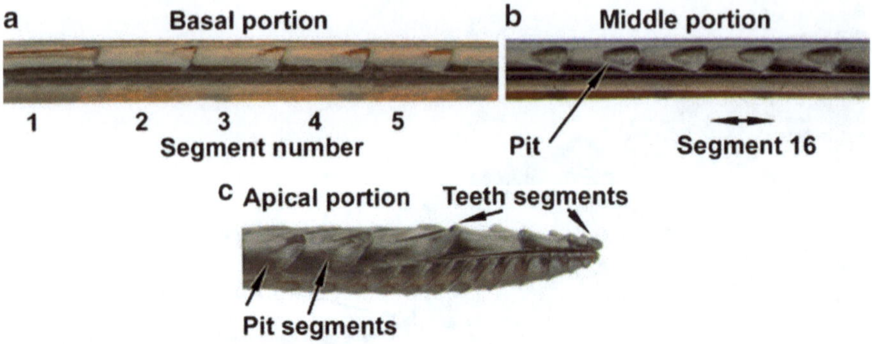

Fig. 1.2 Basal middle and apical sections of ovipositor; (**a**) and (**b**) basal and middle section of *S. noctilio*, (**c**) apical section of *S. longicauda*

are pits found along the lancet from the base to about the middle (Fig. 1.2a, b). The apical teeth segments do not usually show specific characters (Fig. 1.2c). The sheath covers the ovipositor. The sheath consists of two obvious sections, the basal and apical sections separated by a narrow membrane about half way (Fig. 1.1a). Apart from the relative length between the apical and basal sections and its total length relative to fore wing length, we found no structural difference.

There are three ways to study the ovipositor, namely from a dry specimen, a relaxed specimen, and a specimen in liquid preservative.

1.4.1.1 Dry Specimen

The apical half of the ovipositor can been seen by popping out one (either right or left) of the apical sections of the sheath (Fig. 1.1b). This is done by sliding an entomological pin at the end of the sheath between the ovipositor and the sheath. Then, the pin is pushed basally until the apical section breaks off. If the visible portion of the ovipositor is dirty, it can be cleaned with a paintbrush dipped in ethanol (70–95%). This is a practical and rapid method, but it is not possible to observe the ovipositor under the basal section of the sheath.

1.4.1.2 Relaxed Specimen

After 24–48 h in a very humid atmosphere a dry specimen becomes soft and it can be easily dissected. A container with a moist to almost wet cloth will relax the specimen. The specimen is pined on a foam surface put above the moist cloth. The jar is sealed and kept away from the sun for a day or two. An entomological pin can then be used to tease out the ovipositor away from the sheath while the specimen is secured between the fingers of the observer. The pin can then be moved towards the base of

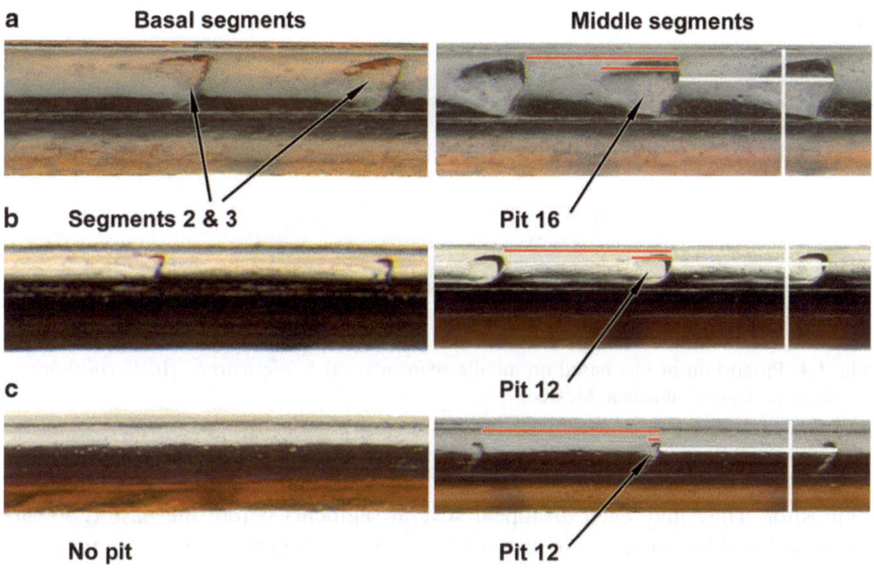

Fig. 1.3 Basal and middle section showing pits on lancet (in middle section, red lines compare the length of pit with that of segment; white lines compare length of segment with width of ovipositor; pit segment number refers to alignment of pit with the junction of basal and apical section of sheath); (**a**) *S. noctilio*, (**b**) *S. nitidus*, (**c**) *S. cyaneus*

the ovipositor without too much effort (the ovipositor could break). At this point, it is possible to observe all of the pits on the ovipositor. If the ovipositor is dirty, it can be cleaned with a paintbrush dipped in ethanol (Fig. 1.1c).

1.4.1.3 Specimen in Preservative

The technique used to study specimens stored in preservative is similar to that for examining relaxed specimens other than the fact that the ovipositor is teased out as soon as the excess of preservative has been removed (Fig. 1.1c).

To study the pits on the ovipositor, a diffuse light source is required. A semi opaque plastic sheet is then placed between the light source and the specimen, positioned about 20 mm from the specimen, or a 13 W desk fluorescent lamp can also be used. The specimen must be shifted until the shadow and lit area reach the middle of pits (as in Fig. 1.4c). The diffuse light renders the pits sharply and clearly visible without glare. The researcher will typically record the size of pits near the middle where the apical and basal section of sheath normally meet. The relative size of pits is a function of the segment length (see red lines in Fig. 1.3a–c). It can then be ascertained whether the size of pits changes little or markedly toward the base of the

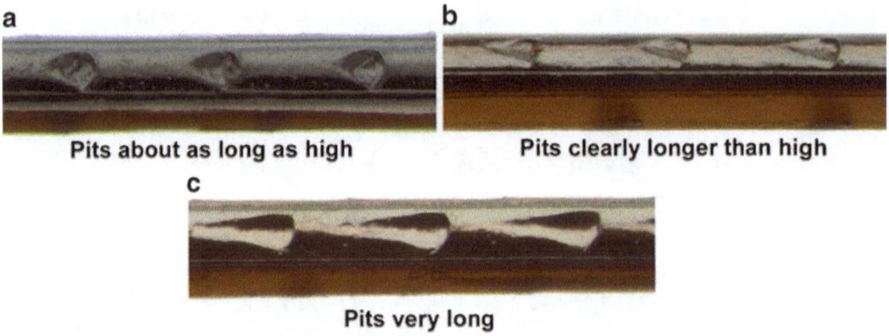

Fig. 1.4 Proportion of pits based on middle segments; (**a**) *S. nigricornis*, (**b**) *S. californicus*, (**c**) *Sirex* sp. (from Chihuahua, Mexico)

ovipositor. They may even disappear several segments before the base (compare basal and middle segments in each of Fig. 1.3a–c). It is then necessary to record the segment number of the ovipositor where the basal and apical section of the sheath would normally meet (as in Fig. 1.3a (pit 16), b (pit 12), c (pit 12)). For some species it is useful to note the segment length relative to ovipositor (lance and lancet) diameter at this segment (see white lines in Fig. 1.3a–c). There have been similar measurements done between segment 1 and 2, 4 and 5, and 9 and 10 (Viitasaari 1984; Viitasaari and Midtgaard 1989). Furthermore, the pit proportions at the middle and base of the ovipositor must be recorded. Pits are commonly about as long as high, but they could be elongate and almost look gutter-like (Fig. 1.4a–c). Finally, for some species the number of pits is exceptionally high (35–43, as opposed to 26–33). All New World species were separated using the above approaches.

1.4.2 *Other Characters*

Once species are recognized using the ovipositor, other characters (e.g., pits on the head and the mesoscutum, proportions of tarsomeres and tarsal pads in females, and color patterns on the legs, abdomen and antennae) are usually found. Moreover, as females become clearly identified, males of most species are quite easily associated and their characters (e.g., structural characters shared usually by both sexes, and color patterns that are often peculiar to each sex) are also discovered.

There are many informative structural characters and significant color patterns. Color patterns that are useful are those on the antennae (Fig. 1.5a–c), the side of the head (Figs. 1.6a, b), the wings after considering the sex of the specimen (Fig. 1.8a–c), the abdomen (Fig. 1.9a–c), and legs (especially femora and tibia of the hind leg) (in males see Fig. 1.7a–d, in females see Fig. 1.10a–d). The main

Fig. 1.5 Antenna in male; (**a**) *S. obesus*, (**b**) *S. californicus*, (**c**) *Sirex* sp. (Chihuahua, Mexico)

Fig. 1.6 Gena behind eye in male; (**a**) *S. behrensii*, (**b**) *S. californicus*

structures are, in both sexes, pits on the vertex (Fig. 1.11a–c) and on the disc of the mesoscutum (Fig. 1.12a–e), and in females, the proportion of the metatarsomere 2 in lateral view (it is the easiest segment to see and measure and the character states are best expressed on this segment) (Fig. 1.13a–c), the relative size of its tarsal pad (Fig. 1.14a, b), and the proportion between the sheath sections.

Fig. 1.7 Hind leg in male; (a) S. nigricornis, (b) S. behrensii, (c) S. nitidus, (d) S. californicus

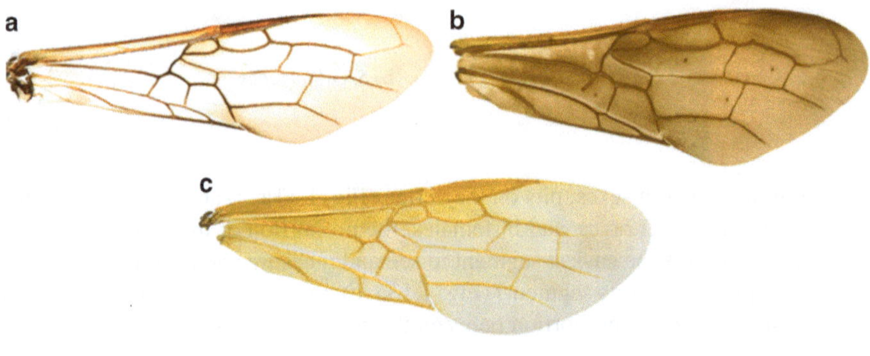

Fig. 1.8 Fore wing; (**a**) female of *S. nitidus*, (**b**) female of *S. nigricornis*, (**c**) male of *S. obesus*

Fig. 1.9 Abdomen in female; (**a**) *S. behrensii*, (**b**) *S. californicus*, (**c**) *S. nigricornis*

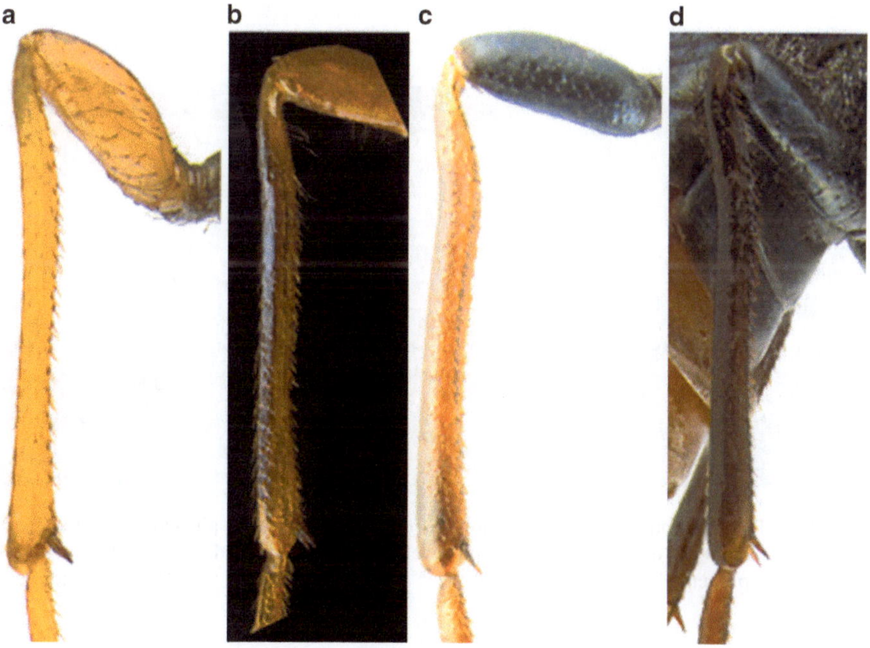

Fig. 1.10 Metafemur and metatibia in female; (**a**) *S. nitidus*, (**b**) *S. varipes*, (**c**) *S. nigricornis* (light color form), (**d**) *S. nigricornis* (dark color form)

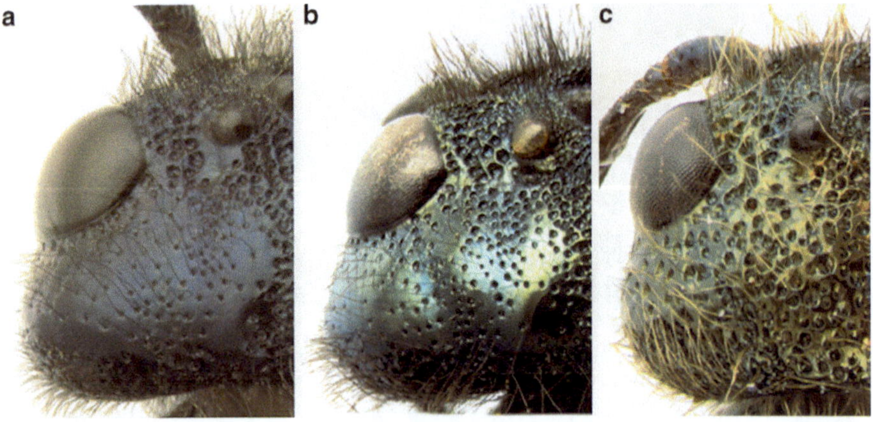

Fig. 1.11 Pit development on dorsal surface of head; (**a**) fine and scattered pits, *S. noctilio*, (**b**) moderate size and dense pits, *S. nitidus*, (**c**) large and dense pits, *S. obesus*

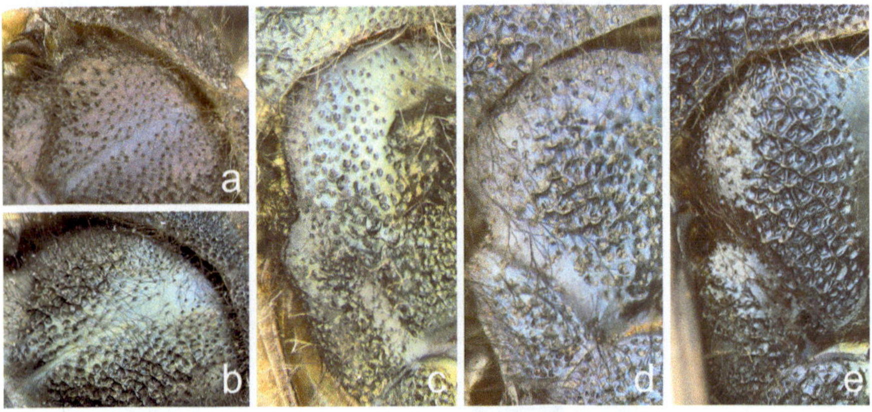

Fig. 1.12 Pit development on mesoscutum; (**a**) dorsolateral view of pit and associated tooth behind, *S. noctilio*, (**b**) dorsolateral view of pit development of net-like ridges, *S. nitidus*, (**c**) dorsal view, *Sirex* sp. (central Mexico), (**d**) dorsal view, *S. obesus*, (**e**) dorsal view, *Sirex* sp. (northern Mexico)

Fig. 1.13 Proportion of metatarsomere 2 length relative to maximum height; (**a**) *S. obesus*, (**b**) *S. noctilio*, (**c**) *S. areolatus*

Fig. 1.14 Proportion of tarsal pad (*blue line*) of metatarsomere 2 relative to maximum ventral length of tarsomere (*red line*); (**a**) *S. nitidus*, (**b**) *S. noctilio*

Acknowledgments I thank Caroline Boudreault for the fine illustrations for this project. I am most grateful to John Huber for scientific editing comments regarding the text.

References

Benson RB (1943) Studies in Siricidae, especially of Europe and southern Asia (Hymenoptera, Symphyta). Bull Entomol Res 34:27–51

Benson RB (1945) *Urocerus californicus* (Norton) and some other interesting Siricidae (Hym., Symphyta) in Britain. Entomol Mon Mag 81:67–68

Benson RB (1962) Holarctic sawflies (Hymenoptera: Symphyta). Bull Br Mus (Nat Hist) Entomol 12:381–409

Benson RB (1963) [1962] A character gradient in *Sirex juvencus* L. (Hym., Siricidae). Entomol Mon Mag 98:252–253

Bradley JC (1913) The Siricidae of North America. J Entomol Zool 5:1–36

Cresson ET (1880) Catalogue of the Tenthredinidae and Uroceridae of North America. Trans Am Entomol Soc 8:53–68

Hoebeke ER, Haugen DA, Haack RA (2005) *Sirex noctilio*: discovery of a palearctic siricid woodwasp in New York. Newsl Mich Entomol Soc 50:24–25

Kjellander E (1945) *Sirex juvencus* L. and *noctilio* Fabr., two good species. Fysiografrafiska Sällskapet i Lund, Förhandlingar 15:121–126

Middlekauff WW (1960) The siricid wood wasps of California (Hymenoptera: Siricidae). Bull Calif Insect Survey 6:59–77

Schiff NM, Valley SA, LaBonte JR, Smith DR (2006) Guide to the siricid woodwasps of North America. USDA Forest Service/FHTET-2006-15, Morgantown

Schiff NM, Goulet AH, Smith DR, Boudreault C, Wilson AD, Scheffler BE (2011) Siricidae (Hymenoptera: Symphyta: Siricoidea) of the Western Hemisphere. Canadian Journal of Arthropod Identification (in press)

Smith DR (1978) Suborder Symphyta (Xyelidae, Parachexyelidae, Parapamphiliidae, Xyelydidae, Karatavitidae, Gigasiricidae, Sepulcidae, Pseudosiricidae, Anaxyelidae, Siricidae, Xiphydriidae, Paroryssidae, Xyelotomidae, Blasticotomidae, Pergidae. In: van der Vecht J, Shenefelt RD (eds) Hymenopterorum catalogus, pars 14. Dr. W. Junk b.v. Publishers, The Hague

Smith DR (1979) Symphyta. In: Krombein KV, Hurd PD Jr, Smith DR, Burks BD (eds) Catalog of Hymenoptera in America North of Mexico, vol 1. Smithsonian Institution Press, Washington, DC

Smith DR (1988) A synopsis of the sawflies (Hymenoptera: Symphyta) of America south of the United States: introduction, Xyelidae, Pamphiliidae, Cimbicidae, Diprionidae, Xiphydriidae, Siricidae, Orussidae, Cephidae. Syst Entomol 13:205–261

Smith DR, Schiff NM (2002) A review of the siricid woodwasps and their Ibaliid parasitoids (Hymenoptera: Siricidae, Ibaliidae) in the eastern United States, with emphasis on the mid-Atlantic region. Proc Entomol Soc Wash 104:174–194

Viitasaari M (1984) Sahapistiäiset 3. Siricicoidea, Orussoidea, ja Cephoidea. University of Helsinki, Department of Agricultural and Forest Zoology Reports 6

Viitasaari M, Midtgaard F (1989) A contribution to the taxonomy of horntails with notes on the genus *Sirex* Linnaeus (Hymenoptera, Siricidae). An Entomol Fen 55:103–110

Wilson AD, Schiff NM (2010) Identification of *Sirex noctilio* and native North American woodwasp larvae using DNA barcode. J Entomol 7:60–79

Chapter 2
Life History and Biology of *Sirex noctilio*

Kathleen Ryan and Brett P. Hurley

Abstract The wood-boring wasp *Sirex noctilio* has an obligate mutualistic relationship with the basidiomycete fungus, *Amylostereum areolatum*. Female wasps carry the fungus in internal mycangia and introduce it into a relatively competition free environment inside the living stems of *Pinus* trees. The fungus and a phytotoxic mucus produced by the wasp act in concert to stress the host tree, creating an ideal environment for the fungus to develop. Fungal development is essential for the initiation of insect development and nutrition of the larvae. Favourable conditions for the fungus produce larger wasps and this influences their longevity, fecundity and flight potential, thus facilitating population growth and expansion. Trees that are physiologically stressed, for example by climatic factors, overcrowding, or other pests and pathogens, are more susceptible to attack. In susceptible stands, populations of the pest were long thought to be regulated by a number of parasites. As the pest complex continues to expand its range, it is becoming evident that a more complex interaction of various biotic and abiotic factors combine to regulate population dynamics. Addressing such knowledge gaps is important if successful control of *S. noctilio* is to be ensured into the future.

K. Ryan (✉)
Faculty of Forestry, University of Toronto, Toronto, Canada
e-mail: kathleen.ryan@utoronto.ca

B.P. Hurley
Forestry and Agricultural Biotechnology Institute, University of Pretoria,
Pretoria, South Africa

B. Slippers et al. (eds.), *The Sirex Woodwasp and its Fungal Symbiont: Research and Management of a Worldwide Invasive Pest*, DOI 10.1007/978-94-007-1960-6_2,

© Springer Science+Business Media B.V. 2012

2.1 Introduction

Sirex noctilio F. (Hymenoptera; Symphyta; Siricidae) is a woodboring wasp that has elicited considerable attention for its ability to kill pines, especially in regions where it has been introduced. This species has a wide host range and most pine species appear to be susceptible (e.g., Morgan and Stewart 1966a; Spradbery and Kirk 1978). *Sirex noctilio* is the only woodwasp species known to be capable of actually killing trees (Spradbery 1973), though it generally favours severely stressed and declining hosts over healthier trees.

Sirex noctilio is endemic to Europe, Asia and North Africa, where it is considered a secondary pest of various pine species (Spradbery and Kirk 1978, Chap. 5) and is of little economic or ecological concern. During the twentieth century, *S. noctilio* became established in several countries in the Southern Hemisphere including New Zealand (1900s), Tasmania (1952), mainland Australia (1961), Uruguay (1980), Argentina (1985), Brazil (1988) and South Africa (1994) (summarized by Ciesla 2003). In regions of this introduced range, the wasp has become a primary pest of some pine species (Rawlings 1948; Hurley et al. 2007). In some situations it has caused extensive mortality in these plantations of introduced pines, especially in those affected by stresses such as drought (Neumann and Minko 1981). In 2004, *S. noctilio* was first discovered in North America (Hoebeke et al. 2005). The extent to which this insect will be of economic concern in this region remains to be determined.

The woodwasp has a cryptic lifestyle (Fig. 2.1). Most of its life cycle is spent as a larva within the sapwood of its host (from which the name woodwasp derives) and this makes it difficult to detect. The development of *S. noctilio,* and thus its reproductive potential, is tied to the vigour of its basidiomycete fungal symbiont, *Amylostereum areolatum* (Fries) Boidin (e.g., Coutts and Dolezal 1965). Interactions among the two biota and a phytotoxin produced by the wasp allow the insect to breach the defences of its living host trees. Woodwasp populations are suppressed by a number of factors, including tree defence, environment and natural enemies. In this chapter, the life history, host preference and factors affecting population dynamics of *S. noctilio* are reviewed.

2.2 Insect-Fungus Complex

The *A. areolatum* symbiont of *S. noctilio* derives considerable benefit from its relationship with the woodwasp. It is protected in the wasp's mycangia before the insect drills through the protective bark of the tree and introduces the fungus directly into a suitable, relatively competition-free, host substrate. *Amylostereum areolatum* primarily spreads vegetatively and clonal lineages over wide geographic areas are demonstrated by a number of authors (e.g., Vasiliauskas et al. 1998; Thomsen and Koch 1999; Vasiliauskas and Stenlid 1999; Slippers et al. 2001). The close association of *A. areolatum* with its woodwasp symbiont is thought to be a reason that it rarely reproduces sexually in some areas of its native range, and is not yet known to fruit in its introduced range (Gilbertson 1984, Chaps. 5, 8).

Fig. 2.1 The life-cycle of *Sirex noctilio* and its fungal mutualist, *Amylostereum areolatum*

Similar to that of *S. noctilio* and *A. areolatum*, other *Amylostereum* species (e.g., *A. chailettii* and *A. laevigatum*) also have symbiotic relationships with woodwasp species (reviewed in Slippers et al. 2003, Chap. 8). The wasps are specialized to ensure the continued relationship with their symbiont. Adult females carry the fungus within paired mycangia that are located at the anterior end of the ovipositor and open into the oviduct (Boros 1968). The mycangia are lined with glands and the secretions produced appear to stimulate fungal growth (reviewed in Morgan 1968). On oviposition, the female woodwasp inoculates arthrospores or fragments of its symbiotic fungus into the host sapwood through the ovipositor and into a separate drill beside the egg. At times no egg is deposited but the fungus and the phytotoxic mucus (discussed below) are (Coutts 1965; Coutts and Dolezal 1969). From the second instar, the fungus is transferred from one larval instar to the next and the adult takes it up into the mycangia when she sheds her pupal skin, thus ensuring the continued association with the fungus (Parkin 1941; Boros 1968). The fungus is thought to be absent in the pupal stage, but is taken up from the wall of the pupal chamber by the adult wasp (Francke-Grosman 1939; Parkin 1942 but see Cartwright 1938).

The wasp produces phytotoxic mucus that is inoculated into the tree, along with the fungus, during oviposition activity (Coutts and Dolezal 1969). The mucus is thought to be a protein-polysaccharide complex. It is known to contain several enzymes including amylase, esterase, phenoloxidase and proteolytic enzymes (Wong and Crowden 1976). The mucus is synthesized in a pair of secretory glands and is stored in a mucus reservoir until use; a duct connects the reservoir to the base of the wasp's oviduct (Boros 1968). *Sirex noctilio* has larger mucus glands and mucus reservoirs than other siricid species and this contributes to its ability to kill trees while its congenerics rarely do so (Spradbery 1977).

The inoculation of *A. areolatum* into a tree on its own has no deleterious effect, but when injected into a tree in combination with the phytotoxic mucus the two function as a pathogen (Vaartaga and King 1964; Coutts 1969a, Chap. 3). In tandem, the fungus and the mucus cause severe physiological stress to the tree, which include impaired water relations and translocation and this is often followed by tree death (Coutts 1969a, b; Fong and Crowden 1973). Signs of toxicity to the tree include foliar chlorosis, often within 2 weeks, or senescence with or without chlorosis (Coutts 1969b). Needle wilting is also described (e.g., Neumann et al. 1987), but this has not been seen in North America (K. Ryan, personal observation 2009). *Sirex noctilio* is the only woodwasp recognized to be capable of triggering this degree of physiological response. This probably relates to both mucus production and behaviour; of all of the woodwasps, *S. noctilio* not only produces the greatest amount of mucus it also shows the greatest oviposition density (Spradbery 1973).

In addition to its role in stressing the tree, *S. noctilio*'s mucus has direct effects on its fungal symbiont. The mucus has been shown to release *A. areolatum* from the wax packet that surrounds it while stored in the mycangia, and stimulates its growth (Boros 1968; Titze and Turnbull 1970).

The role of fungi in the nutrition of woodwasps is not entirely clear. Kukor and Martin (1983) demonstrate the role of fungus-derived cellulases in the digestion of xylem by a congeneric woodwasp species. Other authors report direct consumption of mycelia-impregnated wood in *S. noctilio* (Büchner 1928 in Büchner 1965). Fungal hyphae are digested by gut secretions of the larvae (Francke-Grosman 1939) and in laboratory conditions the siricid larvae can feed on pure fungus (Cartwright 1929). The role of the fungal symbiont in woodwasp nutrition may depend on the life stage of the insect. Madden and Coutts (1979) suggest that first, and some second, instar larvae feed exclusively on the fungus, while later larval stages feed on fungus-colonized wood. Regardless of the specific role of the fungus in larval nutrition, it is essential for the developing larvae. The adult woodwasp is thought not to feed and to survive on stored fat-body reserves (Morgan and Stewart 1966a; Taylor 1981).

The presence of *A. areolatum* is essential throughout the development of the immature woodwasp. Egg eclosion is delayed when conditions in the tree impede fungal growth (Madden 1981) and larvae may starve if symbiont growth is inhibited by the presence of other fungi (Coutts and Dolezal 1965; King 1966). Larval development relates to fungal growth and when conditions are optimal for the fungus, larger adults are produced (Madden 1981). The fungus also modifies environmental conditions. The wood-decaying symbiont dries the wood substrate providing a more suitable micro-environment for egg and larvae development (Coutts and Dolezal 1965) and wood degradation by the fungus facilitates tunneling of the larvae (Gilmour 1965).

2.3 The Life History of *Sirex noctilio*

Sirex noctilio develops in 1–3 years, but in most regions the bulk of emergence occurs after 1 year (Morgan 1968). Development occurs above a 6.8°C threshold and requires 2,500 degree days (Madden 1981), so at a constant 30°C a woodwasp would take approximately 90 days to emerge, while at a constant 10°C it would take about 360 days. This fits with Neumann and Minko's (1981) observation that some wasps may take as little as 3 months to develop in south-eastern Australia. Within a single tree development times may differ; wasps in the upper trunk may develop in 1 year while those in the lower trunk may take 2 years. This difference is thought to relate to lower wood moisture conditions in the upper bole (Morgan 1968).

Sirex noctilio emerges from early or mid-summer to early autumn. These patterns differ drastically across regions in the native and non-native ranges (Fig. 2.2, Chap. 5). Emergence times and lengths vary, even within the same region, and this depends on climate (Neumann et al. 1987). Regardless of variation in emergence, most appears to be concentrated over a period of a few weeks. Emergence patterns vary between regions, some have uni-modal peaks of emergence, others bimodal (Taylor 1981; Neumann et al. 1987). When there is a second peak of emergence it is thought to be a result of the emergence of short-cycle individuals (those with life cycles of a few months) (Neumann et al. 1987). Emergence is favoured by above average temperatures, and falling barometric pressure (Taylor 1981) and this can focus the emergence period. In cool, wet years emergence may be haphazard and without a clear peak (Morgan and Stewart 1966a).

Emergence is protandrous; the males emerge a few days before females (e.g., Rawlings 1948; Morgan and Stewart 1966a). Wasps rest on the bole before moving to the crown of the tree where the males aggregate and are joined by the females for mating (Morgan and Stewart 1966a; Madden 1988). Wasps are sexually mature

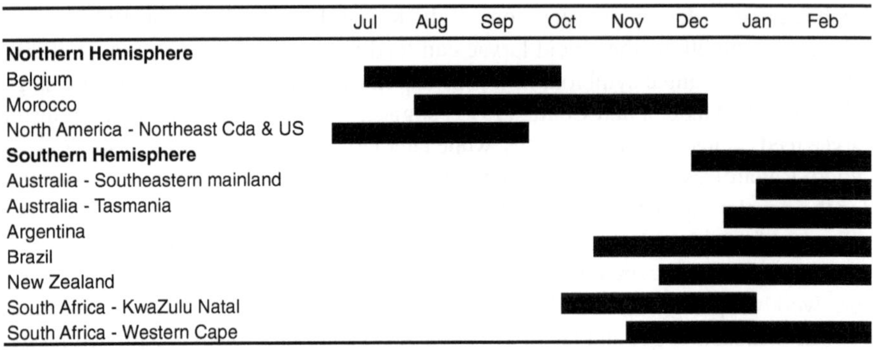

Fig. 2.2 *Sirex noctilio* emergence periods in native and introduced ranges. Data were collated from Morgan and Stewart (1966b), Spradbery and Kirk (1978), Taylor (1978), Neumann and Minko (1981), Iede et al. (1998), Klasmer et al. (2000), Tribe and Cille (2004), Hurley et al. (2008), Long et al. (2009) and Kathleen Ryan (unpublished data from Canada)

at emergence (Neumann et al. 1987). Sunny, warm conditions (>21°C) have been observed to stimulate copulation and cold, wet days to hamper it (Dolezal 1967; Neumann et al. 1987). Life spans are short; males live for up to 12 days and females up to 5 days (Neumann et al.1987). Smaller wasps tend to have shorter adult life spans (Madden 1974). As life-spans are short, prolonged emergence periods in cool rainy seasons may decrease mate-finding, although this may be compensated by greater longevity in cool autumn weather (Neumann et al. 1987).

Following mating, or in the absence of mating after the initial flight, females locate a tree for oviposition (Madden 1988). After a potential host tree is located, the female explores the bole with her antennae on the surface of the bark and then by probing into the sapwood with her ovipositor (Francke-Grosman 1939; Madden 1988). Trees with oleoresin pressures above 18.0×10^5 Pascals are rejected (Madden and Coutts 1979). When a suitable host tree is found, the female typically begins laying eggs near the bottom of the tree and works her way up, drilling single, double, treble or quadruple sets of oviposition chambers every 7.5–50 cm along the stem (Rawlings and Wilson 1949; Morgan and Stewart 1966a). Oviposition generally occurs along the entire stem in young trees and between 3 and 14 m above ground in older trees (Neumann et al. 1982). Eggs are generally laid around the stem (Coutts and Dolezal 1969). When the bark texture is uniform, the distribution of oviposition sites on successfully attacked trees tends to be random (Madden 1974), however, exposed areas on the lower stem may be preferred, especially if they are sun-exposed (Rawlings and Wilson 1949). Several females can attack a single tree (Bedding 1972, K. Ryan, personal observation 2006). Female wasps rarely travel far from the tree in which they are ovipositing (Morgan and Stewart 1966a), perhaps contributing to the clustered pattern of attacked trees that is often evident within a stand (Morgan and Stewart 1966a; Corley et al. 2007, Chap. 4). Oviposition rate is influenced by ambient conditions. Oviposition increases with temperature to a maximum at around 21°C and declines at higher temperatures. Relative humidity also affects egg laying and is most favourable between 49% and 52% and declines above or below this range (Madden 1974). Eggs are laid in sapwood containing between 20% and 100% moisture, but the woodwasp prefers a moisture content of about 60% (Morgan and Stewart 1966a).

Oviposition is up to 19 mm into the sapwood (Gilbertson 1984) depending on the wasp's ovipositor length that ranges from 8 to 20 mm (Coutts 1965). Depending on site quality, the female generally lays up to three eggs per drill site, each in separate tunnels (Madden 1974, 1988). A final, egg-less tunnel at each drill site is inoculated with arthrospores or fragments of *A. areolatum*, as well as the mucus (Coutts and Dolezal 1969). *Sirex noctilio* is also known to make single drills, introducing only the mucus and fungus, perhaps when testing the suitability of the tree for oviposition (Coutts and Dolezal 1969; Spradbery 1977). These test drills are thought to condition the tree and make it prone to further attack (Madden and Irvine 1971).

Egg eclosion is preceded by growth of the fungal symbiont around the egg (Madden 1981). Hatching usually occurs in 16–28 days (Morgan and Stewart 1966a) depending on temperature conditions but ranges between 10 (at 30°C) and 60 (at 10°C) days (Madden 1981). The temperature threshold is 6.2°C, which differs slightly from the developmental threshold of 6.8°C, but Madden (1981) does not consider this difference significant. Eclosion can be delayed, up to 12 months (Waterhouse and Sands 2001), if high moisture conditions or low temperatures, such as those found at tree bases, impede fungal growth (Madden 1981). A wood moisture content of between 40% and 70% is most favourable for egg eclosion and the development of early instar larvae (Morgan and Stewart 1966a).

Larvae begin tunnelling more deeply in the sapwood after the third or fourth instar (Taylor 1981). They bore through the sapwood, excavating tunnels that have been measured up to 26 cm long (Neumann et al. 1987). Galleries are primarily oriented along the wood grain, specific shapes vary and boring patterns depend on wood moisture (Morgan 1968; Madden 1981). There are between 6 and 12 larval instars, depending on aeration and moisture conditions within the sapwood, larger adults tending to have more instars (Madden and Coutts1979, Madden 1981). Adults may differ in size, even within the same area of the tree, depending on wood moisture conditions, smaller adults emerge from drier wood (Coutts 1965; Madden and Coutts 1979). Once larval development is complete, the larva tunnels outward and becomes a pre-pupa within 5 cm of the bark surface (Taylor 1981). The prepupal period lasts up to 4 weeks, while the pupal stage lasts approximately 20–28 days (Morgan and Stewart 1966a). In cooler, wetter conditions, pupation may not occur until the second or third year after eclosion (Taylor 1981).

Natural dispersal of the woodwasp is estimated to be up to 30–50 km/year (Haugen et al. 1990). Recent investigations using a flight mill showed high variability in flight ability between individuals. Female wasps fly between 1.1 and 49.7 km (mean 17.4 km) over a 20 h period (Bruzzone et al. 2009). However, flight ability depends on nematode parasitism; Villacide and Corley (2008, Chap. 4) found that the average flight distance of uninfected females is double that of parasitized ones (30 vs. 16 km/24 h). These authors show that parasitized females are smaller and lose more weight in flight than uninfected wasps. Natural woodwasp spread is easily augmented by movement of infested wood, because all immature life stages can be transported in unprocessed logs, lumber and solid wood packing materials (Haugen 2006).

2.4 Host Selection and Susceptibility

Physiologically stressed trees are more attractive to, and more susceptible to, attack by female *S. noctilio* (Madden 1968). Madden (1977) demonstrated that physiological stress of a tree results in impaired translocation, but enhanced transpiration and phloem respiration. This increases tissue permeability and the rate of monoterpene and water vapour loss through the bark. These physiological changes favour the attraction of the female *S. noctilio*, egg deposition and establishment of *A. areolatum*. The action of the mucus from the wasp and the fungus reinforces the stress condition in the tree (as discussed above), resulting in a feedback loop of increased susceptibility and probability of further attack. In some regions emergence is concomitant with low soil moisture conditions, and therefore with suppressed physiological activity and lowered resistance of host trees (Madden and Coutts 1979). This may contribute to its ability to overcome tree defences (Chap. 5 contains more details on the susceptibility and response of host trees to attack by *S. noctilio*).

Adverse climatic conditions can cause physiological stress in trees and thus increase their susceptibility to attack by *S. noctilio*. Madden (1988) on an examination of infestations by *S. noctilio* in Australia and New Zealand, showed that the incidence of *S. noctilio* is often associated with drought conditions, followed by unusually high rainfall. Extreme differences in rainfall between seasons cause stress in the trees, making them more vulnerable to attack (Madden 1968). Other climatic factors such as hail and snow storms may also result in tree stress and thus increased susceptibility to attack by *S. noctilio*. Prolonged dry conditions also increase the probability of fires. Trees that survive fires will likely be physiologically stressed and thus more susceptible to attack.

Plantation management can greatly influence the susceptibility of trees to attack by *S. noctilio*. Madden (1988) viewed poor silviculture as one of the key reasons for the outbreaks in Australia and New Zealand, and similarly Neumann et al. (1987) stated that outbreaks were largely a management problem and could be prevented by the routine surveillance of plantations and the application of silvicultural measures. Stand density is one of the major factors that can influence tree stress. High-density stands have a greater proportion of suppressed trees that are more susceptible to attack (Neumann et al. 1987). These susceptible trees allow *S. noctilio* to establish in the stand and attack healthier trees in the following seasons as populations grow. The dominance of high-density pulp stands in the KwaZulu-Natal Province of South Africa is thought to be a major factor causing the higher infestations compared to the Western Cape and other countries where lower-density saw timber stands are dominant (Hurley et al. 2007). The timing of silvicultural operations, such as pruning and thinning can also influence the susceptibility of trees to attack by *S. noctilio*. Mechanical damage to trees during these operations can make those trees more susceptible to attack, and thus it is suggested that such operations do not occur during or just before the flight season of *S. noctilio* (Neumann et al. 1987).

Other pest infestations and disease can cause physiological stress to the trees, making them more susceptible to attack by *S. noctilio* (Titze 1965; Spradbery and Kirk 1978). This is especially so when such attacks occur during, or just prior to, the emergence of the wasp. Neumann et al. (1993) showed a positive correlation between levels of attack by *S. noctilio* and infection levels of pine needle blight disease (caused by the pathogen *Dothistroma septosporum*). Spradbery and Kirk (1978) found that infestations of the Lepidopteran larvae *Choristoneura murinana, Lymantria monacha* or *Thaumetopoea pityocampa* made trees more susceptible to attack by *S. noctilio*. In South Africa, infestations of the Lepidopteran larvae *Euprocits terminalis* or the beetles *Hypopholis sommerrii* and *Pissodes nemorensis* are common during the flight season of *S. noctilio* (B.P. Hurley, personal observation), and are also thought to make trees more susceptible to attack. Conversely, infestation by some pathogens, such as blue stain fungi, could prevent *S. noctilio* attack or establishment due to competition of these fungi with *A. areolatum* (King 1966).

In general, trees that are more vigorous and have not sustained mechanical or other damage are more resistant. Madden (1988) summarized resistance to attack by *S. noctilio* in three categories, namely non-preference, anti-biosis and tolerance. Non-preference occurs when an initially attractive tree becomes unattractive after further inspection. Anti-biosis involves the flooding of oviposition holes with resin causing the death of *S. noctilio* eggs and *A. areolatum*, as well as forming a barrier of polyphenols to isolate the growth of *A. areolatum* (Coutts and Dolezal 1966). Tolerance involves suppressing the effect of the mucus by reducing transpiration or the premature senescing of affected needles. Kile et al. (1979) showed that there was some level of mucus resistance in *P. radiata* trees, which was under genetic control. But, even the more resistant trees would be overcome in time and by high infestations.

2.5 Host Records

Recorded hosts of the *S. noctilio – A. areolatum* complex include species of *Pinus, Abies, Larix, Picea, Pseudotsuga* and *Araucaria* (Spradbery and Kirk 1978; Madden 1988, Chap. 5). Of these, *Pinus* species are preferred (Spradbery and Kirk 1978, 1981). *Pinus pinaster* was recorded as the most attacked in its native range (Spradbery and Kirk 1978). In its introduced range in the Southern Hemisphere, the dominant *Pinus* species attacked are *P. radiata* in Australia, New Zealand, Chile and the Western Cape of South Africa, *P. taeda* and *P. elliottii* in Brazil, northern Argentina and Uruguay, *P. elliottii, P. ponderosa, P. radiata* and *P. elliottii contorta* var. *latifolia* in southern Argentina, and *P. patula* in KwaZulu-Natal, South Africa (Rawlings and Wilson 1949; Maderni 1998; Ahumada 2002; Carnegie et al. 2006; Hurley et al. 2007, Chaps. 13–18). In the USA and Canada, *S. noctilio* has only been recently introduced (Hoebeke et al. 2005; de Groot et al. 2007), but has already been recorded on *P. banksiana, P. sylvestris, P. resinosa* and *P. strobus* (Peter de Groot, personal communication, Chap. 19). Besides these, many other *Pinus* sp. have also been attacked by *S. noctilio*, including *P. pinea, P. echinata, P. halepensis, P. palustris, P. brunia, P. nigra, , P. contorta, P. jefferyi, P. chiapensis* and others (Spradbery and Kirk 1978; Tribe 1995; Iede et al. 1998; Maderni 1998; Haugen 2000).

As yet, no *Pinus* sp. have shown resistance against attack, although there have been some reports of certain species being more susceptible or favoured than others.

Maderni (1998) reported that *P. elliottii* appeared to be more resistant than *P. taeda* to attack by *S. noctilio* in Uruguay. Similarly, in South Africa, *S. noctilio* appears to favour *P. patula* over *P. elliottii* (B.P. Hurley, personal observation).

2.6 Factors Influencing Population Dynamics

2.6.1 Potential Fecundity

The potential fecundity of *S. noctilio*, determined by the number of eggs carried by the female wasp, varies greatly between individuals. Madden (1974) calculated that the number of eggs per female was determined by its size, and can be determined by $y = 28.8(52.5)^x$, where y is the number of eggs and x is the width of the female's prothorax (measured in mm). Zondag and Nuttall (1977) determined the number of eggs from wasps collected in New Zealand to range from 50 to 500 eggs. Similarly, Madden (1974) recorded the number of eggs to range from 30 to 450, and Neumann et al. (1987) recorded the number of eggs from wasps collected in Victoria, Australia to range from 21 to 458. On average, female wasps will have about 220 eggs (Neumann et al. 1987).

2.6.2 Reproductive Potential

Neumann and Minko (1981) estimated reproductive potential of *S. noctilio* females as a function of the potential fecundity (F) and the sex-ratio (SR), where RP = FxSR. Thus, in Victoria, Australia, where the average number of eggs per female was 212 and the sex ratio was 4:1 males to females, the estimated reproductive potential for one *S. noctilio* female was 53 females. However, these calculations assumed zero mortality for any life stage of *S. noctilio*. Spradbery and Kirk (1981) calculated about 30% survival of eggs on untreated infested logs. Further, Neumann and Minko (1981) estimated that on average only about 88% of a female's eggs are laid, and of these about 28% are not viable. The development and survival of larvae that do hatch is dependant on adequate temperatures and the growth of *A. areolatum*, which is itself dependant on moisture in the wood and other factors (Madden 1981). Considering these and other mortality factors, the reproductive potential of *S. noctilio* will be considerably less than calculated by Neumann and Minko (1981).

2.6.3 Adult Size

The size of emerging adult *S. noctilio* varies considerably. Female wasps are generally larger than the males, but even within the gender, a great range in size is evident.

Neumann et al. (1987) determined the length of females to range from 12 to 34 mm (mean = 23.2 mm) and that of males to range from 9.3 to 34.9 mm (mean = 25 mm). Similarly, Hurley et al. (2008) found that the length of females range from 10 to 44 mm. The variation in adult size is generally attributed to the conditions for larval development, specifically the availability of the fungal resource (Madden and Coutts 1979). Even larvae from the same oviposition drill, but tunnelling different directions, can encounter different moisture conditions in the tree, which influences fungal growth and potentially determines their size (Madden and Coutts 1979; Madden 1981). The duration that the larva develops in the tree, and thus feeds on the fungus, will also influence its size, which explains the considerable differences in wasp size found by Filho et al. (1998) between short-cycle *S. noctilio*, which emerge earlier (April to June), (mean female length = 13.83 mm; mean male length = 10.95 mm) and long-cycle *S. noctilio*, that emerge later (October to February), (mean female length = 26.35 mm; mean male length = 23.60 mm).

There appears to be a relationship between wasp size, flight ability and probability of parasitism. Villacide and Corley (2008) demonstrated that larger wasps have greater flight velocity and distance, but are also less likely to be parasitized. These findings were supported by those of Eskiviski et al. (2004), but it is in contrast to the findings of Hurley et al. (2008) where parasitism was higher in the larger wasps. Further investigations are needed to clarify this issue, as the outcome has consequences for biological control. As larger wasps have greater flight potential and thus are more likely to infest new areas, it would be hugely beneficial to biological control programmes if large wasps were more likely to be parasitized, but detrimental to these programmes if the opposite were true.

2.6.4 Sex-Ratio

Sex determination in *S. noctilio* is by a haplo-diploid mechanism, where haploid eggs are males and diploid eggs are females (Neumann et al. 1987). Spradbery and Kirk (1978) found the sex-ratio of *S. noctilio* to be 1.8:1 males to females in its native range. The recorded sex-ratio of *S. noctilio* in its introduced range varies greatly. Eskiviski et al. (2004) recorded a sex-ratio of 1.4:1–2:1 from his study in Argentina, similar to that obtained by Spradbery and Kirk (1978). But other recorded sex-ratios have shown a much stronger bias towards males. Zondag and Nuttall (1977) recorded a sex-ratio of 20:1 in New Zealand, Iede et al. (1998) recorded sex-ratios from 1.5:1 to 32: 1 in Brazil and Hurley et al. (2008) recorded a sex-ratio of 12:1 from South Africa. Similarly, Taylor (1981) recorded the sex-ratio in Tasmania to have ranged from 1.5:1 to 16.5:1. The highly male biased sex-ratio often found in the introduced range of *S. noctilio* could result from an expanding population. As female wasps move to new areas to lay their eggs, they are less likely to find a male, resulting in a male biased population occurring at the front of the population. When *S. noctilio* is present throughout this new area, the probability of finding a male increases and thus the sex-ratio becomes less biased.

2.6.5 Natural Enemies

The absence of natural enemies in the introduced range of *S. noctilio* was one of the reasons for the high infestations that have occurred in these areas. In its native range, where natural enemies are present, extensive damage by *S. noctilio* does not occur (Spradbery and Kirk 1978, Chap. 5). Natural enemies of *S. noctilio* include nematodes, parasitoid wasps and woodpeckers. Certain nematodes and parasitoid wasps found in the native range of *S. noctilio* have been used as biological control agents in the Southern Hemisphere, with various levels of success (Hurley et al. 2007).

The parasitoid wasps, *Rhyssa* sp., *Megarhyssa* sp., *Ibalia leucospoides* and *Schletterius cinctipes* have been introduced into various countries as biological control agents. *Ibalia leucospoides* has been recorded to achieve parasitism levels of 40% and higher and even as much as 70% when in combination with the rhyssines (Neumann et al. 1987; Nuttall 1989; Carnegie et al. 2005). For the nematodes, *Deladenus* (=*Beddingia*) *siricidicola* has been widely used, and is considered the primary biological control agent (Bedding and Iede 2005). *Deladenus siricidicola* has been recorded to provide parasitism levels of over 90% (Bedding and Akhurst 1974), although substantial variation exists, with some areas having lower levels of parasitism (Hurley et al. 2007). Predation of siricid larvae by woodpeckers has been recorded to be as high as 29% and 40% by Spradbery (1990) and Marshall (1967) respectively, but they were not considered suitable for biological control agents. The topic of natural enemies is reviewed more extensively in Chaps. 8 and 9.

2.7 Conclusions

Sirex noctilio is the only woodwasp that regularly attacks living trees and overcomes their defences. This is due to the combined effect of its mucus, which appears to differ at least in amount from that of other siricids, and its symbiotic fungus, *A. areolatum*. Pines that are suppressed or stressed by factors such as climate conditions, high stand density, pathogens or pests are particularly vulnerable to *S. noctilio* attack. An abundance of these trees on the landscape, together with insufficient natural enemies, can contribute to high populations of *S. noctilio*, as evident in its introduced range in the Southern Hemisphere. This has been especially evident in introduced environments in the Southern Hemisphere, where the insect and its fungus has caused significant damage.

Amylostereum areolatum is not only essential to help overcome the tree defences, but also for the nutrition and development of the larvae of the wasp. It is, therefore, not surprising that the growth of the fungus is also stimulated by the wasp mucus. Good fungal growth enhances wasp nutrition and leads to larger body sizes. Wasp longevity, fecundity and flight potential all relate to body size, and are all important factors facilitating population growth and expansion. Thus, factors influencing fungal growth will also influence *S. noctilio* populations and spread rates. Climate is an important factor influencing fungal growth and *S. noctilio* populations (as discussed below), but other factors inhibiting fungal growth such as the presence of competing micro-organisms could also play an important role.

Climatic influences on *S. noctilio* populations could contribute significantly to the variability of its activity between different regions and between years. Development times are closely related to climatic conditions and warmer regions where development times are shorter would have more insect activity. In addition, the physiological stress of trees and thus their susceptibility to attack by *S. noctilio* is greatly influenced by climate. Fungal growth is affected by both ambient temperature and wood moisture conditions and when conditions are optimal for the fungus wasps are larger and therefore more fecund. The effects of some climatic factors, such as the seasonality of precipitation and over-wintering temperatures in colder climates, on *S. noctilio* development and survival, and on the population and effectiveness of their natural enemies are less known and are important areas for further research as this insect moves into new ranges.

The research done on *S. noctilio* and its symbiotic fungus, *A. areolatum*, since its arrival in the Southern Hemisphere about a century ago, constitutes a significant portion of knowledge regarding the biology of siricids and their associated fungi. This biological knowledge has also greatly assisted in the management of infestations. However, there are still many gaps in our understanding of the biology of *S. noctilio*, particularly in terms of the complex interaction with different biotic and abiotic variables. As *S. noctilio* continues to expand its range in North America, summer rainfall and subtropical areas of Africa and Australia, and more, these knowledge gaps are becoming more obvious. For that reason, research to further understand the biology and life-history of this insect is needed, in particular for the development and implementation of management strategies.

References

Ahumda R (2002) Diseases in commercial *Eucalyptus* plantations in Chile, with special reference to Mycosphaerella and Botryosphaeria species. MSc dissertation, Faculty of Natural and Agricultural Science, University of Pretoria, Pretoria

Bedding RA (1972) Biology of *Deladenus siricidicola* (Neotylenchidae) entomophagous-mycetophagous nematodes parasitic in siricid woodwasps. Nematologica 18:482–493

Bedding RA, Akhurst RJ (1974) Use of the nematode *Deladenus siricidicola* in the biological control of *Sirex noctilio* in Australia. J Aust Entomol Soc 13:129–135

Bedding RA, Iede ET (2005) Application of *Beddingia siricidicola* for Sirex Woodwasp Control. In: Grewel PS, Ehlers R-U, Shapiro-Ilan DI (eds) Nematodes as biocontrol agents. CAB International, London

Boros CB (1968) The relationship between the woodwasp, *Sirex noctilio* F. and the wood-rot fungus *Amylostereum* sp. M.Ag.Sc dissertation, University of Adelaide, Adelaide

Bruzzone OA, Villacide JM, Bernstein C, Corley JC (2009) Flight variability in the woodwasp *Sirex noctilio* (Hymenoptera: Siricidae): and analysis of light data using wavelets. J Exp Biol 212:731–737

Büchner P (1928) Holznahrung and Symbiose. J. Springer, Berlin

Büchner P (1965) Endosymbiosis of animals with plant microorganisms. Wiley, New York

Carnegie AJ, Elderidge RH, Waterson DG (2005) History and management of Sirex wood wasp in pine plantations in New South Wales, Australia. NZ J For Sci 35:3–24
Carnegie AJ, Matsuki M, Haugen DA, Hurley BP, Ahumada R, Klasmer P, Sun J, Iede ET (2006) Predicting the potential distribution of *Sirex noctilio* (Hymenoptera: Siricidae), a significant exotic pest of *Pinus* plantations. Ann For Sci 63:119–128
Cartwright KStG (1929) Notes on the fungus associated with *Sirex cyaneus*. Ann Appl Biol 16:182–187
Cartwright KStG (1938) A further note on fungus association in the Siricidae. Ann Appl Biol 25:430–432
Ciesla WM (2003) European woodwasp: a potential threat to North America's conifer forests. J For 101:18–23
Corley JC, Villacide JM, Bruzzone OA (2007) Spatial dynamics of a *Sirex noctilio* woodwasp population within a pine plantation in Patagonia, Argentina. Entomol Exp Appl 125:231–236
Coutts MP (1965) *Sirex noctilio* and the physiology of *Pinus radiata,* Canberra bulletin no. 41. Forest Research Institute, Canberra
Coutts MP (1969a) The mechanism of pathogenicity of *Sirex noctilio* on *Pinus radiata* I. Effects of the symbiotic fungus *Amylostereum* sp. (Thelophoraceae). Aust J Biol Sci 22:915–924
Coutts MP (1969b) The mechanism of pathogenicity of *Sirex noctilio* on *Pinus radiata* II. Effects of S. noctilio mucus. Aust J Biol Sci 22:1153–1161
Coutts MP, Dolezal JE (1965) *Sirex noctilio*, its associated fungus and some aspects of wood moisture content. Aust For Res 1:3–13
Coutts MP, Dolezal JE (1966) Polyphenols and resin in the resistance mechanism of *Pinus radiata* attacked by the wood wasp, *Sirex noctilio*, and its associated fungus, leaflet no. 101. Forestry and Timber Bureau, Canberra
Coutts MP, Dolezal JE (1969) Emplacement of fungal spores by the woodwasp, *Sirex noctilio*, during oviposition. For Sci 15:412–416
de Groot P, Nystrom K, Scarr T (2007) Discovery of *Sirex noctilio* (Hymenoptera: Siricidae) in Ontario, Canada. Great Lakes Entomol 39:49–53
Dolezal JE (1967) Some observations on behaviour of *Sirex noctilio* when used for artificial attack on *Pinus radiata*. Aust For Res 2:26–30
Eskiviski E, Nuñez Cresto M, Olmedo D, de Coll OdelR (2004) Biological aspects of *Sirex noctilio* F. and *Ibalia leucospoides* H. parasitism in forest plantations of *Pinus* sp. in Santo Tome, Corrientes. XI Jornadas Tecnicas Forestales y Ambientales – FCF, UNaM – EEA Monetecarlo, INTA
Filho WR, Iede ET, Penteado SdRC (1998) Biological aspects of *Sirex noctilio* F. (Hymenoptera, Siricidae) and its parasitoid *Ibalia leucospoides* (Hymenoptera: Ibaliidae). In: Iede E, Shaitza E, Penteado S, Reardon R, Murphy T (eds) Proceedings of a conference: training in the control of *Sirex noctilio* by use of natural enemies, USDA Forest Service (FHTET 98–13), Morgantown
Fong LK, Crowden RK (1973) Physiological effects of mucus from the wood wasp, *Sirex noctilio* F., on the foliage of *Pinus radiata* D. don. Aust J Biol Sci 29:365–378
Francke-Grosman H (1939) On the symbiosis of woodwasps (Siricinae) with fungi. (Rutland, D translation). Z Angew Entomol 25:647–680
Gilbertson RL (1984) Relationships between insects and wood-rotting basidiomycetes. In: Wheeler Q, Blackwell M (eds) Fungus-insect relationships, perspectives in ecology and evolution. Columbia University Press, New York
Gilmour JW (1965) The life cycle of the fungal symbiont of *Sirex noctilio*. N Z J For 10:80–89
Haugen DA (2000) Sirex woodwasp. In: Pest risk assessment for importation of solid wood packing materials into the United States. Draft August 2000, unpublished report
Haugen DA (2006) *Sirex noctilio*. Provisional record, database of the exotic forest pest information system for North America, North American Forest Commission. http://spfnic.fs.fed.us/exfor/data/pestreports.cfm?pestidval=33andlangdisplay=english. Retrieved 21 May 2008
Haugen DA, Bedding RA, Underdown MG, Neumann FG (winter 1990) National strategy for control of *Sirex noctilio* in Australia. Aust For Grow 13(2): Special liftout section No. 13
Hoebeke ER, Haugen DA, Haack RA (2005) *Sirex noctilio*: discovery of a palearctic siricid woodwasp in New York. Newsl Mich Entomol Soc 50:24–25

Hurley BP, Slippers B, Wingfield MJ (2007) A comparison of control results for the alien invasive woodwasp, *Sirex noctilio*, in the southern hemisphere. Agric For Entomol 9:159–171

Hurley BP, Slippers B, Croft PK, Hatting HJ, van der Linde M, Morris AR, Dyer C, Wingfield MJ (2008) Factors influencing parasitism of *Sirex noctilio* (Hymenoptera: Siricidae) by the nematode *Deladenus siricidcola* (Nematoda: Neotylenchidae) in summer rainfall areas of South Africa. Biol Control 45:450–459

Iede ET, Penteado SRC, Schaitza EG (1998) *Sirex noctilio* problem in Brazil – detection, evaluation and control. In: Iede E, Shaitza E, Penteado S, Reardon R, Murphy T (eds) Proceedings of a conference: training in the control of *Sirex noctilio* by use of natural enemies, USDA Forest Service (FHTET 98–13), Morgantown

Kile GA, Bowling PJ, Dolezal JE, Bird T (1979) The reaction of *Pinus radiata* twigs to the mucus of *Sirex noctilio* in relation to resistance to Sirex attack. Aust For Res 6:25–34

King JM (1966) Some aspects of the biology of the fungal symbiont of *Sirex noctilio* (F.). Aust J Bot 14:25–30

Klasmer P, Botto E, Corley JC, Villacide JM, Arhex F (2000) Avances en el control biológico de *Sirex noctilio* en la región Patagónica Argentina. 1° Simposio do Cone Sul sobre Manejo de Pragas e Doencas de Pinus. Serie técnica IPEF Brasil 13:21–30

Kukor JJ, Martin MM (1983) Acquisition of digestive enzymes by siricid woodwasps from their fungal symbiont. Science 220:1161–1163

Long SJ, Williams DW, Hajek AE (2009) *Sirex* species (Hymenoptera: Siricidae) and their parasitoids in *Pinus sylvestris* in eastern North America. Can Entomol 141:153–157

Madden JL (1968) Physiological aspects of host tree favourability for the woodwasp, *Sirex noctilio* F. P Ecol Soc Aust 3:147–149

Madden JL (1974) Oviposition behaviour of the woodwasp, *Sirex noctilio* F. Aust J Zool 22:341–351

Madden JL (1977) Physiological reactions of *Pinus radiata* to attack by woodwasp, *Sirex noctilio* F. (Hymenoptera: Siricidae). Bull Entomol Res 67:405–426

Madden JL (1981) Egg and larval development in the woodwasp, *Sirex noctilio* F. Aust J Zool 29:493–506

Madden JL (1988) Sirex in Australasia. In: Berryman AA (ed) Dynamics of forest insect populations: patterns, causes, implications. Plenum Press, New York/London

Madden JL, Coutts MP (1979) The role of fungi in the biology and ecology of wood wasps (Hymenoptera: Siricidae). In: Batra LR (ed) Insect-fungus symbiosis. Allanheld, Osmun and Co, Totowa

Madden JL, Irvine CJ (1971) The use of lure trees for the detection of *Sirex noctilio* in the field. Aust For 35:164–166

Maderni JFP (1998) *Sirex noctilio* F.: present status in Uruguay. In: Iede E, Shaitza E, Penteado S, Reardon R, Murphy T (eds) Proceedings of a conference: Training in the control of *Sirex noctilio* by use of natural enemies, USDA Forest Service (FHTET 98–13), Morgantown

Marshall HGW (1967) The effect of woodpecker predation of wood-boring larvae of families Siricidae Hymenoptera) and Melandryidae (Coleoptera). Can Entomol 99:978–985

Morgan FD (1968) Bionomics of Siricidae. Ann Rev Entomol 13:239–256

Morgan FD, Stewart NC (1966a) The biology and behaviour of the woodwasp *Sirex noctilio* (F.) in New Zealand. Trans Roy Soc N Z 7:195–204

Morgan FD, Stewart NC (1966b) The effect of *Rhyssa persuasoria* (L.) on a population of *Sirex noctilio* (F.). Trans R Soc N Z 8:31–38

Neumann FG, Minko G (1981) The Sirex wood wasp in Australian radiata pine plantations. Aust For 44:46–63

Neumann FG, Harris JA, Kassaby FY, Minko G (1982) An improved technique for early detection and control of the Sirex wood wasp in radiata pine plantations. Aust For 45:117–124

Neumann FG, Morey JL, McKimm RJ (1987) The sirex wasp in Victoria, Victoria bulletin no. 29. Lands and Forests Division, Department of Conservation, Forest and Lands, Melbourne

Neumann FG, Collett NG, Smith IW (1993) The Sirex wasp and its biological control in plantations of radiata pine variably defoliated by *Dothiostroma septospora* in north-eastern Victoria. Aust For 56:129–139

Nuttall MJ (1989) *Sirex noctilio* F., sirex wood wasp (Hymenoptera: Siricidae). In: Cameron PJ, Hill RL, Bain J, Thomas WP (eds) Review of biological control of invertebrate pests and weeds in New Zealand from 1874 to 1986, technical communication no. 10. CAB International Institute of Biological Control, Canberra

Parkin WA (1941) Symbiosis in larval Siricidae (Hymenoptera). Nature 147:329

Parkin EA (1942) Symbiosis and siricid woodwasps. Ann Appl Biol 29:268–274

Rawlings GB (1948) Recent observations on the *Sirex noctilio* population in *Pinus radiata* forest in New Zealand. N Z J For 5:411–421

Rawlings GB, Wilson NM (1949) *Sirex noctilio* as a beneficial and destructive insect to *Pinus radiata* in New Zealand. N Z J For 6:20–29

Slippers B, Wingfield MJ, Coutinho TA, Wingfield BD (2001) Population structure and possible origin of *Amylostereum areolatum* in South Africa. Plant Pathol 50:206–210

Slippers B, Coutinho TA, Wingfield BD, Wingfield MJ (2003) A review of the genus *Amylostereum* and its association with woodwasps. S Afr J Sci 99:70–74

Spradbery JP (1973) A comparative study of the phytotoxic effects of siricid woodwasps on conifers. Ann Appl Biol 75:309–320

Spradbery JP (1977) The oviposition biology of siricid woodwasps in Europe. Ecol Entomol 2:225–230

Spradbery JP (1990) Predation of larval siricid woodwasps (Hymenoptera: Siricidae) by woodpeckers in Europe. Entomol 109:67–71

Spradbery JP, Kirk AA (1978) Aspects of the ecology of siricid woodwasps (Hymenoptera: Siricidae) in Europe, North Africa and Turkey with special reference to the biological control of *Sirex noctilio* F. in Australia. Bull Entomol Res 68:341–359

Spradbery JP, Kirk AA (1981) Experimental studies on the responses of European siricid woodwasps to host trees. Ann Appl Biol 98:179–185

Taylor KL (1978) Evaluation of the insect parasitoids of *Sirex noctilio* (Hymenoptera: Siricidae) in Tasmania. Oecologia (Berl) 32:1–10

Taylor KL (1981) The Sirex woodwasp: ecology and control of an introduced forest insect. In: Kitching RL, Jones RE (eds) The ecology of pests. CSIRO, Melbourne

Thomsen IM, Koch J (1999) Somatic compatibility in *Amylostereum areolatum* and *A. chailletii* as a consequence of symbiosis with siricid woodwasps. Mycol Res 103:817–823

Titze JF (1965) Physiological suppression in *Pinus radiata* and its susceptibility to Sirex. Aust For Res 1:51–55

Titze JF, Turnbull CRA (1970) The effect of club gland secretions of *Sirex noctilio* on the growth of the symbiotic fungus *Amylostereum areolatum*. Aust For Res 4:27–29

Tribe GD (1995) The woodwasp *Sirex noctilio* Fabricius (Hymenoptera: Siricidae), a pest of *Pinus* species, now established in South Africa. Afr Entomol 3:215–217

Tribe and Cillié (2004) The spread of *Sirex noctilio* Fabricius (Hymenoptera: Siricidae) in South African pine plantations and the introduction and establishment of its biological control agents. Afr Entomol 12:9–17

Vaartaga O, King JM (1964) Inoculation experiments with *Amylostereum* sp. from a woodwasp. Plant Dis Rep 48:438–440

Vasiliauskas R, Stenlid J (1999) Vegetative compatibility groups of *Amylostereum areolatum* and *A. chailletii* from Sweden and Lithuania. Mycol Res 103:824–829

Vasiliauskas R, Stenlid J, Thomsen IM (1998) Clonality and genetic variation in *Amylostereum areolatum* and *A. chailletii* from northern Europe. New Phytol 139:751–758

Villacide JM, Corley JC (2008) Parasitism and dispersal potential of *Sirex noctilio*: implications for biological control. Agric For Entomol 10:341–345

Waterhouse DF, Sands DP (2001) Classical biological control of arthropods in Australia. Australian Centre for International Agricultural Research, Canberra

Wong LK, Crowden RK (1976) Preliminary studies on the mucus secretion of the wood wasp, Sirex noctilio F. I. Physicochemical and biochemical properties. Aust J Biol Sci 1–2:21–32

Zondag R, Nuttall MJ (1977) *Sirex noctilio* Fabricius (Hymenoptera: Siricidae), forest and timber insects in New Zealand no. 20. New Zealand Forest Service, Rotorua

Chapter 3
Susceptibility and Response of Pines to *Sirex noctilio*

John M. Bordeaux and Jeffrey F.D. Dean

Abstract *Sirex noctilio* uses an unusual, two-component approach to kill pine trees and provide a source of nutrition for its developing larvae. A substantial amount of research was undertaken in the 1960s and 1970s to examine the mechanisms by which these two components – a mucus produced in the woodwasp acid – (venom) gland, and a pathogenic fungus, *Amylostereum areolatum*, both of which are introduced into trees during oviposition – contribute to the death of trees. This chapter reviews the effects that the woodwasp mucus was shown to have when introduced to trees in the absence of the fungus, including resinosis, premature senescence and abscission of needles, and collapse of parenchymal cells with consequent loss of photosynthate translocation, as well as changes in respiration, and reduced growth rates. The effects seen in trees that were artificially inoculated with *A. areolatum*, including increased ethylene production and the induction of polyphenol biosynthesis, are also reviewed. Also discussed are the roles played by tree health and environmental stresses in determining overall susceptibility of host trees to attack by *S. noctilio*. These past observations are placed within the context of our current understanding of plant defense responses to attack by insects and pathogens, and a number of promising areas for future research are highlighted.

3.1 Introduction

Sirex noctilio F. (Hymenoptera: Siricidae) is a woodwasp native to Europe and northern Africa (Spradbery and Kirk 1978). A member of the Symphyta (Sawfly) suborder of the Hymenoptera (Smith and Schiff 2002, Chap. 1), *S. noctilio* attacks

---

J.M. Bordeaux • J.F.D. Dean (✉)
Warnell School of Forestry and Natural Resources, University of Georgia, Athens, USA
e-mail: jeffdean@uga.edu

B. Slippers et al. (eds.), *The Sirex Woodwasp and its Fungal Symbiont: Research and Management of a Worldwide Invasive Pest*, DOI 10.1007/978-94-007-1960-6_3,
© Springer Science+Business Media B.V. 2012

a broad range of pine species, laying eggs in the sapwood of host trees through holes drilled in the bark (Madden 1974). During oviposition, *S. noctilio* inoculates the host tree with a symbiotic fungal pathogen, *Amylostereum areolatum* (Fr.) Boidon, that not only kills the tree, but is also the primary food source for developing wasp larvae (Talbot 1977, Chap. 2). Along with the fungus, the wasp also injects a so-called 'mucus' that affects tree responses and assists the fungus in establishing a productive infection (Coutts 1969b, c). Thus, we must dissect the recognition and response mechanisms of all three organisms if we are to fully understand the infection process.

Sirex noctilio is not typically considered a significant threat to trees in its native range where the forest species have evolved in concert with this pest, and attacks are usually limited to suppressed and dying trees (Borchert et al. 2007a; Ciesla 2003; Gauld et al. 1990). Woodwasps in the Siricidae family, including *Sirex* spp., are successful on conifers generally, but *S. noctilio* shows a marked preference for pines and is the only Siricid in its native range that attacks living trees (Spradbery and Kirk 1981). In its native range (Europe, Turkey, and North Africa) primary host species include Turkish (*Pinus brutia* Ten.), Canary Island (*P. canariensis* Smith), Aleppo (*P. halapensis* Mill.), Austrian (*P. nigra* Arn.), maritime (*P. pinaster* Ait.), stone (*P. pinea* L.), and Scots pines (*P. sylvestris* L.) (Spradbery and Kirk 1978). However, where Sirex populations have become established in the Southern Hemisphere, they have proven a devastating problem for commercial pine plantations (Chaps. 13–18). These commercial forests are frequently planted with North American species, such as Monterey (*P. radiata* D. Don) and loblolly pines (*P. taeda* L.), that have not previously been exposed to *S. noctilio* and in environments that do not contain other Siricids or their associated predators. Under these conditions *S. noctilio* populations can rapidly expand to the point where the woodwasps attack and overwhelm healthy trees (Borchert et al. 2007b; Spradbery and Kirk 1978). Consequently there is great concern about the potential damage this pest may bring to North American forests now that it has established itself in the region around Lake Ontario (Dodds et al. 2007, Chap. 19). Within the current boundaries of this first North American infestation, eastern white pine (*P. strobus* L.) and red pine (*P. resinosa* Ait.) (Fig. 3.1) appear to be the species being hardest hit, although non-native Scots pines in overstocked and abandoned Christmas tree plantations are also suffering heavy mortality (Dodds et al. 2007). Other known North American pine hosts for *S. noctilio* include knobcone (*P. attenuata* Lemm.), jack (*P. banksiana* Lamb.), Caribbean (*P. caribaea* Mor.), lodgepole (*P. contorta* Dougl.), shortleaf (*P. echinata* Mill.), slash (*P. elliottii* Engl.), Jeffrey (*P. jeffreyi* Balf.), Bishop (*P. muricata* D. Don), longleaf (*P. palustris* Mill.), Mexican weeping (*P. patula* Schiede: Schltdl. and Cham.), and ponderosa pines (*P. ponderosa* Dougl. :Laws.) (Borchert et al. 2007b). The economic impact from Sirex outbreaks can be serious when they are left unchecked, and a recent estimate suggested that such an outbreak in U. S. forests might lead to losses as high as \$3–\$17 billion in the affected areas (Rabaglia and Lewis 2006).

In this chapter we review what is known about pine susceptibility and response to attack by *S. noctilio*. Our discussion begins with coverage of the conditions predisposing trees to Sirex attack, and is followed by separate descriptions of tree

Fig. 3.1 *Sirex noctilio* infestations in overstocked pulpwood plantations of *P. patula* in South Africa (*panel a*) and *P. resinosa* in New York, USA (*panel b*). Clear resin beads are visible on a live and freshly oviposited *P. resinosa* (*panel c*), while aged and dried resin remains on dead *P. taeda* in Uruguay (*panel d*). Galleries and larvae are visible in *P. patula* sapwood once bark and overlay is removed using an axe (*panel e, black arrows*), while mature wasp exit holes are distinctly visible on dead *P. patula* stems (*panel f, white arrows*). Note the red patches of discoloration likely due to polyphenol deposition during the defense response (*red arrow*)

responses to woodwasp mucus, responses to the fungal pathogen, and responses for which the inducer remains unclear. We conclude with prescriptions for reducing the severity of Sirex attacks, as well as a discussion of areas needing further research.

3.2 Factors Contributing to Tree Susceptibility

Pines are predisposed to attack by *S. noctilio* by a wide variety of biotic and abiotic factors. Across their native range, *S. noctilio* females favor suppressed or sickly trees, which have not received sufficient light, nutrients, water, or other resources (Coutts and Dolezal 1965a; Madden 1968). Tree age and size at the time of attack also affect susceptibility. Smaller-diameter trees are favored by *S. noctilio* females, although larger trees may be attacked during later years of an outbreak (Madden 1975). Large and healthy dominant trees are least favored for ovipositing, and such trees are most likely to survive attack (Coutts 1965; Talbot 1977). Environmental stresses, particularly drought trigger stress responses in the trees that act as signals for *S. noctilio* attack (Spradbery and Kirk 1978). An outbreak following a severe drought caused losses of 25–33% over 300,000 acres of *P. radiata* plantations New Zealand between 1946 and 1951 (Gilmour 1965). More recently, the implications of declining tree vigor in the face of drought conditions and climate change for expansion of populations of *S. noctilio* and other wood-boring pests was examined for Scots pine forests in Switzerland (Wermelinger et al. 2008). Overstocking and poor silvicultural practices that lead to root exposure, flooding, or wounding of standing trees, as well as poor sanitation practices and monocultures in plantations, can contribute to create conditions conducive to Sirex outbreaks (Fig. 3.1) (Spradbery and Kirk 1978).

Phytophagous insects are often attracted by wounding of potential hosts (Dicke and Hilker 2003; Howe 2004). *Sirex noctilio* females are attracted immediately to trees that have been felled or injured by events such as fire, windthrow, or lightning, and the attraction seems primarily motivated by monoterpenes and possibly other volatiles released from injured tree stems (Madden 1968). Trapping methods to monitor *S. noctilio* populations take advantage of this phenomenon, and the currently recommended lure to attract female wasps to artificial traps is a 70:30 mixture of α- and ß-pinene, respectively (Dodds 2006, Chaps. 11, 12). However, evidence continues to suggest that additional volatile compounds may be involved in signaling since girdled and/or herbicide-treated trap trees (Madden 1971; Morgan and Stewart 1972; Neumann et al. 1982; Simpson 1976; Simpson and McQuilkin 1976) nearly always outperform artificial traps baited with the recommended mixed-pinene lure.

Presumably, forest pests that reduce tree vigor and suppress growth, such as defoliating insects, bark beetles, or root nematodes, would make pines more susceptible to attack by *S. noctilio*. Certainly pines that have been weakened from prior attacks by *S. noctilio* are more attractive and susceptible to subsequent Sirex attacks, and this is the primary means by which once-healthy and dominant trees are eventually killed during large Sirex outbreaks (Madden 1968; Morgan 1968; Rawlings 1948; Vaartaja and King 1964). Although synergisms between Sirex and other pests have not received much attention, it was reported that trees infected by *Heterobasidion annosum* (Fr.) Bref. and consequently suffering annosus root rot were weakened and vulnerable to Sirex attack (Spradbery and Kirk 1978). In contrast, there is some evidence that blue-stain fungi and *A. areolatum* are mutually antagonistic to the growth of one another (Coutts 1965; Coutts and Dolezal 1965b, Chap. 18, BP Hurley, unpublished).

Once a *S. noctilio* female selects a promising tree for oviposition, she drills a series of test bores and, using sensillae in her ovipositor, makes a determination of whether to deposit eggs based on the perceived moisture content and resin pressure of the wood (Coutts 1965; Coutts and Dolezal 1965b; Rawlings 1948). Drought stress typically reduces resin pressure in conifers, and it can also affect resin composition and subsequent resistance to insect pests (Dunn and Lorio 1993; Lombardero et al. 2000; Turtola et al. 2003). When tree phloem sap is under high osmotic pressure, woodwasp females reject the tree (Madden 1974), ostensibly because the resin, which is lethal to Sirex eggs, is more likely to flow into and fill egg galleries in such trees (Coutts and Dolezal 1966a). Small, abundant resin droplets on tree stems (Fig. 3.1c) are often the first symptom noticed in pines under attack from *S. noctilio* (Dodds et al. 2007; Rawlings 1948; Titze and Stahl 1970). Streaming of resin on the trunks of newly attacked healthy trees has been described as a "thin dribble" (Gilmour 1965) and was correlated with resistance, while beading of resin as single droplets was more often the case on trees that were likely to succumb (Coutts 1965). Helpful illustrations of resin beading and streaming are to be found in the report by Titze and Stahl (1970).

As trees weaken and resin pressures drop due to multiple woodwasp attacks, females drill fewer single-bore test holes and more multi-bore galleries into which they lay their eggs (Madden 1988; Rawlings 1948). The galleries typically comprise 2–4 tunnels (0.5 mm diameter) drilled into the sapwood 5–9 mm below the cambium (Coutts and Dolezal 1965a, 1969; Zondag and Nuttall 1977). All of the tunnels drilled by Sirex females receive mucus and inoculation with arthrospores of *A. areolatum,* but fewer than half the tunnels receive an egg (Coutts and Dolezal 1965a; Coutts and Dolezal 1966a, 1969). Although resin is produced constitutively in conifer tissues, it may also be induced by exposure to fungal pathogens (Michelozzi 1999). Thus, resinosis was induced in *P. radiata* by manual inoculation with *A. areolatum* from purified cultures (Titze and Stahl 1970).

The drilling of tunnels through the inner bark (phloem) and into the sapwood (xylem) during oviposition constitutes a wounding event that elicits responses from pine sharing many of the attributes common to wound responses in other plants (DeBruxelles and Roberts 2001; Maffei et al. 2007; Wasternack 2007). When bark above oviposition holes is removed, darkly stained, necrotic lesions are sometimes noted in the cambium (Rawlings 1948) (Fig. 3.2). More commonly, a vertically elongated lens-shaped area of wood stained reddish-brown discoloration quickly becomes apparent around the site of oviposition (Eldridge and Taylor 1989). The discoloration likely stems from polyphenolic compounds induced in response to

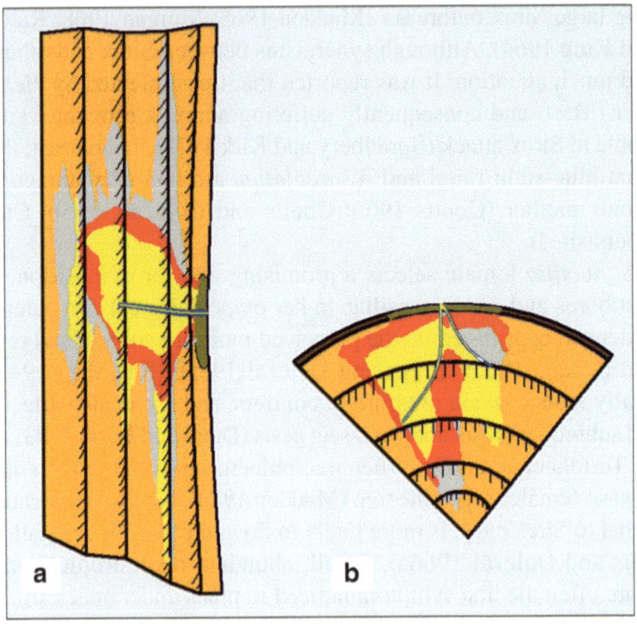

Fig. 3.2 Extent of tree responses to *Sirex noctilio* attack, modeled on figures by Coutts and Dolezal (1966a). *Panel a*, radial section of a resistant tree 3 months after woodwasp oviposition. Fungal drying zones (*blue*), tree-produced polyphenol staining (*red*), resinosis (*yellow*), dying cambium and phloem (*brown*) zones highlighted with color. Tunnels drilled by the wasp for oviposition shown in green, but should be understood to be resin-flooded. *Panel b*, transverse section of a resistant tree 6 weeks after oviposition. Colors are the same as for A. Egg is shown in the shorter of the two tunnel; in a resistant tree, both tunnels would be filled with resin. Oviposition drills are exaggerated in size

wounding or the fungal pathogen, but may also be associated with a polyphenoloxidase activity in the wasp mucus (Coutts 1969b) (Fig. 3.2).

The production of resin and polyphenolic compounds associated with resistance to *S. noctilio* was shown to vary between dominant and suppressed trees (Coutts 1968; Coutts and Dolezal 1966a). In response to oviposition by sawflies in its needles, Scots pine alters the expression of sesquiterpene synthase genes so as to release volatile compounds that attract sawfly parasitoids (Kopke et al. 2008). Recent studies have shown heritable variation in the quantity and quality of secondary compounds produced by pines and other conifers responding to wounding and various pests (Harju et al. 2009; Heijari et al. 2008; Roberds et al. 2003; Rosner and Hannrup 2004). It is conceivable, then, that one means of breeding trees for increased resistance to *S. noctilio* attack would be to select for trees that respond by emitting volatiles that are attractive to *Ibalia, Rhyssa*, and other parasitoid wasps that prey on Siricids. However, our current knowledge of parasitoid attractants is limited, and it remains to be seen whether Scots or other pine species react to *S. noctilio* attack by producing volatile compounds that would attract parasitoid species.

Ethylene is a gaseous hormone that plays a wide variety of roles in plants, including regulation plant defense responses, such as the production of chitinase and the lignification of plant cell walls (Ecker and Davis 1987; Walters and Heil 2007). Dominant *P. radiata* trees attacked by *S. noctilio* were able to increase production of ethylene by more than an order of magnitude compared to suppressed trees, and production of ethylene was more than three times greater in trees attacked by Sirex than in control trees damaged mechanically (Shain and Hillis 1972). Recent studies make clear that ethylene produced in response to wounding or pathogens leads to increased production of resin ducts in conifers and this consequently enhances resin flow, which is the primary line of defense against insects (Nagy et al. 2006; Phillips et al. 2006). It seems likely that the higher levels of ethylene produced in dominant trees may be responsible, at least in part, for the increased resin flows that are also associated with the resistance of pines to attack by *S. noctilio*.

3.3 Tree Responses to *Sirex noctilio* Mucus

In an important series of studies fundamental to our current understanding of the interaction between *S. noctilio* and its hosts, Coutts (Coutts 1968, 1969b, c) demonstrated that the mucus and the fungus injected into the tree during oviposition were responsible for different symptoms in the overall pathology. Neither the mucus nor the fungus alone was sufficient to kill trees, but the combination was lethal. Thus, the mucus was characterized as a 'conditioning' agent necessary for establishing an active and overwhelming infection of the host by *A. areolatum* (Coutts 1969b). It was also noted that in order for it to cause wilting and senescence of needles in the crown of trees, among the most obvious early symptoms of Sirex attack, the mucus must contain elements that can quickly and easily diffuse through the transpirational stream.

The female *S. noctilio* woodwasp produces mucus in an accessory gland that is a cognate to the acid or venom glands found in stinging wasps and bees (Aculeata). The mucus, a clear, sticky and highly viscous material, is stored in a reservoir attached through ducts to the woodwasp reproductive tract near the base of the ovipositor (Boros 1968). An early attempt at biochemical characterization identified non-sulfated acid mucopolysaccharides as a major component of Sirex mucus (Wong and Crowden 1976). The study also reported amylase, esterase, protease and phenoloxidase activities in the mucus. Experiments to isolate and characterize the diffusible active factor(s) responsible for needle wilt and senescence noted that a heat-stable (autoclavable) component dissociated from the mucus in water, initially with an apparent molecular mass of 60–100 kD. This material subsequently dissociated into smaller 2–6 kD subunits with no reduction in physiological activity. These characteristics are consistent with a small, bioactive peptide, but this has not been tested using appropriate procedures, such as protease digestion or exposure to amino acid-modifying agents.

Arthropod venoms are a rich source of biologically active peptides that have provided promising leads for a variety of biotechnological applications

(Pimenta and DeLima 2005). Mastoparan, an antimicrobial peptide commonly found in stinging wasp venoms, is widely utilized to study trimeric G-protein signaling in animal cells (Tomita et al. 1991), however, mastoparan has also been shown to activate mitogen-activated protein (MAP) kinase signaling cascades in plants (Miles et al. 2004). Among other plant responses, MAP kinase cascades have been implicated in the senescence of various plant tissues (Guo and Gan 2005). Whether mastoparan or a related peptide is present in *S. noctilio* mucus where it might play a role in inducing pine needle senescence is as yet unknown. Kuhn-Nentwig (2003) has described how the mélange of peptides, enzymes and low molecular weight compounds contribute to the dual function of arthropod venoms as both antimicrobial agents and toxins that alter host physiology (Kuhn-Nentwig 2003). Kellner (2002) has noted the importance of antimicrobial agents for protecting insect eggs against pathogens, which suggests that examination of Sirex mucus for antimicrobial activities might also be a productive line of inquiry (Kellner 2002). It seems a reasonable evolutionary progression that antimicrobial mucus, originally deposited with eggs to protect them from pathogens, might undergo natural selection for phytotoxic activities that could contribute to the establishment of a fungal food source required by the insect larvae.

Chlorosis or yellowing, especially in older needles, is one of the first visible symptoms of *S. noctilio* attack, and is noticeable on pines as soon as 2 weeks following attack (Rawlings 1948). This effect is inducible in young seedling explants exposed to aqueous solutions of the venom. Premature needle senescence and loss is also common, and these responses are readily reproduced experimentally (Coutts 1968, 1969b). Another early and distinctive symptom of attack, especially noticeable in the crowns of affected trees, is flagging or drooping of young needles at the fascicle sheath (Coutts 1968, 1969b, c). (It should be noted that needle flagging may not be a useful character to assess infection in all pine species as attempts by the authors to reproduce this symptom in a variety other pine species yielded inconsistent results. Field observations of infected red, white and Scots pines in New York also failed to detect noticeable needle flagging). Depending on their age, drooping needles may be retained on the tree for some time, sometimes turning red before falling off (Coutts 1965; Rawlings 1948). These changes in needle color represent distinct physiological changes in the plant. For example, variations in green color typically indicate damage to the photosynthetic machinery, while red or brown suggests accumulation of flavonoids (Gunthardt-Goerg and Vollenweider 2007). However, there have not been any studies to define the specific change in needles of Sirex-attacked trees at the biochemical or molecular levels. At the microscopic level, though, non-lignified cells, including phloem and ray parenchyma cells, in *P. radiata* needle segments were shown to collapse and became necrotic within 7 days of treatment with *S. noctilio* mucus (Fong and Crowden 1973). Such cellular collapse would effectively disrupt transport of carbohydrates and nutrients from the needles to other tissues in the tree.

The effect of *S. noctilio* mucus on needles was replicated in excised pine twigs by several researchers who used it as the basis for bioassays of mucus activity (Fig. 3.3). Thus, genetic and physiological susceptibility of *P. radiata* was gauged

Fig. 3.3 Bioassay of *S. noctilio* venom activity using 4 month-old *P. radiata* seedlings, cut at soil line, 7 days after exposure. *Panel A*, water control. *Panel B*, aqueous solution of venom

by needle flagging and the appearance of chlorosis in detached shoots standing in aqueous solutions of 0.05–0.1 mg/ml mucus (Bowling and Dolezal 1970; Coutts 1969b). Such assays were considered more sensitive than tests performed on live trees, and were used to demonstrate that the mucus from *S. noctilio* was more toxic to *P. radiata* than mucus from several other species of woodwasp (Spradbery 1973). Respiration, as well as changes in amylase and peroxidase activities, was studied in mucus-treated branch tips (Fong and Crowden 1973). These formed the basis of bioassays used to follow activity during biochemical fractionation of the mucus (Wong and Crowden 1976).

Water relations in plants are typically maintained by a delicately balanced process of solute loading and unloading between xylem and phloem tissues (DeBoer and Volkov 2003). Models have suggested that transpiration is particularly sensitive to reduced sugar loading in photosynthetic tissues at the top of the phloem stream (Holtta et al. 2006). *Sirex noctilio* mucus has a rapid effect on water relations in attacked pines, causing water tension in needles to drop in a manner similar to girdling (Madden 1977). Through unknown mechanisms, translocation was inhibited, respiration increased, and both osmotic and turgor pressure were reduced (Madden and Coutts 1979). Declines in osmotic pressure were also reported by Iede and Zanetti (2007), but increased leaf water pressure (defined as total water potential) was noted in trees showing needle chlorosis (Spradbery 1973). Changes in water movement through the transpiration stream following *S. noctilio* attack were assessed by measuring internal tree temperature (Jamieson 1957). *Pinus radiata* trees attacked by *S. noctilio* had internal temperatures 1–5°C above air temperature, whereas healthy control trees maintained internal temperatures 2–5°C below air temperature. This result was attributed to vigorous transpiration in healthy trees in contrast to a "fevered" response due to ineffective transpiration in attacked trees.

A general pattern of starch redistribution, in which needle starch increases while bark and stem starch decreases, has been reported for trees attacked by *S. noctilio* (Gilbertson 1984). Needles began to accumulate starch within 3–4 days of attack, and by 2 weeks, starch grains were abundant in the needles of all *P. radiata* trees attacked by *S. noctilio* (Coutts 1968, 1969b). Treatment with fresh or autoclaved mucus alone reduced the starch in stem samples measured at 4 weeks (Coutts 1969b). Accumulation of starch in needles might fit with reduced sugar loading of the phloem in the model of Holtta et al. (2006). Respiration in *P. radiata* bark tissues, measured as carbon dioxide release, increased significantly within 2–5 days of attack by caged *S. noctilio* females, and was sustained for at least 16 days following the attack (Madden 1977). This increased respiration parallels the reported decline in starch content in stem tissue (Coutts 1969b). Respiration, measured as an increase in oxygen uptake, was shown to peak 10–15 days following treatment of excised stems with *S. noctilio* mucus (Coutts 1970). The quality of respiration was also reported to change in response to mucus, with an increased ratio of carbon dioxide released to oxygen taken up (Fong and Crowden 1973).

Oleoresins are a primary defense against insects and pathogens in conifers (Gijzen et al. 1993; Keeling and Bohlmann 2006), and total resin flow is the measure most frequently correlated with insect resistance (Tisdale et al. 2003). Oleoresin exudation pressure is generally sub-normal in trees that have been attacked by Sirex, but the observations vary with individual trees (Madden 1977). Ovipositing *S. noctilio* females frequently drill multiple tunnels, but tunnels containing eggs remain uniformly free of resin, while tunnels that do not contain eggs are often flooded by resin within 2 or 3 days (Coutts and Dolezal 1969). Reduced resin flooding in tunnels containing eggs was ascribed in most cases to the presence of mucus (Coutts and Dolezal 1965a, 1966a). However, Coutts and Dolezal (1969) mentioned a different substance in egg-containing tunnels that stained intensely with cotton blue (presumably aniline blue used in the lactophenol cotton blue staining technique to detect fungi) and dried with a crazed appearance. There are no corroborating observations of this blue-staining material.

Peroxidase and amylase activities were shown to increase in *P. radiata* needles treated with *S. noctilio* mucus (Fong and Crowden 1973). Increased amylase activity suggests potential mobilization of stored carbohydrate from starch reserves, while peroxidase induction may reflect activation of defensive responses (Kawano 2003). As previously noted, plant cells respond to the G protein activator, mastoparan (Cho et al. 1995; Legendre et al. 1993), and the signaling cascades regulated by G proteins control the production of many of the secondary metabolites involved in plant defense responses (Zhao et al. 2005). If a mastoparan-like peptide is present in *S. noctilio* mucus, it could be a primary inductive agent for such defense responses.

Reductions in stem growth rates were reported as a longer term response of pines to *S. noctilio* attack (Coutts 1968; Coutts and Dolezal 1965a; Gilmour 1965; Iede and Zanetti 2007; Madden 1975). Such growth reductions are to be expected considering the degree of damage incurred during oviposition by the vascular cambium, the meristematic tissue responsible for all radial growth in tree stems (Barbosa and Wagner 1989). Radial growth cessation in *P. radiata* attacked by *S. noctilio* is detectable between 2 and 8 weeks post-attack, but the effect is not permanent in resistant trees, and growth may resume within several weeks to several months (Coutts and Dolezal 1965a; Spradbery 1973). Among seven woodwasps tested, only *S. noctilio* mucus was found to reduce radial growth in *P. radiata* stems (Spradbery 1973). Checking of tree growth was reported for *P. patula* attacked by Sirex (Iede and Zanetti 2007), and checking of growth was correlated with an increased attractiveness for Sirex woodwasps (Madden 1968). Tree mortality from severe *S. noctilio* attacks typically becomes evident within several months, but trees surviving an unsuccessful attack continue to show measurable decreases in growth relative to trees not attacked (Madden 1975).

Several early studies of the effects of *S. noctilio* mucus on pines noted differences in the responses of individual trees (Bowling and Dolezal 1970; Coutts 1969b; Fong and Crowden 1973; Wong and Crowden 1976). However, whereas some of these studies specifically noted that the more responsive trees were suppressed (Coutts 1969b), from which it might be inferred that the variation in sensitivity was a function of environment, others studies suggested that certain trees were genetically predisposed to susceptibility (Fong and Crowden 1973; Wong and Crowden 1976). Spradbery performed one of the few studies that examined the effects of *S. noctilio* mucus across a broad range of *Pinus* species, noting more than a four-fold difference in the rate of response, measured as increased leaf pressure (stomatal closure) over time, across 29 species of conifer. With respect to mucus sensitivity, *P. radiata* was found to be intermediate to such species as *P. nigra* and *P. mugo* (sensitive) and *P. sylvestris* and *P. banksiana* (insensitive) (Spradbery 1973). Perhaps most interestingly, this study reported differences of nearly three-fold in the responsiveness between varieties of the same species, e.g., *P. mugo* versus *P. mugo* var. *rostrata* and *P. nigra* var. *nigra* versus *P. nigra* var. *caramanica* (Spradbery 1973). To the extent that effects of *S. noctilio* mucus are critical to the establishment of productive *A. areolatum* infections, this evidence strongly suggests that an opportunity exists for breeding genetic resistance to *S. noctilio* mucus. Such resistance should, in turn, reduce infectivity by *A. areolatum* and enhance overall resistance of trees to attack from *S. noctilio*.

3.4 Tree Responses to *Amylostereum areolatum*

On its own, *A. areolatum* cannot be considered a particularly aggressive pathogen given the multiple observations of recovery in trees inoculated solely with the fungus (Coutts 1969c; Madden 1971; Titze and Stahl 1970). Woodwasp mucus clearly must alter conifer physiology and defense responses in ways that enable the fungus to gain the upper hand and eventually overwhelm the tree. This is not to say, however, that the fungus does not produce specific factors of its own to facilitate the infection.

Mycelial growth of *A. areolatum* in pines attacked by *S. noctilio* is preceded by a zone of drying plant cells (Coutts and Dolezal 1965b), but the mechanism for this drying effect is unknown. Drying of the wood tissues appears necessary for the infection as optimal moisture content for growth of the fungus is ~70% saturation, far lower than the typical conditions inside a tree (Coutts and Dolezal 1965a; King 1966).

Drying of the tissues precedes mycelial invasion as *A. areolatum* cannot be cultured from either the drying zone or from tissues just beyond it (Coutts and Dolezal 1966a; Kile and Turnbull 1974) (Fig. 3.2). Some experimental evidence suggests that *S. noctilio* mucus and the *A. areolatum* fungus are both involved in the drying phenomenon (Kile and Turnbull 1974). However, drying zones are also observed in conifers infected with the white-rot fungal pathogen, *Heterobasidion annosum,* which is closely related to *A. areolatum* (Coutts 1976; Maijala et al. 2003). *Heterobasidion annosum* produces several low molecular weight compounds that are toxic to a wide variety of conifers (Bassett et al. 1967; Sonnenbichler et al. 1989). Experiments using freshly-cut sections of *P. radiata* stem revealed that ray parenchyma cells were killed in advance of mycelial growth, suggesting that a diffusible agent of fungal origin might be responsible (King 1966). However, it remains to be shown whether toxins of the type produced by *H. annosum* are produced by *A. areolatum.*

Inoculation of *P. taeda* seedlings with *A. areolatum* led to death and collapse of the needles, which subsequently turned gray-green in color and were kinked along the portion of the needle that was still undergoing extension at the time of treatment (Bordeaux 2008). Gray-green coloration has previously been described for *P. radiata* needles, but in those observations it was unclear whether the color change resulted from the combined effect of the fungus and the mucus (Coutts 1969c; Rawlings 1948). Intriguingly, needles in non-resistant trees yellowed and senesced quickly, while in resistant trees the younger needles collapsed, turning from green to gray in color (Coutts 1969c). This color change in young needles of resistant trees is reminiscent of the hypersensitive response sometimes seen in the leaves of other plants. Evidence for hypersensitivity in pines is beginning to accumulate as major resistance genes for fungal pathogens are identified (Kubisiak et al. 2005; Liu and Ekramoddoullah 2008). Whether or not a true hypersensitive response mechanism, premised on gene-for-gene interactions, can be found in the *Pinus/Sirex/Amylostereum* pathosystem remains to be seen (Hammond-Kosack and Jones 1997; Mur et al. 2008).

Drying zones suggest that cell death is taking place ahead of the mycelial front, which is to be expected given the necrotrophic lifestyle of *A. areolatum*, but the exact mechanism the fungus uses to cause cell death remains unclear. It may be that the fungus stimulates pathways related to the tree's own hypersensitive response system, and then takes advantage of the nutrients released by collapsing cells (Levine et al. 1996a; Mayer et al. 2001). Such a system has been described for at least one other necrotrophic fungus that feeds on a woody species (Govrin and Levine 2000). Drying zones might also suggest the involvement of other cellular pathways, such as those that bring about programmed cell death (PCD) (Khurana et al. 2005; Levine et al. 1996b). PCD-type events are presumed to be normal in the development of xylem progenitor cells into water-conductive xylem and wood; however, they must be confirmed by specific tests that distinguish them from other forms of cell death in plants (Reape and McCabe 2008).

The importance of polyphenolic compounds in the resistance of pines to *S. noctilio* attack remains uncertain. *In situ* production of polyphenols at the site of oviposition was described in multiple reports as a pine response to either mechanical wounding or to entry of the fungus (Coutts 1970; Coutts and Dolezal 1966a; Hart 1981; Hillis and Inoue 1968). However, not all experiments detected differences in polyphenol production between controls and wood infected with *A. areolatum* (Coutts 1969a). Polyphenols (visualized using benzidine staining) were produced in the sapwood of infected trees, along the inside edge of the drying zone that precedes advance of the *A. areolatum* mycelium (Coutts and Dolezal 1966a). Accumulation of polyphenolic materials was detectable within 2 weeks of introduction of the fungus, and in resistant trees, accumulation of such materials continued for months after infiltration with the fungus had been stopped (Coutts and Dolezal 1966a). Polyphenolic compounds produced by the tree and deposited at the wound or infection site were fungistatic or fungicidal (Coutts and Dolezal 1969; Hillis and Inoue 1968; Himejima et al. 1992).

Several of the polyphenolic compounds deposited in Sirex-attacked wood were purified and identified as stilbenes (14-carbon diphenylethylenes) or phytoalexins (primarily flavonoids) (Hart 1981). Pinosylvin, a stilbene toxic to *A. areolatum* and many other fungi, was the most abundant polyphenolic compound in the infected zones showing a response (Coutts 1969a; Hillis and Inoue 1968). It may be worth noting that pinosylvin has also been shown to accumulate in pine stems treated with the *H. annosum* toxin, fomannosin (Bassett et al. 1967). The flavonoids, pinobanksin and pinocembrin, were reportedly decreased in the heartwood and knotwood of Sirex-attacked trees (Hillis and Inoue 1968).

There are several intriguing observations in the literature linking enhanced growth of *A. areolatum* to the availability of (undefined) fatty acids from various sources. Titze (1965) noted that histological analyses of sectioned bud tissues from the suppressed trees susceptible to Sirex attack always showed abundant lipid inclusions in the pith compared to dominant, Sirex-resistant trees (Titze 1965). Triglycerides, fatty acids and resin acids produced by pines are a nutritional resource favored by a variety of wood-inhabiting fungi and yeasts, including various *Ophiostoma* (blue-stain) species that are transmitted by bark beetles (Gao et al. 1994). Because they are good competitors and readily establish themselves on fresh conifer wood chips where they preferentially metabolize the fatty acids and resins, albino mutants of *Ophiostoma piliferum* are used as biocontrol agents against blue-stain fungi in pulp mill chip piles (Blanchette et al. 1992). Surprisingly, there has so far been little effort to follow up on early observations and examine the capacity of *A. areolatum* to metabolize the lipophilic extractives in pine wood. Observations that material (fatty acids?) entering the woodwasp reproductive tract from the Dufour's (oil) gland can stimulate growth of *A. areolatum* in culture remain to be further tested (Boros 1968).

3.5 General Responses to *Sirex noctilio* Attack

Plants use a wide variety of mechanisms to deal with pests and disease. Systemic acquired resistance (SAR), induced systemic resistance (ISR), and hypersensitive response (HR) are all possible response paths to which signals from injury or insult may be routed (Walters and Heil 2007). As mentioned previously, *S. noctilio* oviposition induces higher levels of ethylene production in resistant *P. radiata* trees. This suggests that ISR may, in fact, be operative in this pathosystem. However, induced resistance should offer trees enhanced protection from further attacks after an initial encounter with a pest or pathogen, but this has not been the pattern reported for *S. noctilio* attack. While some trees offer far more resistance to Sirex than others, a previously attacked tree is often weakened, and stress-induced increases in the production of α- and β-pinene increase the chances for subsequent attack (Coutts and Dolezal 1966b; Madden 1968, 1988).

Ab initio production of polyphenolic compounds occurs at sites where *A. areolatum* is introduced into trees, suggesting that pathogen challenge induces host tree activation of the phenylpropanoid pathway in part of the defense response (Madden 1988). Since mechanical wounding alone does not elicit the same results, the host tree must have a recognition system for the fungus. What is unclear is whether this system recognizes factors specific to *A. areolatum* or is just part of a general recognition system for intruding fungi. Fungal elicitors have been shown to induce expression of stilbene synthase genes in Scots pine as part of a general defense response (Preisig-Mueller et al. 1999). Suspension-cultured cells and seedlings of loblolly and slash pines expressed pathogen-response (PR) genes when exposed to the necrotrophic fungal pathogen *Fusarium circinatum* (Davis et al. 2002). These genes also responded to the plant defense signaling molecules, salicylic acid and jasmonic acid. Plant resistance gene homologs have been identified in eastern white pine (*P. strobus*) responding to the blister rust pathogen *Cronartium ribicola* (Smith et al. 2006). There is no evidence for specific resistance or avirulence genes in host pines or *A. areolatum*. However, dwelling on this would miss the point that without the mucus produced by *S. noctilio*, *A. areolatum* would fail as a pathogen.

3.6 Reducing Tree Susceptibility

Silvicultural practices that promote the health of individual trees (spacing in the canopy, adequate soil conditioning, removal of unthrifty individuals) clearly promote resistance to attack by *S. noctilio* (Madden 1968). In particular, the combination of overcrowding and drought leaves trees highly vulnerable to Sirex outbreaks (Gilmour 1965). Thinning of stands, removal of smaller suppressed trees, and routine monitoring of stands for evidence of woodwasp attack currently comprise the first line of defense to minimize losses to *S. noctilio* (Dodds et al. 2007). Care should also be taken to avoid mechanical injury to trees during thinning and other silvicultural operations in established stands as wounding is a source of attractants for *S. noctilio* females.

Although evidence in the literature suggests that there might be a genetic basis for resistance to *S. noctilio* in pine, conflicting observations make it unclear whether a breeding program could establish robust resistance to Sirex. Many of the pine responses to *A. areolatum* (e.g., increased production of polyphenols, resin, and ethylene), as well as response to environmental factors predisposing trees to attack (e.g., drought tolerance and in-stand competition), are under a degree of genetic control. Some of these responses have been selected for in tree improvement programs looking to enhance resistance to other fungal pathogens. However, the possibilities for breeding resistance to *S. noctilio* mucus have not yet been examined, most likely because the mucus is in limited supply and the specific active factor(s) are unknown. Current techniques in biochemistry, molecular genetics and genomics provide an opportunity unavailable to researchers in the past to probe in this direction. Given the utility of unexpected results from studies of venoms from Aculeata wasp species, it seems almost certain that detailed studies of *S. noctilio* will yield unique and useful tools for breeding resistance, as well as probing the physiology and metabolism in cells from conifers and likely other plant species. The recent announcement of complete genome sequences for three *Nasonia* species, all parasitoid wasps from the Aculeata (Werren et al. 2010) will no doubt provide insights into aspects of *S. noctilio* biology that will be useful in minimizing this insect's impact on pines.

Acknowledgements The authors acknowledge support from the Georgia Traditional Industries Program for Forest Resources and Products (TIP-FRP) and the USDA/CSREES McIntire-Stennis Program (GEOZ-0154-MS).

References

Barbosa P, Wagner MR (1989) Introduction to forest and shade tree insects. Academic, San Diego

Bassett C, Sherwood RT, Kepler JA, Hamilton PB (1967) Production and biological activity of fomannosin, a toxic sesquiterpene metabolite of *Fomes annosus*. Phytopathol 57:1046–1052

Blanchette RA, Farrell RL, Burnes TA, Wendler PA, Zimmerman W, Brush TS, Snyder RA (1992) Biological control of pitch in pulp and paper-production by *Ophiostoma piliferum*. TAPPI J 75:102–106

Borchert DF, Fowler G, Jackson L (2007a) Organism pest risk analysis: risk to the conterminous United States associated with the woodwasp, *Sirex noctilio* Fabricius, and the symbiotic fungus, *Amylostereum areolatum* (Fries: Fries) Boidin. USDA-APHIS, Raleigh, MD

Borchert DF, Fowler G, Jackson L (2007b) Proposed program for management of the woodwasp *Sirex noctilio* Fabricius (Hymenoptera: Siricidae), Environmental assessment. USDA-APHIS, Riverdale, MD

Bordeaux JM (2008) Characterization of growth conditions for production of a laccase-like phenoloxidase by *Amylostereum areolatum*, a fungal pathogen of pines and other conifers. In: Forest and natural resources. University of Georgia, Athens

Boros CB (1968) The relationship between the woodwasp *Sirex noctilio* F. and the wood-rot fungus *Amylostereum* sp.M Agric Sci thesis. University of Adelaide, Adelaide

Bowling JP, Dolezal JE (1970) Initial note on the development of a mucus test to determine resistance of *Pinus radiata* to attack by *Sirex noctilio*. Aust For Res 5:57–62

Cho MH, Tan Z, Erneux C, Shears SB, Boss WF (1995) The effects of mastoparan on the carrot cell plasma membrane polyphosphoinositide phospholipase-c. Plant Physiol 107:845–856

Ciesla WM (2003) European woodwasp: a potential threat to North Americas conifer forests. J For 101:18–23

Coutts MP (1965) *Sirex noctilio* and the physiology of *Pinus radiata*: some studies of the interactions between the insect, the fungus, and the tree in Tasmania. NDFaT Bureau, Forest Research Institute, Canberra

Coutts MP (1968) Rapid physiological change in *Pinus radiata* following attack by *Sirex noctilio* and its associated fungus, *Amylostereum* sp. Aust J Sci 30:275–277

Coutts MP (1969a) The formation of polyphenols in small blocks of *Pinus radiata* sapwood with and without the fungal symbiont of *Sirex*. Aust For Res 4:29–34

Coutts MP (1969b) Mechanism of pathogenicity of *Sirex noctilio* on *Pinus radiata*. 2. Effects of *S. noctilio* mucus. Aust J Biol Sci 22:1153–1161

Coutts MP (1969c) Mechanism of pathogenicity of *Sirex noctilio* on *Pinus radiata*. I. Effects of symbiotic fungus *Amylostereum* sp. (Thelophoraceae). Aust J Biol Sci 22:915–924

Coutts MP (1970) The physiological effects of the mucus secretion of *Sirex noctilio* on *Pinus radiata*. Aust For Res 4:23–26

Coutts MP (1976) Formation of dry zones in sapwood of conifers. 1. Induction of drying in standing trees and logs by *Fomes annosus* and extracts of infected wood. Eur J For Pathol 6:372–381

Coutts MP, Dolezal JE (1965a) *Sirex noctilio* and the physiology of *Pinus radiata*. Proc 38th A.N.Z.A.A.S. Congress, Hobart

Coutts MP, Dolezal JE (1965b) *Sirex noctilio*, its associated fungus, and some aspects of wood moisture content. Aust For Res 1:3–13

Coutts MP, Dolezal JE (1966a) Polyphenols and resin in the resistance mechanism of *Pinus radiata* attacked by the wood wasp, *Sirex noctilio*, and its associated fungus. NDFaT Bureau, Forest Research Institute, Canberra

Coutts MP, Dolezal JE (1966b) Some effects of bark cincturing on the physiology of *Pinus radiata*, and on *Sirex* attack. Aust For Res 2:17–28

Coutts MP, Dolezal JE (1969) Emplacement of fungal spores by woodwasp, *Sirex noctilio*, during oviposition. For Sci 15:412–416

Davis JM, Wu H, Cooke JEK, Reed JM, Luce KS, Michler CH (2002) Pathogen challenge, salicylic acid, and jasmonic acid regulate expression of chitinase gene homologs in pine. Mol Plant Microbe Interact 15:380–387

DeBoer AH, Volkov V (2003) Logistics of water and salt transport through the plant: structure and functioning of the xylem. Plant Cell Environ 26:87–101

DeBruxelles GL, Roberts MR (2001) Signals regulating multiple responses to wounding and herbivores. Crit Rev Plant Sci 20:487–521

Dicke M, Hilker M (2003) Induced plant defences: from molecular biology to evolutionary ecology. Basic Appl Ecol 4:3–14

Dodds KJ (2006) *Sirex noctilio* trapping protocols. U.S. Forest Service, Durham, NH

Dodds KJ, Cooke RR, Gilmore DW (2007) Silvicultural options to reduce pine susceptibility to attack by a newly detected invasive species, *Sirex noctilio*. North J Appl For 24:165–167

Dunn JP, Lorio PL (1993) Modified water regimes affect photosynthesis, xylem water potential, cambial growth, and resistance of juvenile *Pinus taeda* to *Dendroctonus frontalis* (Coleoptera: Scolytidae). Environ Entomol 22:948–957

Ecker JR, Davis RW (1987) Plant defense genes are regulated by ethylene. Proc Natl Acad Sci USA 84:5202–5206

Eldridge RH, Taylor EE (1989) Sirex woodwasp – a pest of pine. Forestry Commission of New South Wales, Forest Protection Ser. 1, Sydney

Fong LK, Crowden RK (1973) Physiological effects of mucus from wood wasp, *Sirex noctilio* F., on foliage of *Pinus radiata* D. Don. Aust J Biol Sci 26:365–378

Gao Y, Breuil C, Chen T (1994) Utilization of triglycerides, fatty-acids and resin acids in lodgepole pine wood by a sapstaining fungus *Ophiostoma piceae*. Mat Org 28(2):105–118

Gauld ID, Collins NM, Fitton MG (1990) The biological significance and conservation of Hymenoptera in Europe. Council of Europe, Strasbourg

Gijzen M, Lewinsohn E, Savage TJ, Croteau RB (1993) Conifer monoterpenes – biochemistry and bark beetle chemical ecology. In: Bioactive volatile compounds from plants. ACS symposium series 525:8–22

Gilbertson RL (1984) Relationships between insects and wood-rotting basidiomycetes. In: Wheeler Q, Blackwell M (eds) Fungus-insect relationships: perspectives in ecology and evolution. Columbia University Press, New York, pp 130–165

Gilmour JW (1965) The life cycle of the fungal symbiont of *Sirex noctilio*. NZ J For 10:80–89

Govrin EM, Levine A (2000) The hypersensitive response facilitates plant infection by the necrotrophic pathogen *Botrytis cinerea*. Curr Biol 10:751–757

Gunthardt-Goerg MS, Vollenweider P (2007) Linking stress with macroscopic and microscopic leaf response in trees: new diagnostic perspectives. Environ Pollut 147:467–488

Guo YF, Gan SS (2005) Leaf senescence: signals, execution, and regulation. Curr Top Dev Biol 71:83–112

Hammond-Kosack KE, Jones JDG (1997) Plant disease resistance genes. Annu Rev Plant Physiol Plant Mol Biol 48:575–607

Harju AM, Venalainen M, Laakso T, Saranpaa P (2009) Wounding response in xylem of Scots pine seedlings shows wide genetic variation and connection with the constitutive defence of heartwood. Tree Physiol 29:19–25

Hart JH (1981) Role of phytostilbenes in decay and disease resistance. Annu Rev Phytopathol 19:437–458

Heijari J, Nerg AM, Kainulainen P, Vuorinen M, Holopainen JK (2008) Long-term effects of exogenous methyl jasmonate application on Scots pine (*Pinus sylvestris*) needle chemical defence and diprionid sawfly performance. Entomol Exp Appl 128:162–171

Hillis WE, Inoue T (1968) Formation of polyphenols in trees. 4. Polyphenols formed in *Pinus radiata* after Sirex attack. Phytochemistry 7:13–22

Himejima M, Hobson KR, Otsuka T, Wood DL, Kubo I (1992) Antimicrobial terpenes from oleoresin of Ponderosa pine tree *Pinus ponderosa* - a defense-mechanism against microbial invasion. J Chem Ecol 18:1809–1818

Holtta T, Vesala T, Sevanto S, Peramak M, Nikinmaa E (2006) Modeling xylem and phloem water flows in trees according to cohesion theory and munch hypothesis. Trees Struct Funct 20:67–78

Howe GA (2004) Jasmonates as signals in the wound response. J Plant Growth Regul 23:223–237

Iede ET, Zanetti R (2007) Occurrence and management recommendations of *Sirex noctilio* Fabricius (Hymenoptera, Siricidae) on *Pinus patula* (Pinaceae) plantations in the state of Minas Gerais, Brazil. Rev Bras Entomol 51:529–531

Jamieson JPS (1957) Some observations on the relation of the internal temperature of *P. radiata* to infestation by *Sirex*. NZ J For VII:32–34

Kawano T (2003) Roles of the reactive oxygen species-generating peroxidase reactions in plant defense and growth induction. Plant Cell Rep 21:829–837

Keeling CI, Bohlmann J (2006) Diterpene resin acids in conifers. Phytochemistry Rev 67:2415–2423

Kellner RLL (2002) The role of microorganisms for eggs and progeny. In: Hilker M, Meiners T (eds) Chemoecology of insect eggs and egg deposition. Blackwell Publishing Ltd., Oxford

Khurana SMP, Pandey SK, Sarkar D, Chanemougasoundharam A (2005) Apoptosis in plant disease response: a close encounter of the pathogen kind. Curr Sci 88:740–752

Kile GA, Turnbull CRA (1974) Drying in the sapwood of radiata pine after inoculation with *Amylostereum areolatum* and *Sirex* mucus. Aust For Res 6:35–40

King JM (1966) Some aspects of biology of fungal symbiont of *Sirex noctilio*. Aust J Bot 14:25–30

Kopke D, Schroder R, Fischer HM, Gershenzon J, Hilker M, Schmidt A (2008) Does egg deposition by herbivorous pine sawflies affect transcription of sesquiterpene synthases in pine? Planta 228:427–438

Kubisiak TL, Amerson HV, Nelson CD (2005) Genetic interaction of the fusiform rust fungus with resistance gene FR1 in loblolly pine. Phytopathol 95:376–380

Kuhn-Nentwig L (2003) Antimicrobial and cytolytic peptides of venomous arthropods. Cell Mol Life Sci 60:2651–2668

Legendre L, Yueh YG, Crain R, Haddock N, Heinstein PF, Low PS (1993) Phospholipase-C activation during elicitation of the oxidative burst in plant cells. J Biol Chem 268:24559–24563

Levine A, Pennell RI, Alvarez ME, Palmer R, Lamb C (1996) Calcium-mediated apoptosis in a plant hypersensitive disease resistance response. Curr Biol 6:427–437

Liu JJ, Ekramoddoullah AKM (2008) Development of leucine-rich repeat polymorphism, amplified fragment length polymorphism, and sequence characterized amplified region markers to the *Cronartium ribicola* resistance gene CR2 in western white pine (*Pinus monticola*). Tree Gen Genome 4:601–610

Lombardero MJ, Ayres MP, Lorio PL, Ruel JJ (2000) Environmental effects on constitutive and inducible resin defences of *Pinus taeda*. Ecol Lett 3:329–339

Madden JL (1968) Physiological aspects of host tree favourability for the woodwasp, *Sirex noctilio* f. Proc Ecol Soc Aust 3:147–149

Madden JL (1971) Some treatments which render Monterey pine (*Pinus radiata*) attractive to the woodwasp *Sirex noctilio*. Bull Entomol Res 60:467–472

Madden JL (1974) Oviposition behaviour of the woodwasp, *Sirex noctilio* F. Aust J Zool 22:341–351

Madden JL (1975) Analysis of an outbreak of woodwasp, *Sirex noctilio* F. (Hymenoptera, siricidae), in *Pinus radiata*. Bull Entomol Res 65:491–500

Madden JL (1977) Physiological reactions of *Pinus radiata* to attack by woodwasp, *Sirex noctilio* F. (Hymenoptera: Siricidae). Bull Entomol Res 67:405–426

Madden JL (1988) *Sirex* in Australasia. In: Berryman AA (ed) Dynamics of forest insect populations. Plenum Publishing Corporation, New York, pp 407–429

Madden JL, Coutts, MP (1979) The role of fungi in the biology and ecology of woodwasps (Hymenoptera: Siricidae). In: Batra LR (ed) Insect-fungus symbiosis: nutrition, mutualism, and commensalism: proceedings of the symposium. Halsted Press, Montclair, pp 165–174

Maffei ME, Mithofer A, Boland W (2007) Insects feeding on plants: Rapid signals and responses preceding the induction of phytochemical release. Phytochemistry 68:2946–2959

Maijala P, Harrington TC, Raudaskoski M (2003) A peroxidase gene family and gene trees in *Heterobasidion* and related genera. Mycologia 95:209–221

Mayer AM, Staples RC, Gil-ad NL (2001) Mechanisms of survival of necrotrophic fungal plant pathogens in hosts expressing the hypersensitive response. Phytochemistry 58:33–41

Michelozzi M (1999) Defensive roles of terpenoid mixtures in conifers. Acta Bot Gallica 146:73–84

Miles GP, Samuel MA, Jones AM, Ellis B (2004) Mastoparan rapidly activates plant map kinase signaling independent of heterotrimeric G proteins. Plant Physiol Biochem 134: 1332–1336

Morgan FD (1968) Bionomics of Siricidae. Annu Rev Entomol 13:239–256

Morgan FD, Stewart NC (1972) Developing and testing a lure-trap for the woodwasp *Sirex noctilio* f. Aust For 36:38–46

Mur LAJ, Kenton P, Lloyd AJ, Ougham H, Prats E (2008) The hypersensitive response: The centenary is upon us but how much do we know? J Exp Bot 59:501–520

Nagy NE, Krokene P, Solheim H (2006) Anatomical-based defense responses of Scots pine (*Pinus sylvestris*) stems to two fungal pathogens. Tree Physiol 26:159–167

Neumann FG, Harris JA, Kassaby FY, Minko G (1982) An improved technique for early detection and control of the Sirex wood wasp in radiata pine plantations. Aust For 45:117–124

Phillips MA, Bohlmann J, Gershenzon J (2006) Molecular regulation of induced terpenoid biosynthesis in conifers. Phytochemistry Rev 5:179–189

Pimenta AMC, DeLima ME (2005) Small peptides, big world: biotechnological potential in neglected bioactive peptides from arthropod venoms. J Pept Sci 11:670–676

Preisig-Mueller R, Axel S, Ilka B, Hans-Jörg R, Helmut K (1999) Characterization of a pine multigene family containing elicitor-responsive stilbene synthase genes. Plant Mol Biol V39:221–229

Rabaglia R, Lewis J (2006) Economic analysis of the potential impact of *Sirex noctilio*, with emphasis on pines in the southeastern United States: Executive summary. USDA Forest Service, Arlington, pp 1–6

Rawlings GB (1948) Recent observations on the *Sirex noctilio* population in *Pinus radiata* forests in New Zealand. NZ J For 5:411–421
Reape TJ, McCabe PF (2008) Apoptotic-like programmed cell death in plants. New Phytol 180:13–26
Roberds JH, Strom BL, Hain FP, Gwaze DP, McKeand SE, Lott LH (2003) Estimates of genetic parameters for oleoresin and growth traits in juvenile loblolly pine. Can J For Res 33:2469–2476
Rosner S, Hannrup B (2004) Resin canal traits relevant for constitutive resistance of Norway spruce against bark beetles: environmental and genetic variability. For Ecol Manage 200:77–87
Shain L, Hillis WE (1972) Ethylene production in *Pinus radiata* in response to *Sirex-Amylostereum* attack. Phytopathol 62:1407–1409
Simpson R (1976) Bioassay of pine oil components as attractants for *Sirex noctilio* (Hymenoptera: Siricidae) using electroantennogram techniques. Entomol Exp Appl 19:11–18
Simpson R, McQuilkin R (1976) Identification of volatiles from felled *Pinus radiata* and the electroantennograms they elicit from *Sirex noctilio*. Entomol Exp Appl 19:205–213
Smith DR, Schiff NM (2002) A review of the siricid woodwasps and their Ibaliid parasitoids (Hymenoptera: Siricidae, Ibaliidae) in the eastern united states, with emphasis on the mid-atlantic region. Proc Entomol Soc Wash 104:174–194
Smith JA, Blanchette RA, Burnes TA, Jacobs JJ, Higgins L, Witthuhn BA, David AJ, Gillman JH (2006) Proteomic comparison of needles from blister rust-resistant and susceptible *Pinus strobus* seedlings reveals upregulation of putative disease resistance proteins. Mol Plant Microbe Interact 19:150–160
Sonnenbichler J, Bliestle IM, Peipp H, Holdenrieder O (1989) Secondary fungal metabolites and their biological activities I. Isolation of antibiotic compounds from cultures of *Heterobasidion annosum* synthesized in the presence of antagonistic fungi or host plant cells. Biol Chem Hoppe Seyler 370:1295–1303
Spradbery JP (1973) A comparative study of the phytotoxic effects of Siricid woodwasps on conifers. Ann Appl Biol 75:309–320
Spradbery JP, Kirk AA (1978) Aspects of the ecology of Siricid woodwasps (Hymenoptera: Siricidae) in Europe, North Africa and Turkey with special reference to the biological control of *Sirex noctilio* F. in Australia. Bull Entomol Res 68:341–359
Spradbery JP, Kirk AA (1981) Experimental studies on the responses of European Siricid woodwasps to host trees. Ann Appl Biol 98:179–185
Talbot PHB (1977) *Sirex-Amylostereum-Pinus* association. Annu Rev Phytopathol 15:41–54
Tisdale RA, Nebeker TE, Hodges JD (2003) The role of oleoresin flow in the induced response of loblolly pine to a Southern Pine Beetle associated fungus. Can J Bot 81:368–374
Titze JF (1965) Physiological suppression in *Pinus radiata* and its susceptibility to *Sirex*. Aust For Res 1:51–55
Titze JF, Stahl W (1970) Inoculations of *Pinus radiata* trees with *Amylostereum areolatum*, the symbiotic fungus of *Sirex noctilio*. Aust For Res 4:3–14
Tomita U, Inanobe A, Kobayashi I, Takahashi K, Ui M, Katada T (1991) Direct interactions of mastoparan and compound 48/80 with GTP-binding proteins. J Biochem, Tokyo 109:184–189
Turtola S, Manninen AM, Rikala R, Kainulainen P (2003) Drought stress alters the concentration of wood terpenoids in Scots pine and Norway spruce seedlings. J Chem Ecol 29:1981–1995
Vaartaja O, King J (1964) Inoculation experiments with *Amylostereum* sp. from a woodwasp. Plant Dis Rep 48:438–440
Walters D, Heil M (2007) Costs and trade-offs associated with induced resistance. Physiol Mol Plant Pathol 71:3–17
Wasternack C (2007) Jasmonates: an update on biosynthesis, signal transduction and action in plant stress response, growth and development. Ann Bot 100:681–697
Wermelinger B, Rigling A, Schneider-Mathis D, Dobbertin M (2008) Assessing the role of bark- and wood-boring insects in the decline of Scots Pine (*Pinus sylvestris*) in the Swiss Rhone valley. Ecol Entomol 33:239–249

Werren JH, Richards S, Desjardins CA, Niehuis O, Gadau J, Colbourne JK (2010) Functional and evolutionary insights from the genomes of three parasitoid *Nasonia* species. Science 327:343–348

Wong LK, Crowden RK (1976) Preliminary studies on mucus secretion of wood wasp, *Sirex noctilio* F.1. Physicochemical and biochemical properties. Aust J Biol Sci 29:21–32

Zhao J, Davis LC, Verpoorte R (2005) Elicitor signal transduction leading to production of plant secondary metabolites. Biotechnol Adv 23:283–333

Zondag R, Nuttall MJ (1977) *Sirex noctilio* Fabricius (Hymenoptera: Siricidae). In: Service NZF (ed) N.Z. Forest Service. Forest Research Institute, Rotorua, pp 1–8

Chapter 4
Population Dynamics of *Sirex noctilio*: Influence of Diapause, Spatial Aggregation and Flight Potential on Outbreaks and Spread

Juan C. Corley and José M. Villacide

Abstract The most significant ecological feature of invasive populations of the wood-boring wasp, *Sirex noctilio*, is the occurrence of pulse-like eruptive population outbreaks. Pulse-like outbreaks are rapid, local increases in density and are usually terminated by natural enemies or resource defenses or depletion. While *S. noctilio* populations may remain at low, endemic levels for lengthy periods killing stressed tree specimens, during outbreaks more generalized damage, through the attack of large numbers of healthy trees, can be of great economic significance. Outbreak phases can also favor geographical spread. In this chapter, we present an overview of recent ecological and behavioral studies of *S. noctilio* in Patagonia (Argentina). We outline the patterns of adult emergence, where delayed adult emergence is noted and describe strong spatial aggregation and redistribution of *S. noctilio* attacks within a pine plantation. Finally, we report recent work on the potential dispersal capacities of *S. noctilio* males and females. We discuss how these ecological features observed for *S. noctilio* populations in Patagonia may influence or are influenced by the introduction of natural enemies, during the implementation of biological control programmes. The general aim is to stimulate a critical understanding of the population ecology of *S. noctilio*, and consequently improve our abilities to manage established and expanding populations.

J.C. Corley (✉)
CONICET-Laboratorio de Ecología de Insectos,
INTA-EEA Bariloche, CC 277 (8400) Bariloche,
Río Negro, Argentina
e-mail: jcorley@bariloche.inta.gov.ar

J.M. Villacide
Laboratorio de Ecología de Insectos, INTA-EEA Bariloche, CC 277 (8400) Bariloche,
Río Negro, Argentina

B. Slippers et al. (eds.), *The Sirex Woodwasp and its Fungal Symbiont: Research and Management of a Worldwide Invasive Pest*, DOI 10.1007/978-94-007-1960-6_4,
© Springer Science+Business Media B.V. 2012

4.1 Introduction

For pine tree plantations in South America, the woodwasp *Sirex noctilio* F. (Hymenoptera: Siricidae) is probably the most important pest problem. Native to Eurasia, this wood-boring solitary wasp, was first recorded in the region in the early eighties (Iede et al. 1988) and has since invaded all areas were pine trees are cultivated. *Sirex noctilio* populations are established within a broad eco-climatic range, from Southern Brazil and Northern Argentina to South Western Argentina and Chile (Chaps. 15–17).

Among an array of studies on the ecology, behavior and, more importantly, biological control and management strategies (e.g., Coutts 1969; Madden 1974, 1988; Spradbery and Kirk 1978; Haugen 1990; Carnegie et al. 2006), the dynamics of *S. noctilio* populations is amongst the least understood aspects (Madden 1988; Corley et al. 2007). It is known however, that characteristic of the species is the occurrence of pulse-like eruptive population outbreaks. Pulse-like outbreaks are described by a rapid, unpredictable increase in densities and are usually terminated by natural enemies or resource defenses or depletion (Berryman 1987).

Long term studies carried out since the accidental introduction of *S. noctilio* into Australia and New Zealand, have allowed noting that woodwasp populations can remain at relatively low levels (i.e. killing 5–10 trees/ha/year) during many seasons, where attacks are focused on stressed trees. However, when outbreaks occur, these may last between 4 and 10 years (Madden 1988). It is during outbreaks that an otherwise unimportant forest insect may cause extensive tree mortality and severe economic damage as woodwasps then also attack healthy trees (Morgan 1968; Berryman 1987; Madden 1988).

In an attempt to understand the observed *S. noctilio* outbreaks, Madden (1988) based on field and experimental data, proposed the *"intermittent drought hypothesis"*, which emphasizes the role of density-independent factors on outbreaks. According to this hypothesis, discontinuous drought levels occurring during the woodwasp emergence season increase tree attractiveness and susceptibility to *S. noctilio* attacks. However more recently, Corley et al. (2007) proposed that strong spatial aggregation, as consequence of limited female wasp spatial redistribution, could lead to increased insect performance through the concentration of attacks on trees. Still, the ecological mechanisms underlying outbreaks of invasive *S. noctilio* populations, remains an important open question. Because *S. noctilio* is probably the most damaging pest of cultivated pine-trees in several countries of the world (Ciesla 2003), knowledge on the ecology of outbreaks is an important part of predictive modeling of spread and population growth, and essential to successful population management and control.

In this chapter we review recent ecological and behavioral studies of this forest pest carried out in its southernmost distribution (Patagonia, Argentina). By discussing their significance to wasp population dynamics and spread, we present information on the patterns of adult emergence and describe the spatial arrangement of attacked trees within a *Pinus* plantation. We then report work on the potential, long-range dispersal capacities of *S. noctilio* males and females, and on how these are influenced by the introduction of biological control measures. Our emphasis throughout is on outlining recent research of wasp population ecology as related to prior information, to increase our understanding of outbreaks. Consequently, we hope to improve opportunities as to manage established and expanding populations of this damaging forest insect.

4.2 Pine Plantations and *Sirex noctilio* in Patagonia

In Patagonia, in the south of Argentina, pine trees are increasingly being planted. While several species have been cultivated, existing effort is concentrated on planting the North American species, Ponderosa pine (*Pinus ponderosa*), mainly in the arid steppe areas. Currently, plantations in this region cover nearly 68,000 ha of which most bear established populations of *S. noctilio*. The first recordings of this pest in the region date back to 1993 (Klasmer et al. 1998, Chap. 15). Since, woodwasp populations have spread to most pine plantations, at an approximate rate of 15 km/year, much slower than that observed for other regions where *S. noctilio* has invaded (J.C. Corley, J.M. Villacide and A. Liebhold, unpublished data).

The area where pines are planted is arid (600 mm mean annual rainfall) and cold (8.5°C annual mean temperature). Scrub vegetation and short grasses dominate the steppe, and native trees are absent, with the exception of a few marginal populations of *Austrocedrus chilensis* (Cupresaceae) in some sites. Most plantations have been established on the less productive sites within private land, fostered by governmental incentive policies. Thus, pine plantations appear as a slow-growing, conspicuous patchwork of trees widely distributed within the region (Fig. 4.1).

4.3 Life Cycle and Adult Emergence Patterns

The woodwasp flight season in Patagonia begins by late December, when summer starts, and can last until May, when most rainfall and eventually some snowfall, occur. Peak adult emergence, as observed from 1 m long logs, cut-down and caged in the field, is usually observed in January, the hottest month of the year (mean temperature 14.1°C), but a smaller peak can be also seen in March (Klasmer et al. 2000). During the cold, winter months (mean temperature 3.7°C), woodwasps display arrested development, and diapause deep inside the trees.

Sirex noctilio is described as a univoltine species, but it has been noted that not all larvae enter pupation the year in which eggs have hatched (Taylor 1981). Such arrested development in part of the population leads to an extension of adult emergence for additional time. In Patagonia, approximately 75% of the *S. noctilio* population emerges 1 year after oviposition, 24% after 2 years and the remainder after burrowing inside the wood for 3 years (Corley 2001). Very similar patterns

Fig. 4.1 A view of typical pine tree plantations, established on steppe habitats of Patagonia (Argentina)

have been reported for this species in the colder and wetter regions of New Zealand and Australia (Hanson 1939; Rawlings and Wilson 1949; Coutts 1965; Taylor 1981). Delayed adult emergence is also a well-known fact of several other Siricidae species (Morgan 1968). In warmer climates, it has been observed that short (within a season) and long (between seasons) cycles may occur (Neumann and Minko 1981, Chap. 2).

Prolonged or extended diapause is an adaptive feature of forest insects living in unpredictable environments (Hanski 1988). For instance, it is common in seed and cone feeding species where mast-seeding patterns can affect resource availability and predictability (Turgeon et al. 1994). Although the timing (duration) and frequency of diapause within a population is determined by environmental factors, such as temperature and photoperiod (Tauber et al. 1986), prolonged diapause leading to delayed emergence is a life history characteristic. While there is little evidence on the evolutionary background for the delayed diapause patterns observed in Siricidae in general, extended developmental cycles in *S. noctilio* have been associated with current environmental cues such as the size of the tree attacked, wood moisture and tree specimen exposition to sunlight or shading (Morgan and Stewart 1966, Morgan 1968).

Delayed woodwasp emergence can have important consequences on population dynamics and bio-control. For example, Corley et al. (2004) have shown through theoretical modeling how prolonged diapause can affect the stability properties of the interaction of hosts with parasitoid species. In subsequent studies, Corley and

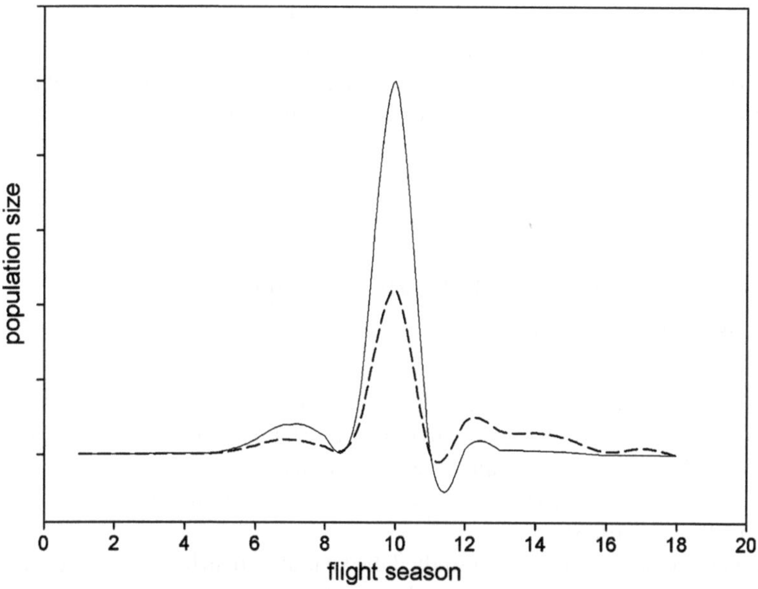

Fig. 4.2 Schematic representation of changes of the population size of *Sirex noctilio* with time, for a fully univoltine population (*full line*) and one undergoing spread-out adult emergence (*dashed line*). Population rates of increase have been established arbitrarily to vary as simulating good and bad wasp years. The prolonged diapause curve was drawn starting from the same seed-number of individuals and using identical rates of increase per time interval, but considering the temporal displacement of reproductive adults. We took as the first year adult fraction emerging to be 75%, the second year fraction 25% and the reminder 1% is emerging in year three since oviposition, throughout. Note that delayed emergence does not affect any individual wasp characteristics, such as size or fecundity

Bruzzone (2009) suggest that the variable fractions of *S. noctilio* undergoing extended diapause and how this relates to the different physiological interactions between its parasitoids, partly explains the also variable success of Rhyssine parasitoids (*Rhyssa persuasoria* L. and *Megarhyssa nortoni* (Cresson)) and *Ibalia leucospoides* (Hochenw.) at two climatically different Australian localities (Taylor 1978). While *I. leucospoides* attacks eggs and first instar larvae and thus has tightly synchronized development to that of its host, the Rhyssineae species attack older larvae, interrupting host development (Hanson 1939). Note that both these parasitoid species have been introduced into most regions were invasive populations of *S. noctilio* are established, as a part of elaborate biological control programmes (Neumann et al. 1987).

Delayed adult emergence may also affect population outbreak dynamics. Extension of adult emergence can depress outbreak intensity and extend its duration, through a slower, time-displaced, population build-up process (Fig. 4.2). Note that because delayed emergence can also affect adult size (Morgan 1968), in turn affecting longevity, fecundity and dispersal potential of female wasps (see below), the consequences of extended diapause on population dynamics are likely to be more complex. It is possible that delayed adult emergence patterns, again by changing outbreak frequency, duration or intensity, influence the spread rates of established populations as well. From a practical perspective, the fact that a number of individuals remain deep inside the wood for up to 3 years, can increase the chance for accidental human transport into new areas and this reinforces the need for efforts to control wood movement and treatment.

4.4 Spatial Dynamics of Woodwasp Populations at a Stand Scale

Spatial aggregation of forest insects has been often associated with population outbreaks. This is because organisms that feed in groups may enjoy better protection from natural enemies or else can increase their effectiveness against plant defenses, which can lead to increased individual performance and a consequent population growth (Rhoades 1985).

Recently, Corley et al. (2007) studied the spatial re-distribution of wasp-attacked pine trees, within two stands located close to the city of Bariloche (Patagonia, Argentina, 71°15′W–41°03′S) that had been recently invaded by *S. noctilio*. Through a census of trees and with the aid of GPS, they were able to count and map all trees attacked by woodwasps within the 70 ha, during three consecutive years. Dating of attacks was possible because recently attacked trees (the season where field work took place) are easily recognizable through characteristic resin droplets associated with tree wilting and evidence of the presence of the wasp fungal symbiont *Amylostereum areolatum*. In turn, clean wasp exit holes are observed when adult emergence has occurred recently (and hence most attacks have occurred the previous season) and old, dirty (and often bearing spider cobwebs) on partly bark-less trees, suggests that attacks have occurred at some earlier point in time (Villacide and Corley 2006; Corley et al. 2007).

Most newly attacked trees were found within a limited range from trees with preceding attacks. While half of attacked trees found in a given season were within 45 m radius from trees attacked the previous season, 90% were found to lie within 130 m (Fig. 4.3). These findings indicated that the spatial aggregation of *S. noctilio* attacks, within a pine tree plantation during the early stages of pest colonization, was strong. Attacked trees were concentrated during the three wasp seasons, with aggregation becoming stronger with time despite a significant increase in the number of attacked trees (Corley et al. 2007). Such reported aggregation is in line with findings on the chemical ecology of *S. noctilio*. It has been observed that through the repeated inoculations of fungal spores and mucus on trees (even if no eggs are laid on them) healthy trees may rapidly (i.e. within the same flight season) become attractive to other wasps, as the liberation of mono-terpenes is induced by the attacks (Coutts and Dolezal 1969; Madden 1974; Spradbery 1977; Böröczky 2008).

Spatial aggregation of woodwasp attacks can also help explain observed population dynamics (Corley et al. 2007). The spatial concentration of attacks by *S. noctilio*

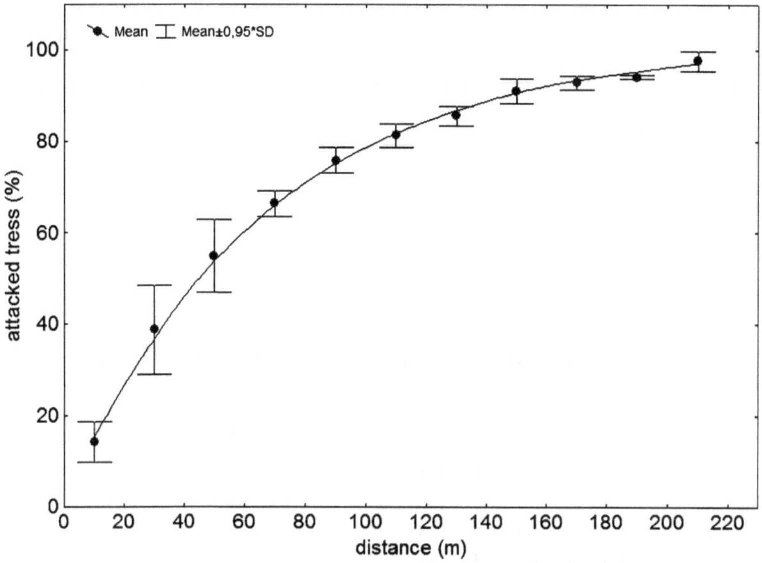

Fig. 4.3 Accumulated percent of newly attacked trees respective to previous season attacks as a function of radial distance, within a 70 ha pine plantation of NW Patagonia (Argentina). Dots and whiskers indicate means and SD of 2 inter-annual comparisons. Note that 50% of new attacks are found within 30 m and 90% within 130 m from any previous attacks (Corley et al. 2007)

females, could contribute to improved larval performance in the absence of climatically induced tree stresses (Madden 1974). Through repeated inoculations of spores and mucus, wood conditions that favor fungal growth can be created. In this regard, it is relevant that feeding by woodwasp immature stages depends on fungal decomposition of wood. In this sense, improved individual performance may lead to more fecund females, and through a population build-up process enhances the probability of outbreak occurrence (Coutts and Dolezal 1969; Madden 1974; Spradbery 1977).

Recently, through a spatially explicit model of *S. noctilio* - pine tree interactions, Bruzzone and Corley (2004) noted that spatial aggregation of attacked trees increases the probability of outbreak occurrence. Their model also showed that *S. noctilio* population outbreaks may occur even in the absence of density independent factors, such as droughts, intermittent droughts or other disturbances. The relationship between strong localization of insect attacks and population build-up is again, through the potential increase in individual performance, associated with aggregation.

Woodwasp spatial aggregation can also affect biological and mechanical control measures. For example, sampling and monitoring programmes of *S. noctilio* for classification purposes, or else density estimations, are fed existing information of the spatial distribution of attacks (e.g., Penteado 1995). Also, control measures that imply finding attacked individual host trees, such as sanitary thinning (Villacide and Corley 2006), can benefit from knowledge on the spatial dynamics of pests, through a reduced searching effort.

Theoretical work has shown that host spatial aggregation can play an important role in the dynamics of host-parasitoid interactions (Hassell and Wilson 1997). Here it is important to note that existing *S. noctilio* IPM programmes include several parasitoid species as biological control agents (Chap. 8). While the dynamic consequences of these specific interactions remain largely unknown, the introduction procedures of these and of the nematode *Deladenus* (=*Beddingia*) *siricidicola* may benefit from knowledge of the spatial distribution of woodwasp populations. Spatial aggregation should also be taken into account during the development of predictive population models.

4.5 Flight Potential of *Sirex noctilio* Males and Females

While geographical spread of many invasive pests is closely related to human transportation, natural dispersal capabilities of flying insects can play an important role (Liebhold et al. 1995; Lavandero et al. 2004). How far an insect can fly can also determine the spatial structure of populations and can play an important role in species interactions (Commins et al. 1992).

Villacide and Corley (2008) and Bruzzone et al. (2009) have recently studied the flight potential of *S. noctilio* tethered individuals in flight mill devices. Additional studies of male flight, using the same flight mills, have also recently been addressed (J.M. Villacide et al. unpublished data). Flight mills allow estimation of potential flight characteristics such as duration, speed and distance covered in given time periods. With the exception of a single mention of *S. noctilio* flight distance, obtained from an unpublished flight mill study (Taylor 1981), there are no data on potential flight capacity for this species.

Individuals used for the flight mill assays (Villacide and Corley 2008; Bruzzone et al. 2009, J.M. Villacide et al. unpublished data) were recently emerged adults, collected from cages holding 1 m long billets obtained from several *Pinus contorta* var. *latifolia* trees recently attacked by *S. noctilio*, from plantations located in NW Patagonia (Argentina). Before and after flight, each individual was weighed (*Scientech SA210*; d: 0.0001 g). In all experiments, recorded variables were: the accumulated flight distance (in kilometers) and mean flight speed (in meters/second) during a period of 23 h, for 28 healthy females. The effects of infection by the nematode on flight parameters were studied separately by flying 46 (22 infected and 24 uninfected) female wasps and 56 males (42 infected and 14 uninfected). Infection status was determined after flight, by dissecting wasps and inspecting their abdomens under a stereo-microscope.

The results of these studies showed that *S. noctilio* females are capable of long, sustained flights. The average distance flown by a healthy female wasp during the 1 day-long trial was 17.4 km with a maximum of 49.7 km and a minimum of 1.1 km. The average speed for all female wasps studied was 0.37 m/s, although individuals remained at rest most of the time. However, it is noteworthy that the estimated flight parameters of wasps were highly variable between individuals, with some females being able of performing very long, uninterrupted flight spells.

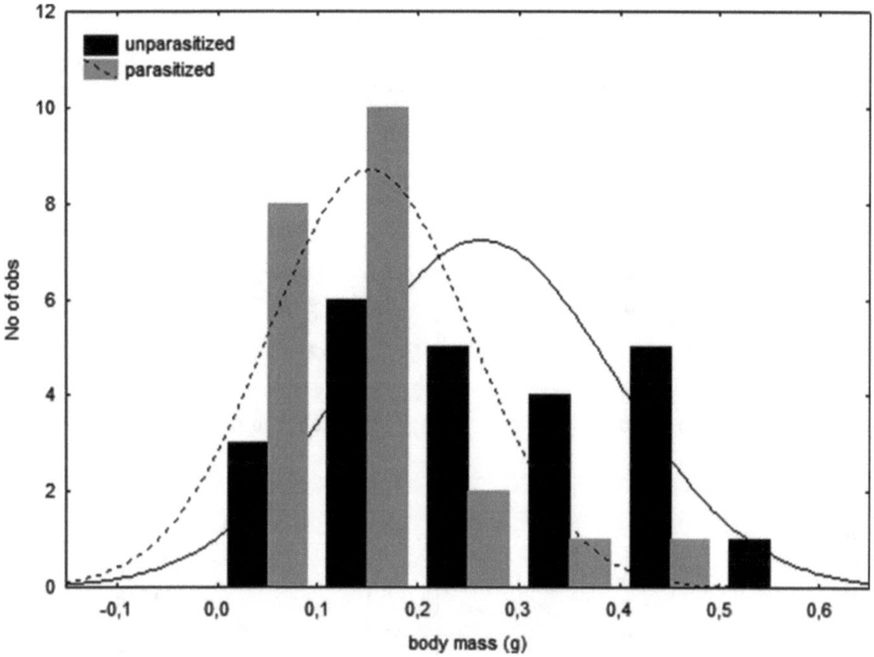

Fig. 4.4 Frequency distribution of the initial body mass of healthy and nematode-bearing females flown in flight mills. Grey bars represent nematode-infected females (n = 22), while black bars are healthy wasps (n = 24). Parasitized *S. noctilio* were smaller and consequently flew for shorter distances at slower mean velocity (Villacide and Corley 2008)

Nematode-infected *S. noctilio* males showed no differences in their flight performance compared to healthy males (J.M. Villacide et al. unpublished data). However, parasitized female wasps showed a marked reduction in the total flight distances (infected wasps flew 16.1 km, while uninfected wasps flew 30.5 km; Villacide and Corley 2008) compared to control wasps. Villacide and Corley (2008) analyzed body mass of flown female wasps and concluded that the reduced flight capabilities are probably a result of the effects of parasitism on wasp size. As a consequence of competition for fungus resources between nematodes and developing wasp larvae or else through the direct effect of the parasites on wasp development, parasitized wasps are significantly smaller than their healthy counterparts (Fig. 4.4, see also Bedding 1984).

Both speed and distance flown are closely dependent on initial body mass for males or females (Fig. 4.5). Larger individuals fly faster, probably because of their larger wings, and may also fly longer distances. Long distance flights must relate to body fat reserves, acquired during larval development. In their paper, Bruzzone et al. (2009) also show that wasps display different flight performances in relation to initial body size.

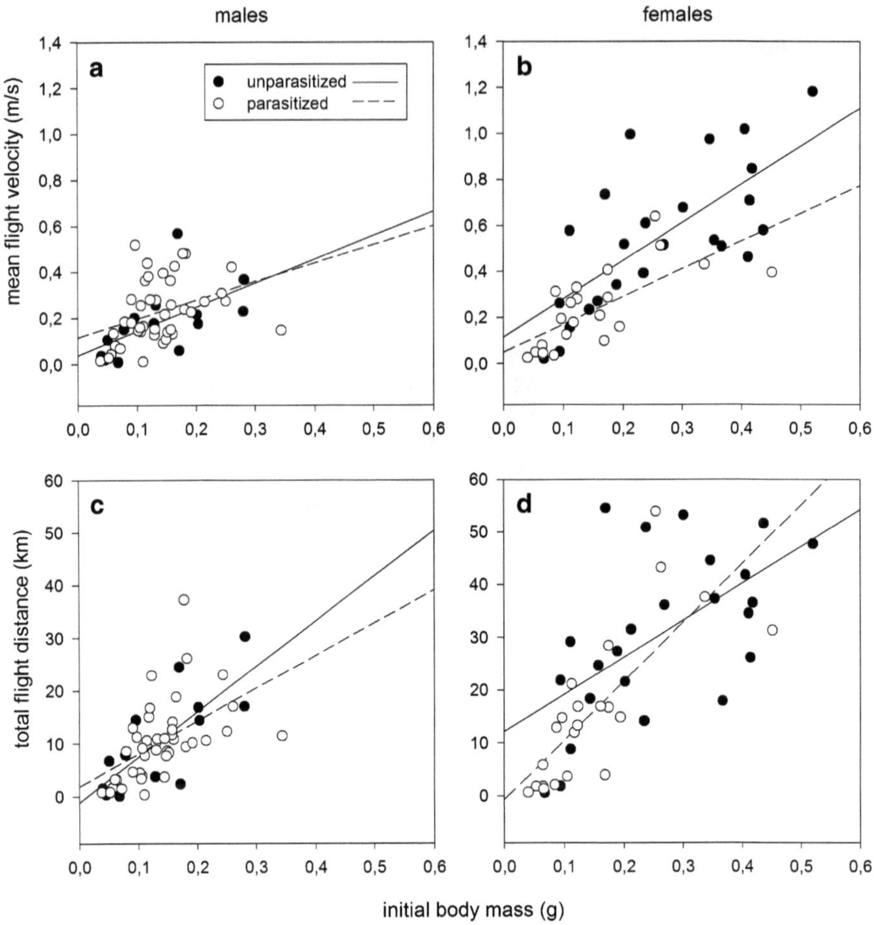

Fig. 4.5 Relationship between flight velocity (*upper plates*) and flight distance (*lower plates*) recorded for *Sirex noctilio* males (**a** and **c**) and females (**b** and **d**) flown in tethered flight devices during a 23 h long period, at ambient temperature. *Open circles* show nematode bearing individuals, while healthy ones are depicted by full circles. Males are typically smaller than females, which are capable of faster and longer flights. In turn, infected individuals tend to be smaller, also affecting flight characteristics (Villacide and Corley 2008; J.M. Villacide et al. unpublished data)

The larger females, while flying nearly 50 km in day through long, uninterrupted flight spells, are potentially capable of colonizing new pine stands. However, geographical spread rates, even though aided by human transportation of infested wood, are much smaller than may be expected. On the one hand, it is known that flight mills notably overestimate flight potential as insects are forced to fly once attached to the spinning axle (Blackmer et al. 2004). Their use is more appropriate for comparison purposes. On the other hand, as observed from spatial aggregation studies (Corley et al. 2007), most wasps should remain in the vicinity of the tree they have emerged from, following chemical cues that favor mating and oviposition success.

The contribution of natural dispersal of wasps to invasion spread is still unclear. The colonization and successful establishment of new populations requires that a number of individuals arrive at a new pine plantation. It is possible that during endemic population phases, wasp dispersal outwards from a pine plantation is minimal. During this phase, population growth is limited and competition for resources (appropriate pine trees in which to oviposit) unlikely. However, more attention should be paid to wasp dispersal behavior during outbreaks, when populations may grow exponentially and resources are rapidly depleted.

Different flight capacities displayed by nematode infected and healthy female wasps could affect biological control success. Here it is relevant that the nematode *D. siricidicola*, the most important component of woodwasp population management plans, relies on wasp movement for its own dispersal. Moreover, to prevent wasp outbreaks at the same time as optimizing introductions, healthy wasps should attack trees also attacked by infected wasps, thus ensuring that nematode infection naturally spreads throughout the population (Villacide and Corley 2008).

4.6 Conclusions

We have outlined aspects of the ecology and behavior of *S. noctilio* that are important to understand population dynamics and manage invasive populations. The main objective of pest management plans for established populations of this forest insect, consist in reducing the frequency and/or magnitude of outbreaks. This is because *S. noctilio* is an economically significant problem to commercial pine forestry when populations reach epidemic levels.

Delayed adult emergence, a consequence of extended diapause that is commonly observed in the colder areas of *S. noctilio* distribution, can alter the magnitude and duration of outbreaks and affect geographical spread rates. The duration of the life cycle and the frequency of extended cycles within a population should attract further detailed attention. Prolonged diapause may also influence the relative success of parasitoids used in biological control programmes and quarantine measures (Corley and Bruzzone 2009).

The strong spatial aggregation of woodwasp attacks is probably essential to outbreaks. While clumping, through enhanced individual performance may favor population build-up, it may also facilitate the efficiency of sanitary thinning of forests and the introduction of biological control agents such as the parasitic nematode *Deladenus siricidicola*, both well-known methods to minimize the damage caused by *S. noctilio* (Neumann and Minko 1981; Haugen et al. 1990; Haugen and Underdown 1993; Villacide and Corley 2006).

While the contribution of *S. noctilio* dispersal potential to regional spread is probably small if compared to human-assisted introduction, it should not be disregarded. The association of flight capacity with body size, as well as a yet unclear relationship of dispersal with local population outbreaks, suggests that natural dispersal could be important during colonization and establishment processes. From an applied perspective, the importance of knowledge regarding wasp flight potential in the design of wood inspection processes and the execution of biological control practices should be emphasized.

In this chapter, we have attempted to illustrate the significance of understanding the spatio-temporal dynamics of *S. noctilio* populations. By gaining an insight into some ecological and behavioral features of this pest, we expect to improve an ability to predict local and regional dynamics and to improve the management of established populations. Further research, with an emphasis on population studies in different regions of the world, is likely to draw a more complete picture of our understanding of *S. noctilio* population dynamics and geographical spread.

Acknowledgments Research reported in this chapter was partially funded by grants PICT 2006 # 01200 (ANPCyT), PNFOR 2212 (INTA) and 07-JV-11242300-062 (Northern Research Station, Forest Service, U.S. Department of Agriculture).

References

Bedding RA (1984) Nematode parasites of Hymenoptera. In: Nickle WR (ed) Plant and insect nematodes. Marcel Dekker Inc., New York

Berryman AA (1987) The theory and classification of outbreaks. In: Barbosa P, Schultz J (eds) Insect outbreaks. Academic, San Diego

Blackmer JL, Naranjo SN, Williams L (2004) Comparative study of tethered and untethered flight by *Lygus hesperus* and *Lygus lineolaris* (Heteroptera: Miridae). Environ Entomol 33:1389–1400

Böröczky K (2008) Chemical ecology of in *Sirex noctilio*. In: 19th USDA interagency research forum on invasive species, Annapolis

Bruzzone OA, Corley JC (2004) Influencia de la heterogeneidad espacial en la dinámica poblacional de *Sirex noctilio*. 2° Reuni6n Binacional de Ecología Argentino-Chilena y 21° Reuni6n Argentina de Ecología. Mendoza

Bruzzone OA, Villacide J, Bernstein C, Corley JC (2009) Flight polymorphism in the woodwasp *Sirex noctilio* (Hymenoptera: Siricidae): an analysis of tethered flight data using wavelets. J Exp Biol 212:731–737

Carnegie AJ, Matsuki M, Haugen DA et al (2006) Predicting the potential distribution of *Sirex noctilio* (Hymenoptera: Siricidae), a significant exotic pest of *Pinus* plantations. Ann For Sci 63:119–128

Ciesla WM (2003) European woodwasp: a potential threat to North America's conifer forests. J For 101:18–23

Commins HN, Hassell M, May RM (1992) The spatial dynamics of host-parasitoid systems. J Anim Ecol 61:735–748

Corley JC (2001) La diapausa prolongada y su papel en la dinámica poblacional de sistemas huesped-parasitoide. PhD thesis, University of Buenos Aires, Argentina

Corley JC, Bruzzone OA (2009) Delayed emergence and the success of parasitoids in biological control. Biol Contr 51:471–474

Corley JC, Capurro AF, Bernstein C (2004) Prolonged diapause and the stability of host-parasitoid interactions. Theor Popul Biol 65:193–203

Corley JC, Villacide JM, Bruzzone OA (2007) Spatial dynamics of a *Sirex noctilio* woodwasp population within a pine plantation in Patagonia, Argentina. Entomol Exp Appl 125:231–236

Coutts MP (1965) *Sirex noctilio* and the physiology of *Pinus radiata*. Some studies of interactions between the insect, the fungus, and the tree in Tasmania. Australia bulletin no 41. Forestry and Timber Bureau, Canberra

Coutts MP (1969) The mechanism of pathogenicity of *Sirex noctilio* on *Pinus radiata*. I. Effects of the symbiotic fungus *Amylostereum sp.* (Thelophoraceae). Aust J Biol Sci 22:915–924

Coutts MP, Dolezal JE (1969) Emplacement of fungal spores by the woodwasp, *Sirex noctilio*, during oviposition. For Sci 15:412–416

Hanski I (1988) Four kinds of extra long diapause in insects: a review of theory and observations. Annu Zool Fenn 25:37–53

Hanson HS (1939) Ecological notes on the *Sirex* wood wasps and their parasites. Bull Ento Res 30:27–65

Hassell P, Wilson H (1997) The dynamics of spatially distributed host--parasitoid systems. In: Tilman D, Kareiva P (eds) Spatial ecology. The role of space in population dynamics and interspecific interactions. Princeton University Press, Princeton

Haugen DA (1990) Control procedures for *Sirex noctilio* in the Green Triangle: review from detection to severe outbreak (1977–1987). Aust For 53:24–32

Haugen DA, Underdown MG (1993) Reduced parasitism of *Sirex noctilio* in radiata pines inoculated with the nematode *Beddingia siricidicola* during 1974–1989. Aust For 56:45–48

Haugen DA, Bedding RA, Underdown MG, et al (1990) National strategy for control of *Sirex noctilio* in Australia. Aust Fort Grower, No 13 (special liftout section)

Iede ET, Penteado SR, Bisol JL (1988) Primeiro registro de ataque de *Sirex noctilio* em *Pinus taeda* no Brasil. EMBRAPA-CNPF, Circular técnica 20

Klasmer P, Fritz G, Corley J, Botto E (1998) Current status of research on *Sirex noctilio* F in the Andean-Patagonian region in Argentina. In: Iede E, Shaitza E, Penteado S, et al (eds) Proceedings of a conference: training in the control of *Sirex noctilio* by use of natural enemies, FHTET 98–13. USDA Forest Service, Morgantown

Klasmer P, Botto E, Corley JC, Villacide J, Fernandez-Arhex V (2000) Avances en el control biológico de *Sirex noctilio* en la región Patagónica Argentina. Serie técnica IPEF Brasil 13:21–30

Lavandero B, Wratten S, Hagler J, Jervis M (2004) The need for effective marking and tracking techniques for monitoring the movements of insect predators and parasitoids. Int J Pest Manag 50:147–151

Liebhold AM, MacDonald D, Bergdail WL, Mastro VC (1995) Invasion by exotic forest pests: a threat to forest ecosystems. For Sci Monogr 30:1–49

Madden JL (1974) Oviposition behaviour of the woodwasp *Sirex noctilio* F. Aust J Zool 22:341–351

Madden JL (1988) *Sirex* in Australasia. In: Berryman AA (ed) Dynamics of forest insect populations: patterns, causes and implications. Plenum Press, New York

Morgan FD (1968) Bionomics of Siricidae. Annu Rev Entomol 13:239–256

Morgan FD, Stewart NC (1966) The biology and behavior of the woodwasp *Sirex noctilio* F. in New Zealand. Trans R Soc N Z Zool 7:195–204

Neumann F, Minko G (1981) The *Sirex* wood wasp in Australian radiata pine plantations. Aust For 44:46–63

Neumann FG, Morey NC, McKimm RJ (1987) The sirex wasp in Victoria, Bulletin no 29. Department of Conservation, Forests and Lands, Melbourne, pp 1–41

Penteado SR (1995) Métodos de amostragem para avaliação populacional de *Sirex noctilio* F. (Hymenoptera: Siricidae) e de seus inimigos naturais, em *Pinus taeda* L. e aspectos do controle biológico. Msc thesis, Universidade Federal do Paraná, Brazil

Rawlings GB, Wilson NM (1949) *Sirex noctilio* as a beneficial and destructive insect of *Pinus radiata* in New Zealand. N Z J For 6:20–29

Rhoades DF (1985) Offensive–defensive interactions between herbivores and plants: their relevance in herbivore population communities. Ecol Monogr 58:39–55

Spradbery JP (1977) The oviposition biology of siricid wasps in Europe. Ecol Entomol 2:224–230

Spradbery JP, Kirk AA (1978) Aspects of the ecology of siricid woodwasps (Hymenoptera: Siricidae) in Europe, North Africa and Turkey with special reference to the biological control of *Sirex noctilio* F. in Australia. Bull Entomol Res 68:341–359

Tauber MJ, Tauber CA, Masaki I (1986) Seasonal adaptations of insects. Oxford University Press, Oxford

Taylor KL (1978) Evaluation of the insect parasitoids of *Sirex noctilio* (Hymenoptera: Siricidae) in Tasmania. Oecologica 32:1–10

Taylor KL (1981) The sirex woodwasp: ecology and control of an introduced forest insect. In: Kitching RL, Jones RE (eds) The ecology of pests – some Australian case histories. CSIRO, Melbourne

Turgeon JJ, Roques A, Groot P (1994) Insect fauna of coniferous seed cones: diversity, host plant interactions and management. Annu Rev Entomol 39:179–212

Villacide JM, Corley JC (2006) Control de niveles poblacionales endémicos de la avispa de los pinos *Sirex noctilio* (Hymenoptera: Siricidae) mediante el raleo sanitario de hospederos atacados. Rev Invest Agrop 35:121–134

Villacide JM, Corley JC (2008) Parasitism and dispersal potential of *Sirex noctilio* implications for biological control. Agr For Entomol 10:341–345

Chapter 5
The Woodwasp *Sirex noctilio* and Its Associated Fungus *Amylostereum areolatum* in Europe

Beat Wermelinger and Iben M. Thomsen

Abstract The current knowledge about the biology and ecology of siricid woodwasps and their *Amylostereum* mutualistic fungi, with a specific focus on *Sirex noctilio* and *Amylostereum areolatum*, are summarized from a European perspective. The woodwasp females deposit their eggs together with fungal spores and toxic mucus in single or multiple drills. The symbiotic fungus supports the development of the larvae, which usually takes 2 years. The fungus propagates vegetatively through the formation of asexual spores in the teneral females and is stored in the wasps' mycangia. Both *A. areolatum* and the related species *A. chailletii* are known to have vegetative compatibility groups as a consequence of the symbiosis with woodwasps. Among the natural enemies of the woodwasps, woodpeckers have a minor significance in Sirex mortality compared to parasitoids. Particularly the parasitic *Ibalia* species can account for 40–70% larval mortality. Like all European woodwasps, *S. noctilio* preferably colonizes newly dead, damaged or greatly weakened trees with considerable needle loss and low moisture content. Thus, unlike in the countries where it has been accidentally introduced, *S. noctilio* causes hardly any pine mortality in Europe and is economically irrelevant.

B. Wermelinger (✉)
Swiss Federal Institute for Forest, Snow and Landscape Research WSL,
8903 Birmensdorf, Switzerland
e-mail: beat.wermelinger@wsl.ch

I.M. Thomsen
Forest & Landscape Denmark, University of Copenhagen,
1958 Frederiksberg, Denmark

B. Slippers et al. (eds.), *The Sirex Woodwasp and its Fungal Symbiont: Research and Management of a Worldwide Invasive Pest*, DOI 10.1007/978-94-007-1960-6_5,
© Springer Science+Business Media B.V. 2012

5.1 Introduction

Most woodwasp species (Siricidae) are native to forests in the northern hemisphere. Nineteen European species are currently listed in the Fauna Europaea database (Taeger and Blank 2008) of which ten are native to Central Europe (Bachmaier 1966). Siricids do not have a major economic impact in Europe, because they most often colonize trees that have recently died. This is in contrast to the situation in New Zealand, Australia, various countries in South America, and South Africa, where *Sirex noctilio* was accidentally introduced into plantations of *Pinus radiata* and other *Pinus* spp. In these regions, the woodwasps attack vigorous pine trees and thus cause considerable damage and cost to the local economies (e.g., Madden 1988; Hurley et al. 2007).

The three most common *Sirex* (=*Paururus*) species present in Europe, i.e. *Sirex cyaneus* F., *S. juvencus* (L.), and *S. noctilio* F., colonize coniferous trees, mostly *Pinus* and *Abies* species. *Sirex noctilio* is endemic to Eurasia and North Africa and has been found mostly on pines (Spradbery and Kirk 1978; Eichhorn 1982). Occasionally, fir (*Abies alba* Mill.) or spruce (*Picea abies* (L.) Karst) may be attacked as well (Browne 1968; Schimitschek 1968; Spradbery and Kirk 1978). The woodwasp is most common on the Iberian Peninsula and in western France (Hall 1968). Other European woodwasps that colonize conifers include the genera *Urocerus* and *Xeris*, while the two *Tremex* species develop in broadleaved trees. *Xeris spectrum* is the only species not associated with symbiotic fungi. For a successful development it therefore needs wood that has been colonized by other siricids and their fungi (Morgan 1968).

The symbiotic fungi associated with the genus *Sirex* are the basidiomycetes *Amylostereum areolatum* (Fr.) Boid. and *A. chailletii* (Pers.: Fr.) Boid. These two fungi have been described and studied in their own right as wood decay fungi, both before and after the association with woodwasps became known (Boidin 1958; Pechmann et al. 1967; Siepmann and Zycha 1968; Jahn 1971; Stalpers 1978; Thomsen 1998). In relation to mycology or forest pathology, articles and data on both fungi are numerous, but *A. areolatum* has received more attention as it is a symbiont of the invasive *S. noctilio* wasp. There is consequently a need for more studies of the biology and ecology of other *Amylostereum* spp. in relation to their roles a symbionts of woodwasps.

There is an extensive literature on almost every conceivable aspect of *S. noctilio*'s ecology and significance, but most originates from non-European studies. This paper summarizes the most important aspects of the woodwasp and its associated fungus *A. areolatum* with a focus on European sources. Where appropriate, this synopsis is complemented with new data from our own studies, and comparisons with other European woodwasps and with *A. chailletii* are made.

5.2 Biology of *Sirex noctilio*

The biology and ecology of *S. noctilio* have been studied in great detail (e.g., Morgan 1968; Spradbery and Kirk 1978; Madden 1981) because it has become a pest organism in the countries where it was accidentally introduced. Like siricids in general,

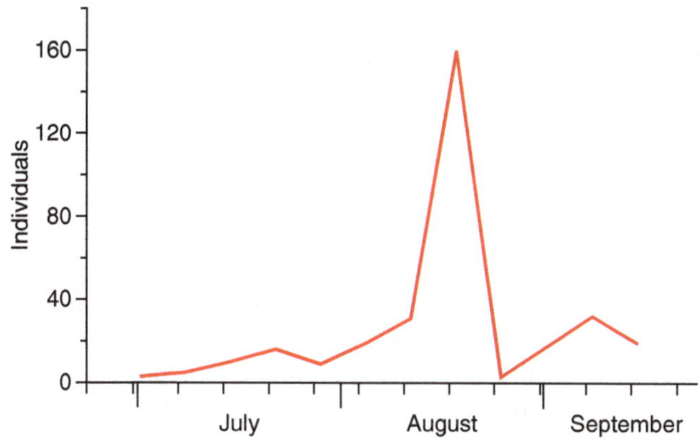

Fig. 5.1 Emergence phenology of *S. noctilio* from *Pinus sylvestris* logs collected in Valais, a dry inner-Alpine valley in Switzerland

S. noctilio emerges and flies in bright sunshine in summer and autumn, mainly between July and August (Benson 1951; Schimitschek 1968). At an altitude of 1,200 m a.s.l. in South Tyrol (Italy), it has been observed in flight up to mid October (Schimitschek 1968). In a study on phloeo- and xylophagous insects attacking pines in southern Switzerland (Wermelinger et al. 2008), a total of 320 siricid woodwasps were collected from 16 pine logs incubated in emergence traps between July and September, with a maximum collected in August (Fig. 5.1).

Females are attracted to suitable hosts by tree volatiles (Simpson and McQuilkin 1976, see Chap. 11). Males in particular usually stay in the tree canopy, where mating takes place as well. Similar to other siricids, females may oviposit prior to mating and these eggs will develop into male wasps (Stillwell 1966). After mating, the female descends to the tree stem and starts oviposition by walking up the stem in a counter clockwise movement, drilling holes every 7–50 cm. When she reaches the tree top, she flies down and begins ovipositing again. The drillings are more frequent at the crown base. No information is available on the oviposition behaviour of the wasps on fresh logs. The functional anatomy of the siricid ovipositor and the drilling process have been described for example in Escherich (1942). During drilling, the sheath of the ovipositor points backwards and the thin ovipositor penetrates the wood. The ovipositor is composed of a middle dorsal part and two ventral parts thus forming a tube (see Chap. 1). During penetration, the middle part and both ventral parts are forced alternately into the wood. Their surfaces possess ridges and depressions and the alternating action of the ovipositor parts transports the boring dust to the wood surface. Upon completion of the drill, an egg may be pressed through the ovipositor tube. Each egg is supplied with fungal spores and toxic mucus. Up to five drills, originating from one entrance hole on the surface, radiate out into the wood, with at least one of them containing only spores of the symbiotic fungus without any eggs (Coutts and Dolezal 1969; Spradbery 1977). The spores are stored in specialised organs, the mycangia, in the female's body. Females may lay between 200 and 400 or more eggs in approximately 90 drills during their life times (Eichhorn 1982). The length of the drills varies between 6 and 10 mm (Escherich 1942). Madden (1974) undertook detailed studies of the drilling activity and egg deposition as well as their relationship to tree and environmental conditions. Most drills were single drills and contained no egg while multiple drills increased the frequency of egg deposition. In resistant trees, multiple drills were rare and confined to discrete areas, whereas on susceptible trees, drills were randomly distributed. Drilling type also depended on the osmotic pressure of the phloem sap with high pressure favouring single drills. During the oviposition period, multiple drills became more frequent and were mostly found in the lower tree parts. Drilling activity was found to be highest at 20–22°C and 40–54% relative humidity which conditions are usually encountered in early afternoon.

The spores germinate after they have been placed in the tree and the symbiotic fungus proceeds to colonize the wood, growing rapidly in the vertical direction and more slowly radially and circumferentially. The larvae hatch from the eggs after a short delay, presumably in order to ensure that the fungus has spread first, as suppression of fungal growth could prolong embryogenesis (Madden 1981). The larvae excavate galleries that steadily grow larger with increasing body diameter and the larvae pack these galleries tightly with fine wood powder.

While feeding in the wood, the larvae ingest mycelium of the symbiotic fungus, thus acquiring enzymes from the fungus and using these to degrade the wood components (Kukor and Martin 1983; Martin 1987). In Europe, the larval development usually lasts 2 years with an average of seven instars. It may be accelerated to 1 year in southern parts or in exceptionally warm years, or delayed by several years due to adverse conditions (Eichhorn 1982). The fully-grown larvae pupate at the end of the 5–20 cm long tunnels in the outermost 50 mm of the stems. After a prepupal and a pupal phase of 3–4 weeks each, the adult woodwasps emerge and gnaw their way through the remaining wood and bark. After emerging through the typical circular holes 3–7 mm in diameter, they rest for some minutes on the bark surface before they take flight.

The sex ratio is usually markedly shifted in favour of the males (5–36% females), but it also depends on the gallery position within the tree (Spradbery and Kirk 1978; Eichhorn 1982). Females are more frequent at the bottom 2 m of the stem, and males dominate at the tree tops. Upon emergence, males usually head for the crown of their host tree, while females first travel some 100 m through the stand and then start their drilling attempts at the upper stem parts. Once they begin egg-laying, they hardly disperse further. For this reason attacked trees are often found in clusters (also see Chap. 4).

5.3 Natural Enemies

Among the generalist natural enemies of xylophagous insects, woodpeckers are known to be predators of siricid larvae (Escherich 1942; Spradbery 1990). While woodpeckers are quite efficient in excavating the larvae, the mortality rate of siricid populations due to woodpeckers was only 6% in one survey in Europe (Spradbery 1990). Thus, they do not seem to have as much impact as the more specific parasitoids of siricids. The most important parasitoids are *Ibalia leucospoides* and *I. rufipes* (Ibaliidae), as well as the ichneumonid *Rhyssa persuasoria* (Spradbery and Kirk 1978). In a European survey insect parasitoids were found to account for at least 40% of siricid mortality (Spradbery and Kirk 1978) and to be important agents for controlling the Sirex populations (Hall 1968). In Australasia, the mortality of *S. noctilio* by total parasitism was estimated at 70% in Tasmania (Taylor 1976), and that by *R. persuasoria* alone at 10% in New Zealand (Morgan and Stewart 1966). The parasitoid woodwasps are olfactorily attracted to the larvae, their frass and, in the case of *S. noctilio*, to the symbiotic fungus *A. areolatum* (Morgan and Stewart 1966; Morgan 1968).

Rhyssa-females do not rely on the drillings of the woodwasp females for their own oviposition, but they are able to penetrate the wood with their thin and flexible ovipositor (Escherich 1942). However, most penetrations do not result in the deposition of an egg. *Rhyssa* acts, according to the summary by Escherich (1942), as an ectoparasite on older larvae and pupae. Young host larvae would not support complete parasitoid development, and medium-aged hosts may be out of reach of the parasitoid female's ovipositor. The development of *R. persuasoria* lasts 5 weeks, but the parasitoids do not emerge before the next spring (Escherich 1942). In contrast to *Rhyssa* spp., *I. leucospoides* females access their hosts through the woodwasp drillings and deposit their eggs directly into the host eggs or in the newly hatched larvae. Their total development lasts 2–3 years. Sirex larvae parasitized by *I. leucospoides* do not intrude as deeply into the wood as healthy ones and, therefore, the parasitoids pupate close to the surface. The two parasitoid genera are considered equally effective (Escherich 1942).

In addition to parasitoid wasps, the parasitic nematode *Deladenus* (=*Beddingia*) *siricidicola*, has been employed as a biocontrol agent throughout the Southern Hemisphere (Bedding and Akhurst 1974, see Chap. 9, 13–18). It is the cornerstone of the control programs in those regions, albeit not uniformly effective, and is considered for its potential use in North America (see Chap. 10). Unfortunately, little is known about its distribution, diversity and effect on *S. noctilio* populations in Europe.

5.4 Ecology of the *Amylostereum* Fungal Symbionts

Regarding mutualism between *Sirex* spp. and *Amylostereum* spp., it could be said that the woodwasps are unable to proliferate without the fungi, but the latter may exist apart from the insects. The fruitbodies of *A. areolatum* (Fig. 5.2) and *A. chailletii* (Fig. 5.3) can be found on stumps or logs of various conifers in Europe. Both fungi are tetrapolar heterothallic, and basidiospore derived heterokaryons are, therefore, separate genets. The basidiospores of *A. areolatum* are smaller (4.5×2.4 μm) than those of *A. chailletii* (6.1×3.1 μm) (Figs. 5.4 and 5.5), and the difference is visible without measuring when looking at spore prints through the microscope (Thomsen 1998). This allows the fungus to be identified rapidly without culturing on artificial media, via spore prints from fruitbodies.

Fig. 5.2 Fruitbodies of *A. areolatum* from Denmark. Note the purple brown (**a**) or umbra-brown (**b**) colour and white fruitbody margins characteristic of this species. (Photos J. Koch)

Fig. 5.3 *A. chailletii* fruitbodies from Denmark. Note the variable colouring, from pale leather brown to brick red. (Photos J. Koch)

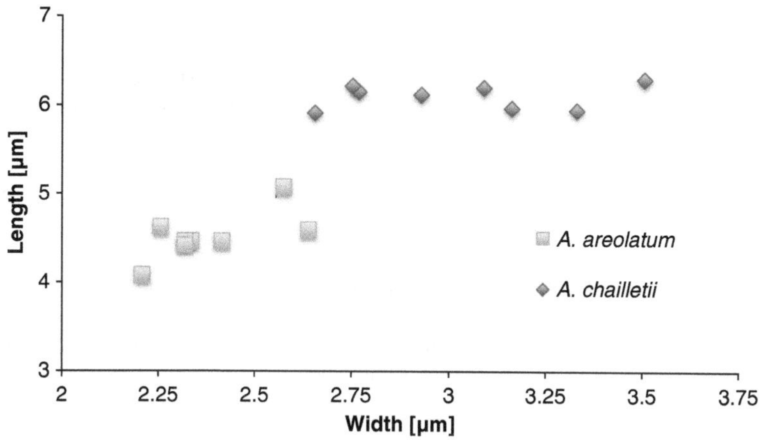

Fig. 5.4 Measurements of basidiospores from *A. areolatum* and *A. chailletii* in lactophenol. Each marker represents the average of 25 basidiospores from one fruitbody. On average, *A. areolatum* basidiospores are smaller than those of *A. chailletii*

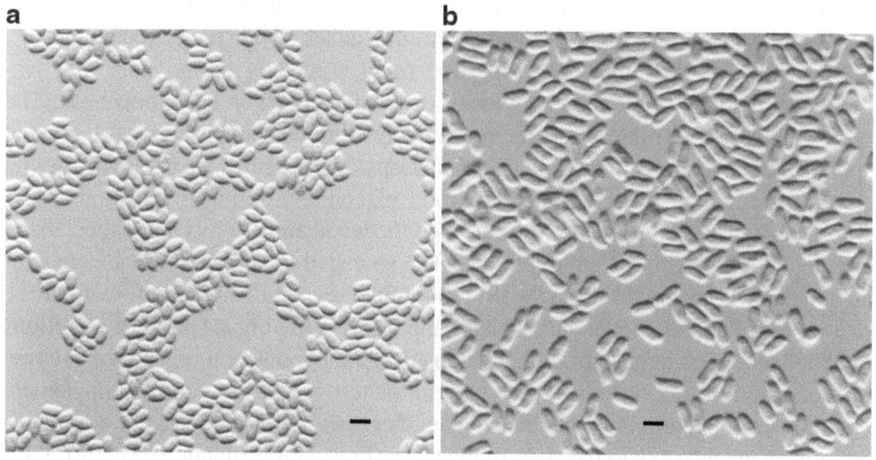

Fig. 5.5 Basidiospores from *A. areolatum* (**a**) and *A. chailletii* (**b**). Scale bars = 5 µm (Photos I.M. Thomsen)

In Europe, *A. chailletii* fruitbodies have been found in most countries, but *A. areolatum* fruitbodies are rare and only reliably documented from Austria, Czech Republic, Denmark, France, Germany, Lithuania, Switzerland and maybe Slovenia (Table 5.1). In the Southern Hemisphere, where *A. areolatum* has been introduced with the woodwasp *S. noctilio*, fruitbodies have never been found. Although *S. noctilio* and *S. juvencus*, the other European vector of *A. areolatum*, have quite wide host ranges, most records of *A. areolatum* fruitbodies are from *Picea abies*. In Europe, most isolations of this fungus from wood have also been made from *P. abies*. As a consequence

Table 5.1 Host species recorded for *Amylostereum areolatum* in Europe by various authors

Tree host species		Source
Fir, exotic conifers, silver fir	Fruitbodies	Bourdot and Galzin 1921 + 1927
Picea sp.	Isolations	Pechmann et al. 1967
Picea abies	Fruitbodies	Jahn 1968
Picea abies	Isolations	Schönhar 1969
Picea sp.	Isolations	Pechmann and Aufsess 1971
Picea abies	Isolations	Siepmann 1971
Picea abies	Fruitbodies and isolations?	Jahn 1971 + 1979
Picea abies, *Pinus* sp., *Abies holophylla*, *Pinus radiata*	Isolations and fruitbodies?	Boidin and Lanquetin 1984
Picea sp.	Fruitbodies	Breitenbach and Kränzlin 1986
Picea abies	Fuitbodies and isolations	Thomsen 1996
Picea abies	Isolations	Solheim 2006

some European mycological books state that *A. areolatum* may be found almost exclusively on spruce (e.g., Jahn 1971, 1979; Eriksson et al. 1978; Breitenbach and Kränzlin 1986). But the meagre list of hosts may also be due to the relative rareness of *A. areolatum* fruitbodies, or that fructification occurs for some reason more easily on *P. abies* substrate, as both *S. noctilio* and *S. juvencus* attack a wide range of conifers.

As symbionts of siricid woodwasps, both fungi, especially *A. areolatum*, are striking and provide exceptional examples of dispersive clones in basidiomycetes. The propagation of the fungi happens vegetatively through the formation of asexual spores in the mycangia in the female woodwasps. These spores, also known as oidia or arthrospores, are inserted into the wood of conifers during oviposition of the woodwasps. The fungus then grows through the wood and also occupies specialized structures in the skin of (female) larvae. The so-called hypopleural organs are situated in the hypopleural folds on the first abdominal segment. Shortly before moulting that precedes pupation, glandular cells in the hypopleural organs produce secretions in abundance. The hyphae or oidia contained in the hypopleural organs are encased in the secretion that turns waxy and forms plates. When the young adult female emerges from pupation and begins chewing her way out of the tree, the wax plates are broken up and stick to the body of the wasp (Francke-Grosmann 1957, 1967). By the movement of her body and perhaps especially of the ovipositor, wax pieces containing fungus travel up the outside of the ovipositor, through the genital opening, and into the intersegmental sacs (Gilmour 1965). After the retractive movements of the ovipositor have transported the wax plates to the mycangia, the oidia are set free, start growing and produce mycelium, which subsequently breaks up to oidia again and are ready to be transferred together with the eggs (Francke-Grosmann 1967).

Amylostereum areolatum and *A. chailletii* may, therefore, be spread by a purely vegetative process without going through sexual recombination. However, the existence of fruitbodies of both *A. areolatum* and *A. chailletii* shows that sexual recombination and production of basidiospores is another dispersive method besides woodwasps as vectors. Pechmann et al. (1967) found that the basidiospores of a species, which they identified as *A. areolatum* but probably was *A. chailletii*, were able to colonize freshly cut disks of spruce wood.

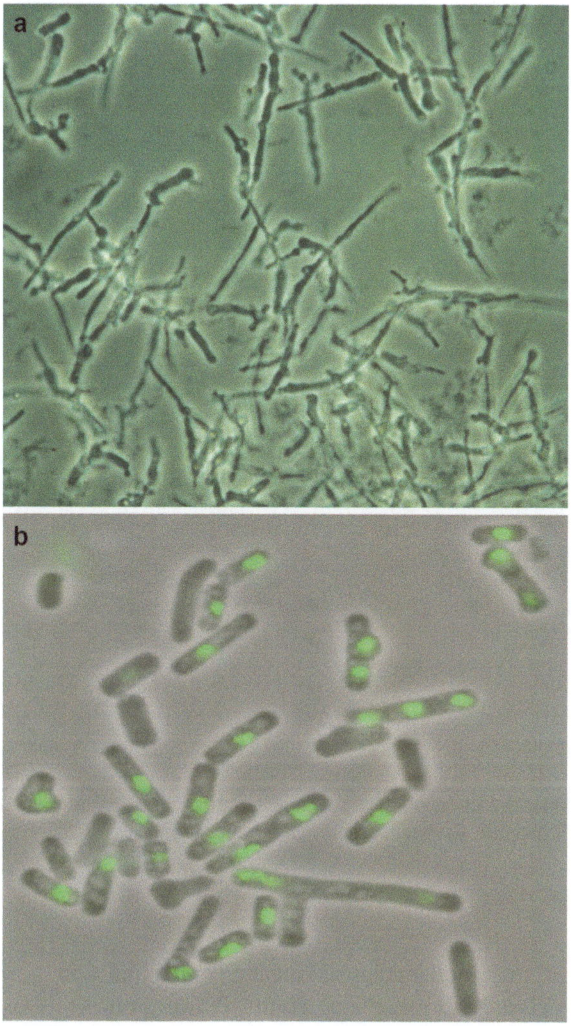

Fig. 5.6 (**a**) Hyphal fragments (arthrospores), often called oidia, of *A. areolatum* from the mycangia of a female *S. noctilio* woodwasp and (**b**) from a monokaryotic isolate in culture on malt yeast agar with DAPI stained nuclei. Notice the oidia from mycelium have a more regular shape of than those seen in the mycangia of the woodwasp. (Photos B. Slippers)

In culture, both *A. areolatum* and *A. chailletii* have numerous clamp connections on their hyphae, and later the mycelium develops cystidia, i.e. hyphal ends that are pointed and encrusted (Stalpers 1978; Slippers 1998). In some cases, the colour of the mycelium grown on potato dextrose agar (PDA) can be used to differentiate *A. areolatum* and *A. chailletii* (Thomsen 1998; Thomsen and Harding 2011). However, distinguishing *A. areolatum* from *A. chailletii* is most easily achieved by looking for oidia in culture (Siepmann and Zycha 1968; Stalpers 1978). Although both fungi break up into hyphal fragments reminiscent of oidia in the mycangia of the woodwasps, only *A. areolatum* produces oidia in culture. The oidia produced in the mycelium of *A. areolatum* grown on agar are more regularly shaped than those found in the mycangia of *S. noctilio* (Fig. 5.6) (Slippers 1998; Thomsen and Harding 2011).

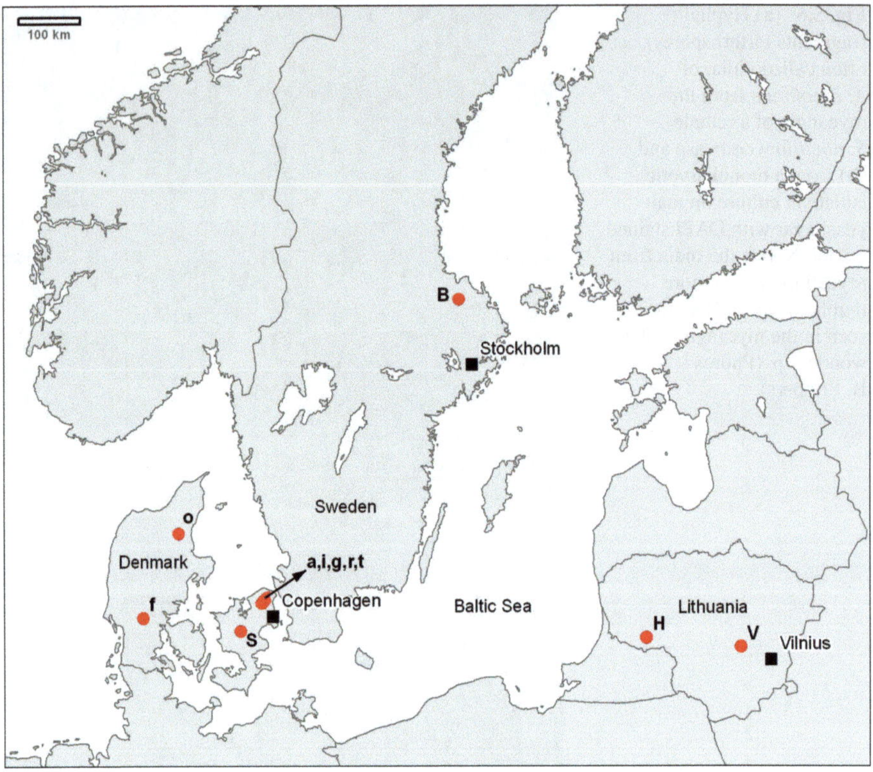

Fig. 5.7 *A. areolatum* vegetative compatibility group (VCG) found in Denmark, Sweden and Lithuania. Each letter represents one location

In several studies of somatic compatibility in *A. areolatum* and *A. chailletii*, the existence of vegetative compatibility groups (VCGs) has been demonstrated for both fungi (Vasiliauskas et al. 1998; Thomsen and Koch 1999; Vasiliauskas and Stenlid 1999; Slippers et al. 2001, 2002). There was a high degree of somatic compatibility between isolates of *A. areolatum*, and large, widespread clones have been found around the Baltic Sea (Fig. 5.7). In *A. chailletii*, clones were smaller and occurred geographically closer but more rarely, especially amongst fruitbody isolates. However, for both fungi, somatic compatibility could be found between isolates from woodwasps, fruitbodies and wood obtained over a period of more than 10 years (1983–1994). The rarity of *A. areolatum* fruitbodies is consistent with the total dominance of clones, with low genetic variation found in this species. Its rarity may also be related to the fact that *A. areolatum* is transferred by only two woodwasp species in Europe.

Fig. 5.8 Drought-mediated die-back of pine forests near Visp (Valais), Switzerland

5.5 Relationships Between *S. noctilio* Attack and Host Quality

The preferred substrates of siricids for oviposition are trees that have recently died or freshly produced stumps. Living trees are rarely attacked in Europe, and attacked trees most often suffer from some kind of stress that renders them susceptible to infestation (see Chap. 5.6). Colonization density in damaged living trees does not differ from that in wind-thrown timber (Spradbery and Kirk 1978).

In recent years, a die-back of pine forests has been observed in several inner-Alpine valleys in Europe (Fig. 5.8). A common feature of these valleys is their xerothermic environment. The role of insects in the pine (*Pinus sylvestris* L.) decline in the Swiss Rhone valley has recently been investigated (Wermelinger et al. 2008). In this study, some 200 pines with different levels of crown transparency (i.e. needle loss), which is an estimator of tree vigour, were felled between 2001 and 2005, and sections of their stems and branches were collected and incubated in insect emergence traps. Colonization densities were then related to the transparency level of each host tree at the time of attack.

Out of 209 logs, only 16 were colonized by *S. noctilio*. At the time of emergence, colonized trees exhibited a crown transparency between 75% and 100%, 80% of the trees suffered from 90% or more needle loss (Fig. 5.9a). Woodwasps were found in densities of up to 70 emerging adults per meter stem length. The crown transparency of the colonized trees at the time of woodwasp attack was identified based on a 1 year

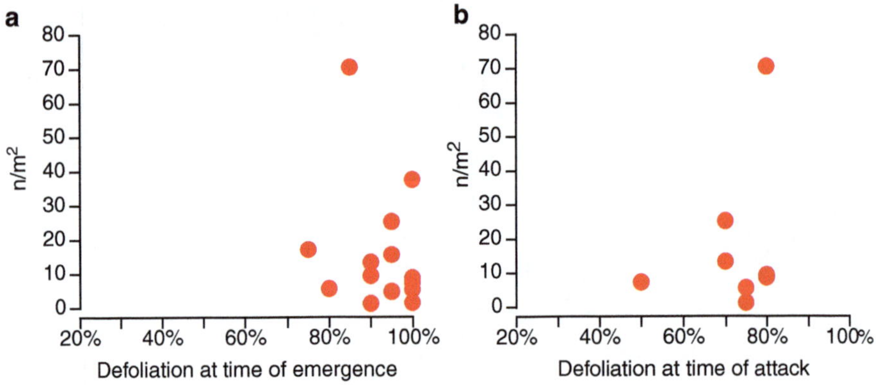

Fig. 5.9 (**a**) Offspring production of *S. noctilio* per m^2 bark surface in Scots pines with different crown transparencies (= levels of needle loss) at the time of Sirex emergence. (**b**) The same at the time of Sirex attack

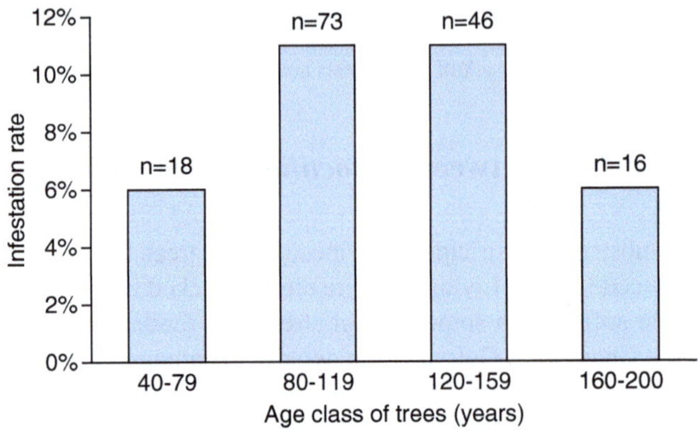

Fig. 5.10 Proportion of weakened pines infested with *S. noctilio* in different age classes

developmental time of *S. noctilio* and on the transparency records. The infestation densities showed that, with one exception, the ovipositing females preferred heavily defoliated trees with 70–80% needle loss (Fig. 5.9b). These trees had been suffering from drought stress and obviously had little vigour. Moribund trees were also reported to be optimal for woodwasp development in the countries of introduction (Morgan 1968). Infestation densities did not depend on changes in transparency shortly before the attack, i.e. trees were infested regardless of whether their needle mass had increased or decreased in the year before attack (B. Wermelinger, unpublished data). The woodwasps preferably attacked trees aged 80–160 years (Fig. 5.10). The highest breeding densities were found in trees with diameters of 15–30 cm dbh.

Once a tree is accepted as a host, its physiological condition affects both oviposition and larval development. The number of drills branching off from the entrance hole depends on the osmotic pressure of the phloem sap. In healthy trees with higher osmotic pressure, fewer tunnels are drilled and, therefore, fewer eggs are deposited (Coutts and Dolezal 1969). Wood moisture affects several of the biological traits of the progeny. In trees with a high moisture content, drillings often remain sterile, but the woodwasps inject fungal spores, which predisposes the tree for future colonization (Coutts and Dolezal 1969). The highest rate of egg deposition was found on wood with 60% humidity based on oven dry weight (Morgan 1968; Eichhorn 1982). The moisture content and lipid levels of the wood are also crucial for the establishment and growth of *A. areolatum* (Morgan 1968). Furthermore, it influences the larval boring behaviour and the size of the emerging adults, even from offspring originating from the same egg batch. The length and shape of the galleries also depend on the growth of the fungal symbiont and on larval competition (Viitasaari and Heliövaara 2004).

5.6 Economic Significance of *S. noctilio* in Europe

Unlike in the countries where it has been accidentally introduced (New Zealand, Australia, various countries in South America and South Africa), *S. noctilio* and woodwasps in general are considered an irrelevant factor for tree mortality in Europe. A review on detrimental bark and wood boring insects in Europe lists three siricid species, but not *S. noctilio* (Långström et al. 2004). This indicates that the species has been largely insignificant as yet in European silviculture. The woodwasps preferably colonize trees that have been weakened or already killed by primary insect attack such as leaf feeders and bark beetles or wounded by deer browsing, pruning or harvest machinery (Escherich 1942; Schimitschek 1968; Spradbery and Kirk 1978; Wermelinger et al. 2008). Trees 1–2 weeks after felling and fresh stumps are particularly attractive to siricids (Simpson and McQuilkin 1976). In a survey in Greece on pines impaired by wildfires, *S. noctilio* attack was observed at only one location out of five. European *Pinus radiata* plantations seem not to be at risk; in south-western Europe and England most infestations were recorded from scattered and suppressed trees (Hall 1968; Spradbery 1973). Often *S. noctilio* appeared to first colonize mechanically damaged trees and then attack the least vigorous or newly dead trees in the vicinity. Moreover, vigorous trees that have been successfully attacked are rarely killed (Spradbery and Kirk 1978).

Woodwasp infestations may have some effect on timber quality, especially in windfalls supplying abundant breeding substrates. Because the larval galleries are tightly packed with boring dust, they are difficult to detect in cut timber. Adult woodwasps may emerge even several years after timber processing and thus reduce the aesthetic value or the stability of woody products.

There is an evident difference in the aggressiveness of *S. noctilio* in European forests and in exotic plantations in the Southern Hemisphere. Most pines in Europe are native and adapted to the regional environment. In the Southern Hemisphere *P. radiata* has been introduced and is mostly grown in large and dense monospecific plantations. They suffer from various stressors such as intraspecific competition and drought (Morgan 1968). In addition, attacks in summer occur approximately 1 month earlier than in Europe. At this point in a tree's development its susceptibility

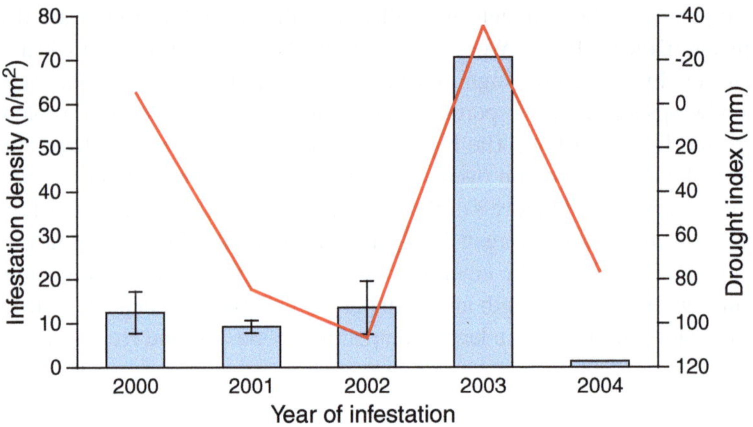

Fig. 5.11 Yearly density of *S. noctilio* infestations of Scots pines (columns) and development of the drought index (difference between precipitation and potential evaporation from January to August; solid line)

seems to be higher, thus increasing the risk of attack by *S. noctilio* (Spradbery and Kirk 1978).

As outlined above, *S. noctilio* is not at present of major economic significance in Europe, but it may have greater impact in the future in view of the ongoing climate change. In Europe, the woodwasp is more abundant in regions with low precipitation levels and high temperatures, i.e. in Mediterranean climates (Spradbery and Kirk 1978). In the Swiss Rhone valley a clear relationship between drought intensity and attack by the buprestid *Phaenops cyanea* could be demonstrated (Wermelinger et al. 2008). A similar pattern can be found for *S. noctilio* (Fig. 5.11). The exceptionally warm and dry year 2003 appeared to favour a woodwasp infestation of weakened pines. Global warming is expected to increase average and extreme temperatures and to alter the precipitation regimes (e.g., Fuhrer et al. 2006). If this happens, woodwasp development will accelerate, generation times will probably decrease, and the abundance of susceptible host trees suffering from drought stress will most likely increase. Therefore, *S. noctilio* may become more significant in Europe in the future but it will have to compete with numerous other insect species for the weakened hosts.

References

Bachmaier F (1966) Übersicht und Bestimmungstabelle der europäischen nadelholzbewohnenden Siriciden (Hymenoptera, Symphyta). Anz Schädl kde 39:129–132

Bedding RA, Akhurst RJ (1974) Use of the nematode *Deladenus siridicola* in the biological control of *Sirex noctilio* in Australia. J Aust Entomol Soc 13:129–135

Benson RB (1951) Handbook for the identification of British insects; vol VI Hymenoptera, 2. Symphyta, section (a). Royal Entomological Society of London, London

Boidin J (1958) Hétérobasidiomycètes saprophytes et Homobasidiomycètes résupinés. V. Essai sur le Genre *Stereum* Pers. ex S.F.Gray. Rev Myc 23:318–346

Boidin J, Lanquetin P (1984) Le genre *Amylostereum* (Basidiomycetes), intercompatibilités partielles entre espèces allopatriques. Bull Soc Mycol Fr 100:211–236
Bourdot H, Galzin A (1921) Hyménomycètes de France, VII. *Stereum*. Bull Soc Mycol Fr 37:117–130
Bourdot H, Galzin A (1927) Hyménomycètes de France. Hetérobasidiés - Homobasidiés gymnocarpes. Soc Mycol Fr, Paris, pp 378–379
Breitenbach J, Kränzlin F (1986) Fungi of Switzerland. vol 2 Fungi of Switzerland Heterobasidiomycetes, Aphyllophorales, Gasteromycetes. Mykologia, Lucerne
Browne FG (1968) Pest and diseases of forest plantation trees. An annotated list of the principal species occurring in the British commonwealth. Clarendon Press, Oxford
Coutts MP, Dolezal JE (1969) Emplacement of fungal spores by woodwasp, *Sirex noctilio*, during oviposition. For Sci 15:412–416
Eichhorn O (1982) Siricoidea. In: Schwenke W (ed) Die Forstschädlinge Europas, vol 4 Hautflügler und Zweiflügler. Paul Parey, Hamburg, pp 196–231
Eriksson J, Hjortstam K, Ryvarden L (1978) The Corticiaceae of North Europe, vol 5. Fungiflora, Oslo, pp 890–893
Escherich K (1942) Die Forstinsekten Mitteleuropas. Ein Lehr- und Handbuch. vol 5 Hymenoptera (Hautflügler) und Diptera (Zweiflügler). Paul Parey, Berlin
Francke-Grosmann H (1957) Über das Schicksal der Siricidenpilze während der Metamorphose. Bericht 8. Wandervers deutsch Entomol 11:37–43
Francke-Grosmann H (1967) Ectosymbiosis in wood-inhabiting insects. V. Symbiosis in Siricidae and Xiphydriidae. In: Henry SM (ed) Symbiosis, vol 2 – Associations of invertebrates, birds, ruminants and other genera. Academic Press, London, pp 141–205
Fuhrer J, Beniston M, Fischlin A et al (2006) Climate risks and their impact on agriculture and forests in Switzerland. Clim Change 79:79–102
Gilmour JW (1965) The life cycle of the fungal symbiont of *Sirex noctilio*. NZ J For 10:80–89
Hall MJ (1968) A survey of siricid attack on radiata pine in Europe. Aust For 32:155–162
Hurley BP, Slippers B, Wingfield MJ (2007) A comparison of control results for the alien invasive woodwasp, *Sirex noctilio*, in the southern hemisphere. Agric For Entomol 9:159–171
Jahn H (1968) Pilze an Weißtanne (*Abies alba*). Westf Pilzbr 7:17–40
Jahn H (1971) Stereoide Pilze in Europa (Stereaceae Pil. emend. Parm. u.a., Hymenochaete) mit besonderer Berücksichtigung ihres Vorkommens in der Bundesrepublik Deutschland. Westf Pilzbr 8:69–176
Jahn H (1979) Pilze die an Holz wachsen. Busse, Herford
Kukor JJ, Martin MM (1983) Acquisition of digestive enzymes by siricid woodwasps from their fungal symbiont. Science 220:1161–1163
Långström B, Heliövaara K, Moraal LG et al (2004) Non-coleopteran insects. In: Lieutier F, Day KR, Battisti A et al (eds) Bark and wood boring insects in living trees in Europe – a synthesis. Kluwer Academic Publishers, Dordrecht, pp 501–538
Madden JL (1974) Oviposition behavior of woodwasp, *Sirex noctilio* F. Aust J Zool 22:341–351
Madden JL (1981) Egg and larval development in the woodwasp, *Sirex noctilio* F. Aust J Zool 29:493–506
Madden JL (1988) *Sirex* in Australasia. In: Berryman AA (ed) Dynamics of forest insect populations. Plenum Press, New York, pp 407–429
Martin MM (1987) Acquired enzymes in the siricid woodwasp *Sirex cyaneus*. In: Martin MM (ed) Invertebrate-microbial interactions. Cornell University Press, Ithaca, pp 37–48
Morgan FD (1968) Bionomics of Siricidae. Annu Rev Entomol 13:239–256
Morgan D, Stewart NC (1966) Effect of *Rhyssa persuasoria* (L) (Ichneumonidae) on a population of *Sirex noctilio* F. (Siricidae). Trans R Soc NZ Zool 8:31–38
Pechmann H von, Aufsess H (1971) Untersuchungen über die Erreger von Stammfäulen in Fichtenbeständen. Forstwiss Centralbl 90:259–284
Pechmann H von, Aufsess H, Liese W et al (1967) Untersuchungen über die Rotstreifigkeit des Fichtenholzes. Forstwiss Forsch, Beihefte Forstwiss Centrbl 27
Schimitschek E (1968) Ursachen der Befallsbereitschaft für Holzwespen, Ausmass der Schäden und Vorbeugungsmassnahmen. Z ang Entomol 61:45–60

Schönhar S (1969) Untersuchungen über das Vorkommen von Rotfäulepilzen in Fichtenbeständen der Schwäbischen Alb. Mitt Ver Forstl Stand Forstpfl 19:20–28

Siepmann R (1971) *Über Odontia* bicolor (Alb. & Schw) Quél. und *Amylostereum areolatum* (Fr.) Boidin, zwei stammfäuleerregende Basidiomyceten in lebenden Fichten (*Picea abies*). Forstwiss Centralbl 90:337–340

Siepmann R, Zycha H (1968) Artdiagnose einiger holzzerstörender Hymenomyzeten an Hand von Reinkulturen. Nova Hedwigia 15:559–569

Simpson RF, McQuilkin RM (1976) Identification of volatiles from felled *Pinus radiata* and electroantennograms they elicit from *Sirex noctilio*. Entomol Exp Appl 19:205–213

Slippers B (1998) The *Amylostereum* symbiont of *Sirex noctilio* in South Africa. MS thesis University Orange Free State, South Africa

Slippers B, Wingfield MJ, Coutinho TA et al (2001) Population structure and possible origin of *Amylostereum areolatum* in South Africa. Plant Pathol 50:206–210

Slippers B, Wingfield BD, Coutinho TA et al (2002) DNA sequence and RFLP data reflect geographical spread and relationships of *Amylostereum areolatum* and its insect vectors. Mol Ecol 11:1845–1854

Solheim H (2006) Treveps og assosierte sopper i Norge [Woodwasps and associated fungi in Norway]. Agarica 26:87–95

Spradbery JP (1973) A comparative study of the phytotoxic effects of siricid woodwasps on conifers. Ann Appl Biol 75:309–320

Spradbery JP (1977) The oviposition biology of siricid woodwasps in Europe. Ecol Entomol 2:225–230

Spradbery JP (1990) Predation of larval siricid woodwasps (Hymenoptera: Siricidae) by woodpeckers in Europe. Entomol 109:67–71

Spradbery JP, Kirk AA (1978) Aspects of the ecology of siricid woodwasps (*Hymenoptera: Siricidae*) in Europe, North Africa and Turkey with special reference to the biological control of *Sirex noctilio* F. in Australia. Bull Entomol Res 68:341–359

Stalpers JA (1978) Identification of wood-inhabiting *Aphyllophorales* in pure culture. Studies in mycology 16. Centraalbureau voor Schimmelculture, Baarn

Stillwell MA (1966) Woodwasps (Siricidae) in conifers and the associated fungus, *Stereum chailletii*, in eastern Canada. For Sci 12:121–128

Taeger A, Blank SM (2008) Fauna Europaea: Siricidae. In: van Achterberg K (ed) Fauna Europaea: Symphyta. Fauna Europaea version v. 1.3, http://www.faunaeur.org

Taylor KL (1976) The introduction and the establishment of insect parasitoids to control *Sirex noctilio* in Australia. Entomophaga 21:429–440

Thomsen IM (1996) *Amylostereum areolatum & Amylostereum chailletii*, symbiotic fungi of woodwasps (*Sirex* sp. and *Urocerus* sp.). Forest & Landscape, Ph.D. thesis, 130 pp

Thomsen IM (1998) Fruitbody characters and cultural characteristics useful for recognizing *Amylostereum areolatum* and *A. chailletii*. Mycotaxon 69:419–428 (note that photos of basidiospores are switched)

Thomsen IM, Harding S (2011) Fungal symbionts of siricid woodwasps: isolation techniques and identification. For Pathol 41:325–333

Thomsen IM, Koch J (1999) Somatic compatibility in *Amylostereum areolatum* and *A. chailletii* as a consequence of symbiosis with woodwasps. Mycol Res 103:817–823

Vasiliauskas R, Stenlid J (1999) Vegetative compatibility groups of *Amylostereum areolatum* and *A. chailletii* from Sweden and Lithuania. Mycol Res 103:824–829

Vasiliauskas R, Stenlid J, Thomsen IM (1998) Clonality and genetic variation in *Amylostereum areolatum* and *A. chailletii* from northern Europe. New Phytol 139:751–758

Viitasaari M, Heliövaara K (2004) Siricidae (horntails). In: Lieutier F, Day KR, Battisti A et al (eds) Bark and wood boring insects in living trees in Europe – a synthesis. Kluwer Academic Publishers, Dordrecht, pp 529–534

Wermelinger B, Rigling A, Schneider Mathis D et al (2008) Assessing the role of bark- and wood-boring insects in the decline of Scots pine (*Pinus sylvestris*) in the Swiss Rhone valley. Ecol Entomol 33:239–249

Chapter 6
Genetics of *Amylostereum* Species Associated with Siricidae Woodwasps

Magriet A. van der Nest, Brenda D. Wingfield, Michael J. Wingfield, Jan Stenlid, Rimvydas Vasaitis, and Bernard Slippers

Abstract The basidiomycete fungi, *Amylostereum areolatum*, *A. chailletii* and *A. laevigatum*, are closely associated with various Siricid woodwasps. DNA-based studies have shown that specific Siricid species always carry the same *Amylostereum* species, suggesting that the association between the fungus and woodwasp is specific. Of these associations, the *A. areolatum* and *Sirex noctilio* complex represent an important threat to pine based forestry in countries where it has been introduced. For this reason, various molecular tools have been used to study the phylogenetics, population genetics and genomics of *Amylostereum* spp., particularly *A. areolatum*. The phylogenetic studies revealed an evolutionary relationship between *Amylostereum* species and *Echinodontium* species, as well as with *Russula* species, *Heterobasidion* species and *Peniophora* species. Studies making use of vegetative compatibility groups (VCGs) and molecular markers investigating the population biology of *A. areolatum* and *A. chailletii* have increased our knowledge regarding the diversity and spread of these fungi. Molecular studies have also shed light on how the association with the wasp has influenced the ecology, biology and evolution of the fungal partner. Due to the association between the fungus and the woodwasp the fungus is asexually distributed, resulting in the presence of dispersive clonal lineages which are preserved over time. Despite the asexual spread by the wasp, sexual reproduction continues to play an important part in the biology of these fungi. *Amylostereum* spp. are heterothallic and have a tetrapolar mating system. Recent studies suggest that

M.A. van der Nest (✉) • B.D. Wingfield • M.J. Wingfield • B. Slippers
Department of Genetics, Forestry and Agricultural Biotechnology Institute, University of Pretoria, Pretoria, South Africa
e-mail: magriet.vandernest@fabi.up.ac.za

J. Stenlid • R. Vasaitis
Department of Forest Mycology and Pathology, Swedish University of Agricultural Sciences, Uppsala, Sweden

B. Slippers et al. (eds.), *The Sirex Woodwasp and its Fungal Symbiont: Research and Management of a Worldwide Invasive Pest*, DOI 10.1007/978-94-007-1960-6_6,
© Springer Science+Business Media B.V. 2012

these mating type genes appear to be similar in structure and function to that of other Homobasidiomycete fungi. Here we review these recent developments and opportunities for future research.

6.1 Introduction

Amylostereum areolatum (Basidiomycotina) lives in an obligate mutualistic symbiosis with various Siricid woodwasps, including *Sirex noctilio* (Hymenoptera: Siricidae) (Morgan 1968; Slippers 1998; Slippers et al. 2003, Chap. 2). The *A. areolatum* – *S. noctilio* complex is native to Eurasia and North Africa (Morgan 1968; Spradbery and Kirk 1978, Chap. 5). During the last 100 years it has been introduced throughout the Southern Hemisphere (Hurley et al. 2007, Chaps. 13–19). The wasp/fungus complex poses no substantial economic threat in native environments, however it represents one of the most serious sources of damage to pine-based forestry in countries where it has been introduced (Carnegie et al. 2006; Hurley et al. 2007).

Amylostereum areolatum is considered a weak pathogen of many of *Pinus* spp., but the action of the fungus and the woodwasp collectively, can kill the host tree (Coutts and Dolezal 1969; Fong and Crowden 1973; Madden 1981, Chaps. 2, 3). Aside from carrying *A. areolatum* spores, the wasps produce a phytotoxic mucus that further reduces the resistance of the host tree. This enables the fungus to establish and eventually cause a dry white rot in the wood, which disrupts the vascular system of the host. The larvae of the wasp spend 1–3 years in the tree boring tunnels as they feed, resulting in limited damage which does not contribute markedly to tree death.

In countries were *S. noctilio* and its fungal symbiont have been introduced, the nematode *Deladenus (=Beddingia) siricidicola* is used as the primary control agent (Bedding 1995; Haugen 1990; Hurley et al. 2007, Chap. 9). The nematode sterilizes the female wasps during a parasitic phase of its life cycle when its juveniles enter the eggs of the wasp. During a second, mycetophagous phase of the *D. siricidicola* life cycle it feeds on *A. areolatum*. This latter phase of the life cycle is exploited to mass-rear the fungus for biocontrol purposes (Bedding 1995). The fungus is consequently a critical factor in understanding the biology of the nematode, and in control programs to reduce the impact of the wasp.

The success of biological control programs varies significantly in different regions (Bedding 1995; Hurley et al. 2008, Chaps. 13–18). Despite years of research, the reasons for the variable levels of success are poorly understood (Hurley et al. 2007, 2008). As a consequence, the pest complex continues to spread to previously unaffected areas and still kills significant numbers of trees in many areas (Hoebeke et al. 2005; Hurley et al. 2007, Chaps. 9, 18–19).

A thorough understanding of the biology of *A. areolatum* is required in order to understand the factors that affect the mutualism as a whole, as well as from the perspective of developing control strategies. In recent years, various molecular tools have been applied to study the phylogenetics, population genetics and genomics of

Amylostereum spp. The aim of this chapter is to review these recent studies and consider how they have influenced the understanding of the systematics, ecology and evolution, as well as control of the fungus and its woodwasp host.

6.2 Systematics and Identification

Based on morphological similarities of the sexual fruiting structures and vegetative mycelia *Amylostereum* has historically been linked to *Stereum* and later *Peniophora* (Boidin and Lanquetin 1984; Slippers et al. 2003). However, phylogenetic studies have revealed a relationship between *Amylostereum* spp. and *Echinodontium, Russula, Heterobasidion* and *Bondarzewia* in addition to *Peniophora* and *Stereum* spp. (Hsiau 1996; Hibbett et al. 1997; Kim and Jung 2000; Slippers et al. 2002; Maijala et al. 2003). Studies based on sequence data for the nuclear and mitochondrial ribosomal RNA internal transcribed spacer (nuc-ITS) and 16S small subunit gene (mtSSU) demonstrated that the genus *Amylostereum* is more closely related to *Echinodontium tinctorium* than to other fungi (Tabata et al. 2000; Maijala et al. 2003; Miller et al. 2006). *Amylostereum* and *Echinodontium* species also share micro-morphological characteristics, such as their amyloid basidiospores and thick-walled and heavily encrusted cystidia. These genera are currently both accommodated in the family Echinodontiaceae (Slippers et al. 2003). However, the phylogenetic relationships of *Amylostereum* spp. are still not strongly supported in molecular analyses and need further attention (Miller et al. 2006).

The genus *Amylostereum* includes four species, namely the type species *A. chailletii* (Pers.:Fr.) Boid. (=*Stereum chailletii*), *A. areolatum* (Fr.) Boid. (=*S. areolatum*), *A. laevigatum* (Fr.) Boid (=*Peniophora laevigata*) and *A. ferreum* (Berk. & Curt.) Boid. & Lanq. (=*S. ferreum*). All of these species have smooth amyloid basidiospores, hyaline-encrusted cystidia, and resupinate to effuso-reflexed fruiting bodies (Boidin 1958). These species can be distinguished based on morphology of the sexual fruiting structures, using characters such as spore size and the colour and texture of the basidiocarps (Thomsen 1998, Chap. 5). However, an important limitation of this approach is that basidiocarps of these fungi are rarely found in nature and in some areas, such as in the Southern Hemisphere, they have never been observed (Slippers et al. 2003, B. Slippers, unpublished data). For this reason, the identification of *Amylostereum* spp. has in the past also been done based on morphology in culture (e.g., mono- or dimyctic hyphae and the formation of arthrospores) or asexual spores found in the mycangia of the woodwasp (Thomsen 1996; Slippers et al. 2003). These characters, however, have limited variation and substantial experience is required to apply them accurately.

The phylogenetic relationship amongst species of *Amylostereum* was first determined using mating experiments and the Buller phenomenon (Buller 1931; Boidin and Lanquetin 1984). *Amylostereum areolatum* was not compatible with any of the other species, and was thus considered to be the most divergent species in the genus, despite its similarity to *A. chailletii*. *Amylostereum laevigatum* and *A. chailletii* were considered to be both more closely related to *A. ferreum* than to each other, as *A. chailletii* and *A. laevigatum* were completely incompatible while they were both partially compatible with *A. ferreum*. These relationships were later confirmed using comparisons of sequence data for nuc-ITS, the ribosomal RNA intergenic spacer (IGS) region, mtSSU and the partial manganese-dependent peroxidase gene. These studies also showed that *A. ferreum* and *A. laevigatum* were more closely related to each other, than to *A. chailletii* and *A. areolatum* (Vasiliauskas et al. 1999; Slippers et al. 2000, 2002; Tabata et al. 2000; Nielsen et al. 2009). Sequence data for the IGS region could consequently be used to develop a rapid and accurate PCR-RFLP (restriction fragment length polymorphism) method to distinguish between the species of *Amylostereum* (Slippers et al. 2002). However, phylogenetic studies have also shown that there is at least one undescribed species of *Amylostereum* in North America (Slippers et al. 2000).

6.3 Symbiotic Specificity

The results of DNA sequence comparisons have confirmed previous reports that specific Siricid species always carry the same *Amylostereum* species, i.e. the association between the fungus and woodwasp appear to be specific for the wasp host (Tabata and Abe 1997, 1999; Slippers et al. 2003, B. Slippers et al. unpublished). Recent work on the Xiphydridae, a group of woodwasps closely related to *S. noctilio* that infest hardwoods, has shown that a species of these wasps can be associated with more than one *Daldinia* spp. (Paźoutavá et al. 2010). For example, while a single *Daldinia* spp. normally dominated the association with a specific *Xiphidria* spp. (e.g., *X. longicollis* associated with *D. childiae*), a small number of wasps associated with different symbiont species (e.g., *X. camelus* associated with *D. decipiens* and *D. petriniae*). This was observed in a study of fungal associates of a large number of wasps (n=1389), using molecular identification tools. Such large scale studies have not been done for Siricidae and this begs the question as to whether the current views regarding symbiont specificity in the Siricidae are not based on inordinately few samples and incorrect identifications. Also, studies of smaller numbers of wasps possibly would have overlooked these rare cases of symbiont switching.

It is believed that transmission of *Amylostereum* is uniparental and vertical between host generations, i.e. offspring obtain the fungal mutualist directly from the female parent. Strict vertical transmission of the symbiont would, however, predict a tight pattern of co-evolution of species of the fungus and wasp. On the contrary, however, one *Amylostereum* species can be carried by more than one wasp species and even different wasp genera. For example, *A. areolatum* is the symbiont of *S. noctilio*, *S. juvencus* Linn., *S. nitobei* Mats., *Sirex* sp. "*nitidus*" and *S. edwardsii* Brullé (=*S. nigricornis*; Chap. 1), while *A. chailletii* is carried by *S. cyaneus* Fabr., *S. imperialis* Kirby., *S. areolatus* Cress., *S. californicus* Nort., *Urocerus gigas* Linn., *U. augur augur* Klug. and *U. augur sah* Mocs. (Gaut 1969, 1970; Tabata et al. 2000; Slippers et al. 2003; Nielsen et al. 2009). These patterns suggest that there is not a strict co-evolution of species. Rather that historically, species of

Amyostereum have been exchanged horizontally between the different species and genera of the Siricidae. However, the specificity of the wasp/fungus association is still under debate (Nielsen et al. 2009). Also, not all of the wasp/fungus associations have fully been discovered, especially in North America.

Nielsen et al. (2009) found identical genotypes of *A. areolatum* associated with *S. noctilio* and *S. edwardsii* (=*S. nigricornis*; Chap. 1), which was thought to carry *A. chailletii*, when they emerged from the same logs. Identical genotypes of *A. areolatum* have also been shown to be associated with *S. juvencus* and *S. noctilio* (B. Slippers, J. Stenlid and R. Vasaitis, unpublished data). The latter authors have also found identical genotypes of *A. chailletii* to occur in association with *U. augar* and *U. gigas*. These results suggest that horizontal transmission of fungal genotypes between wasp generations, and even species, might occur fairly commonly, given that none of the studies considered particularly large numbers of specimens. Larger sample sizes from local populations containing different wasp species are needed to clarify the question of how frequently horizontal transmission of fungal genotypes occur between different wasp generations and different species.

Amylostereum laevigatum occurs throughout Europe, but is not known to be associated with woodwasps in this region. In Japan, *A. laevigatum* has, however, been shown to be associated with *U. japonicas* and *U. antennatus* (Tabata and Abe 1997, 1999). *Amylostereum ferreum* is not known to be associated with any woodwasps.

6.4 Mode of Reproduction

Typical of a homobasidiomycete fungus, *Amylostereum* spp. has a life cycle (Fig. 6.1) consisting of a short homokaryotic phase (i.e. having hyphae with one single type of genetically distinct nucleus), followed by a predominant heterokaryotic phase (i.e. hyphae with more than one type of genetically distinct nucleus) (Kües 2000). The fertile heterokaryotic phase arises when the hyphae of two sexually compatible homokaryons fuse (mate). Karyogamy (nuclear fusion) and meiosis occur later within specialized cells of the basidiocarp. This temporal and spatial separation of hyphal fusion and karyogamy is unique to *A. areolatum* and other Basidiomycetes.

Amylostereum basidiocarps are typically resupinate and smooth macroscopically (Boidin 1958; Thomsen 1996). The fruiting structures appear infrequently and in unique distribution patterns for each of the species. Basidiocarps of *A. chailletii* are, for example, more common than *A. laevigatum* and *A. areolatum* in Europe (Vasiliauskas and Stenlid 1999; Slippers et al. 2003, B. Slippers, R. Vasaitis and J. Stenlid, personal observation, Chap. 5). Fruiting bodies of *A. areolatum* are much rarer and geographically restricted, occurring most frequently in central parts of Europe (Thomsen 1996; Solheim 2006, B. Slippers, R. Vasaitis and J. Stenlid, personal observation, Chap. 5). The fruiting structures of the latter species have never been reported from North America, the Southern Hemisphere or Asia, where it is known to occur (Slippers et al. 2003; Nielsen et al. 2009). This is despite the fact that *A. areolatum* isolates from the Southern Hemisphere have been shown to be

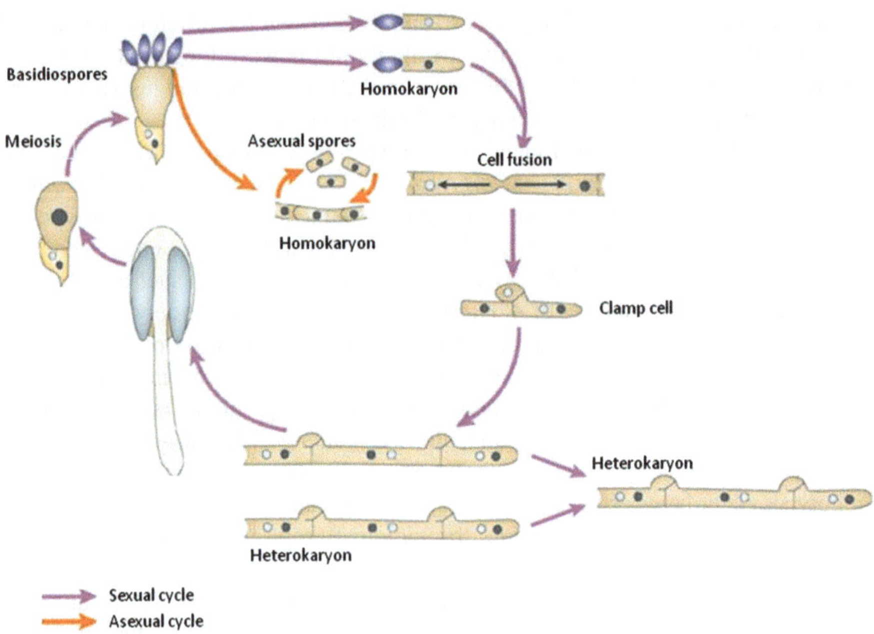

Fig. 6.1 A typical basidiomycete life-cycle consists of a short homokaryotic phase followed by a predominant fertile heterokaryotic phase (Kothe 2003) (Figure adapted from Casselton and Zolan (2002)

able to fruit in the laboratory (Talbot 1977). The patterns of fruiting of *A. laevigatum* are less well documented, but are known to be rare.

Amylostereum areolatum is a heterothallic basidiomycete that has a tetrapolar mating system (Fig. 6.2). In this mating system the alleles present on two unlinked mating type (*mat*) loci (*mat-A* and *mat-B*) must be different for two homokaryons to be sexually compatible (Boidin and Lanquetin 1984; van der Nest et al. 2008, 2009). The *mat-A* locus harbours genes that encode homeodomain transcription factors, while the *mat-B* locus harbours genes encoding pheromones and pheromone receptors that control separate but complementary pathways involved in sexual compatibility (Casselton 2002). Only a gene encoding a putative pheromone receptor (*RAB1*) at the *mat-B* locus and a gene encoding a mitochondrial intermediate peptidase (*MIP*), that is closely linked to the *mat-A* locus, have been identified for *A. areolatum* (van der Nest et al. 2008). Future studies could utilise this information to obtain a larger portion of the *mat* loci, as well as to investigate the structure and evolution of these loci.

The *mat* loci of basidiomycete fungi that have tetrapolar mating systems are usually multiallelic (Kothe et al. 1999, 2003; Schirawski et al. 2005; James et al. 2006). This is also true for the *mat* loci of *A. areolatum* (van der Nest et al. 2008, 2009). Despite the unique life history of the mutualistic association with the woodwasps, these latter studies have shown that the *mat* loci of this fungus appear to be subject to similar evolutionary forces as those acting on the recognition loci of other

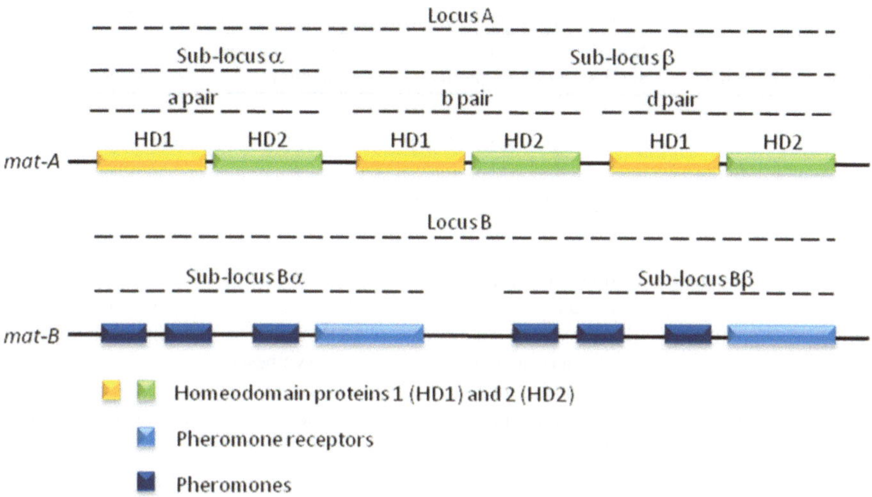

Fig. 6.2 The genes located at the *mat-A* locus of basidiomycetes encode homeodomain proteins, while the genes present at the *mat-B* locus encode pheromones and pheromone receptors (Kronstad and Staben 1997; Casselton and Olesnicky 1998; Casselton 2002)

eukaryotes (van der Nest et al. 2008, 2009). These forces may include balancing selection and the selection for rare alleles, as well as suppressed recombination that retain the polymorphic nature of these loci (Awadalla and Charlesworth 1999; May et al. 1999; van der Nest et al. 2009).

Amylostereum spp. can reproduce asexually through a vegetative mycelium that fragments into oidia or arthrospores. This mode of reproduction is associated with the symbiosis between *Amylostereum* and Siricid woodwasps. Asexual spores of the fungus are carried in specialized mycangia situated near the base of the ovipositor in the female wasps from where they are inoculated into the wood during oviposition (Coutts 1969; Coutts and Dolezal 1969; Madden 1981). This ensures that not only the same species, but the same genotype of the fungus is transmitted between generations. Such clonal, vertical transmission of a symbiont between host generations is common in mutualistic systems and is thought to align the reproductive interests of the different partners (Herre et al. 1999). As has been shown in other mutualistic systems, and discussed above, a certain level of horizontal acquisition of the symbiont (either from asexually or sexually produced lineages) also appears to occur.

6.5 Vegetative Incompatibility and Population Diversity Studies

A self/nonself recognition mechanism known as somatic or vegetative incompatibility that distinguishes individual genotypes from one another has been widely used to differentiate genotypes of *A. areolatum* and *A. chailletii* (Vasiliauskas et al. 1998;

Table 6.1 The *Neuropora crassa* and *P. anserina het* genes characterized at the DNA level

Gene	Class of gene	Nature of the protein and its functional motifs	Other cellular function	References
Neurospora crassa:				
mat A-1	Allelic *het* gene	Mating-type transcriptional regulator; α-domain	Mating	Glass et al. (1990)
mat a-1	Allelic *het* gene	Mating-type transcriptional regulator; HMG box	Mating	Staben and Yanofsky (1990)
het-c	Allelic *het* gene	Signal peptide, variable domain, glycine-rich region	None	Saupe et al. (1996), Saupe and Glass (1997)
un-24	Nonallelic *het* gene	Ribonucleotide reductase, allosteric activity site, variable domain	DNA synthesis	Smith et al. (2000)
het-6	Nonallelic *het* gene	Region with similarity to TOL and HET-E	None	Smith et al. (2000)
Podospora anserina:				
het-s	Allelic *het* gene	Prion analog; single amino acid differences alters allelic specificity	None	Coustou et al. (1997), Turcq et al. (1991)
het-e	Nonallelic *het* gene	GTP-binding domain, WD repeat, region with similarity to TOL and HET-6	None	Saupe et al. (1995)
het-d	Nonallelic *het* gene	GTP-binding domain, WD repeat, region with similarity to TOL and HET-6	None	Espagne et al. (2002)
het-c	Nonallelic het gene	Glycolipid transfer protein; amphipathic α-helix, single amino acid differences alters allelic specificity	Ascospore formation	Saupe et al. (1994)
het-r	Nonallelic het gene	GTP-binding domain, WD repeat, region with similarity to TOL and HET-6	None	Chevanne et al. (2009)

Thomsen and Koch 1999; Vasiliauskas and Stenlid 1999; Slippers et al. 2001). Fungal isolates, including those of *Amylostereum* spp., are considered to be vegetatively compatible when their interacting hyphae are able to merge (anastomose) and where hyphal anastomoses persists. In contrast, where cell death prevents hyphal anastomosis from persisting, isolates are considered incompatible. Evidence of incompatibility is usually characterized by the formation of a zone of sparse aerial hyphae between the interacting heterokaryons (Thomsen and Koch 1999; Worrall 1997; van der Nest 2008).

Heterokaryons of basidiomycete fungi are vegetatively compatible when they are genetically similar and share identical alleles at all of their *het* loci (Table 6.1). Conversely, they are vegetatively incompatible when they are genetically different with different alleles at some or all of their *het* loci (Rayner 1991; Worrall 1997).

Amylostereum areolatum has a relatively low number of *het* loci (at least two) (van der Nest et al. 2008), which is similar to the number of *het* loci present in other Homobasidiomycetes. For example, *Phellinus weirii* has a single *het* locus, *Serpula lacrymans* has two *het* loci and *Heterobasidion annosum* has three to four *het* loci (Hansen et al. 1994; Kauserud et al. 2006; Lind et al. 2007). The two *het* loci identified in *A. areolatum* appear to be multiallelic, similar to the other homobasidiomycetes mentioned above (Stenlid and Vasiliauskas 1998; Lind et al. 2007).

The limited number of *het* loci in *Amylostereum* spp. may limit the ability of vegetative compatibility group (VCG) assays to distinguish between genetically different individuals, and so lead to an underestimation of the genetic diversity. This is despite the fact that in some cases different genets with identical DNA fingerprinting profiles have been distinguished using VCGs (one case in *A. areolatum*, and two cases in *A. chailletii*) (Vasiliauskas et al. 1998; Kauserud 2004; van der Nest et al. 2008). It is also important to recognise that VCG assays are not selectively neutral. They should therefore ideally be used in combination with selectively neutral genetic markers such as neutral DNA based markers (van der Nest et al. 2008). This has been achieved in *A. areolatum* and *A. chailletii* by combining data from VCGs and DNA fingerprinting using the M13 core sequence as a primer, RFLP and DNA sequencing data (Vasiliauskas et al. 1998; Thomsen and Koch 1999; Vasiliauskas and Stenlid 1999; Slippers et al. 2001).

VCGs, DNA sequencing and DNA fingerprinting studies with *A. areolatum* have shown the presence of large clonal populations that persist over time and space (Vasiliauskas et al. 1998; Thomsen and Koch 1999; Vasiliauskas and Stenlid 1999; Slippers et al. 2001). For example, *A. areolatum* isolates that belong to the same VCG were isolated in Denmark, Sweden and Lithuania (Thomsen and Koch 1999). This pattern differs from the diversity observed in central parts of Europe (Austria and the Czech Republic), where 55 isolates from wood represented 50 distinct VCGs (B. Slippers, T. Kirisits and R. Vasaitis, unpublished data). These patterns of diversity correlate with levels of sexual sporulation of the fungus, which is relatively common in central Europe and rare in Scandinavia (as discussed above). These results suggest that genotypic diversity is likely to be strongly influenced by the ability of the fungus to reproduce and spread sexually in an area, or predominantly asexually via the wasp in areas where the fungus fruits less successfully.

An even more extreme example of the presence of clonal territorial lineages is found in isolates of *A. areolatum* from Brazil, Uruguay and South Africa, which belong to the same VCG. This is in contrast to isolates from New Zealand and Tasmania that displayed only partial vegetative compatibility (Slippers et al. 2001). The genetic similarity of these populations was confirmed based on the fact *A. areolatum* isolates from South Africa, Brazil, New Zealand and Tasmania contained the same heterogenic combination of sequences for the IGS region (Slippers et al. 2001). These data indicate that the wasp/fungus complex was most likely introduced into the Southern Hemisphere only once or a limited number of times from the same origin, after which they spread between countries in the Southern Hemisphere (Slippers et al. 2001, 2002).

Clonal lineages (groups of isolates with identical VCG and DNA fingerprinting profiles) are present in *A. chailletii*, but to a much lesser degree than in *A. areolatum* (Vasiliauskas et al. 1998). The proportion of compatible isolates of *A. chailletii* was significantly lower than those for *A. areolatum* from Sweden and Lithuania indicating that *A. chailletii* spreads less frequently through asexual spores than *A. areolatum* in this region (Vasiliauskas et al. 1998). In fact, the population structure of *A. chailletii* in these regions is similar to that of natural populations of other airborne wood decaying Basidiomycetes that are not associated with insects (Stenlid et al. 1994; Högberg and Stenlid 1994; Vasiliauskas et al. 1998). Not only are there fewer clonal lineages present in *A. chailletii*, but clonal lineages are also not as widely distributed as those of *A. areolatum*. For example, in Denmark compatible isolates of *A. chailletii* were never separated by more than 75 km (Thomsen 1996; Vasiliauskas et al. 1998). These findings are concordant with the fact that the basidiocarps of *A. chailletii* are more commonly observed in forests of Scandinavia than those of *A. areolatum* (Thomsen 1998, Chap. 5).

6.6 Conclusions and Future Prospects

DNA-based studies during the course of the last decade have significantly increased our understanding of the relationships between *Amylostereum* spp. and between other Basidiomycetes. However, to determine the exact phylogenetic relationship among *Amylostereum* spp. and their relatives, multigene phylogenies that include all the representative taxa is needed. DNA based technologies have also provided the opportunity to accurately identify the different species in the genus. These studies have for example, led to the discovery of an undescribed species of *Amylostereum* in North America.

The ability to accurately identify *Amylostereum* spp., as well as genotypes of these fungi has increased our knowledge regarding the co-evolution between the Siricidae and their associated symbionts. For example, the identity of a number of *Amylostereum* spp. associated with different wasp species has been accurately determined. These data have confirmed that the same fungal species, and even genotype, can be carried by more than one wasp genus or species, suggesting that horizontal exchange of the fungal symbiont occurs between host species. It is thus possible that previous conclusions regarding the specificity of the wasp and fungus represent artefacts of restricted experimental sample size as well as incorrect identifications. Identification studies utilising DNA sequence data on a large number of isolates from native areas where diverse wasp species/populations co-exist would be valuable. Such studies would make it possible to determine the extent of the levels of specificity between the symbiont and host.

Studies based on VCGs and molecular markers have revealed the dramatic effect that the spread of asexual spores of *A. areolatum* by *S. noctilio* has had on populations of this fungus. This has given rise to clones that span across extensive geographic areas, and even across continents. Sexual reproduction, however, clearly continues to play an important role in the ecology of *Amylostereum* spp. The interaction between sexually spread individuals and populations associated with woodwasps is, however, poorly understood. This is likely to differ between regions where sexual sporulation in the fungus occurs at different frequencies.The latter is a phenomenon which deserves to be characterized further.

The contemporary project to sequence the genome of *A. areolatum* (C. Currie, B. Slippers, M. van der Nest and co-workers, personal communication) holds much promise to significantly elucidate questions regarding the evolution of the wasp-fungus mutualism and its effects on the fungus. Comparative genomics including the genome of *Heterobasidion annosum*, a close relative of *A. areolatum* that has recently been sequenced (J. Stenlid and co-workers, unpublished), and other Basidiomycetes will prove to be a powerful tool in the analysis of the genome. At the time of writing this manuscript the genomes of at least nine basidiomycetes (i.e., *Cryptococcus neoformans*, *C. neoformans* var *neoformans*, *Coprinopsis cinerea*, *Schizophyllum commune*, *Moniliophthora perniciosa*, *Malassezia globosa*, *Ustilago maydis*, *Postia maydis* and *Laccaria bicolor*) have been sequenced completely (http://www.ncbi.nlm.nih.gov) and numerous other basidiomycete genome projects are under way (http://www.genomesonline.org). These data will add substantial value to the results emerging from the *A. areolatum* genome-sequencing project. These data will also provide a rich source of markers to study ecological and evolutionary questions. For example, the work will entail characterization of recognition loci (mating and vegetative), as well as the intriguing linkage between these regions and those that influence growth rate. Importantly, genes regulating the interaction between the fungus and the tree host can also be studied with much greater accuracy. Finally, these data will provide a valuable source of markers to study ecological and evolutionary questions regarding the relationship between Siricidae and their fungal symbionts.

It is expected that the continued use of genetic tools and the increased use of genomic resources in coming years will further increase our understanding of the ecology and evolution of *Amylostereum* spp. They should also promote a much deeper understanding of the Siricid-*Amylostereum* mutualism as a whole. It is anticipated that these advances will also enhance our understanding of the natural history of the mutualism to a level where it is on par with other better studied systems (such as leaf cutting ants and termites). Such knowledge will be necessary for a meaningful comparison and will make it possible to address fundamental questions regarding the evolution of fungal-insect mutualism in general.

References

Awadalla P, Charlesworth D (1999) Recombination and selection at *Brassica* self-incompatibility loci. Genetics 152:413–425

Bedding RA (1995) Biological control of *Sirex noctilio* using the nematode *Deladenus siricidicola*. In: Bedding RA, Akhurst RJ, Kaya H (eds) Nematodes and biological control of insect pests. CSIRO, Melbourne

Boidin J (1958) Heterobasidiomycetes saprophytes et Homobasidiomycetes resupines: V. – Essai sur le genre *Stereum* Pers. ex S.F. Gray. Rev Mycol 23:318–346

Boidin J, Lanquentin P (1984) Le genre *Amylostereum* (Basidiomycetes) intercompatibilités partielles entre espèces allopartriques. Bull Soc Mycol France 100:211–236

Buller AHR (1931) Researches on fungi, vol IV. Longmans, Green, London

Carnegie AJ, Mamoru M, Hurley BP, Ahumada R, Haugen DA, Klasmer P, Sun J, Lede ET (2006) Predicting the potential distribution of *Sirex noctilio* (Hymenoptera: Siricidae), a significant exotic pest of *Pinus* plantations. Ann For Sci 63:119–128

Casselton LA (2002) Mate recognition in fungi. Heredity 88:142–147

Casselton L, Zolan M (2002) The art and design of genetic screens: filamentous fungi. Nature Rev Genet 9:683–97

Casselton LA, Olesnicky NS (1998) Molecular genetics of mating recognition in basidiomycete fungi. Microbiol Mol Biol Rev 62:55–70

Chevanne D, Bastiaans E, Debets A, Saupe SJ, Clave C, Paoletti M (2009) Identification of the *het-r* vegetative incompatibility gene of *Podospora anserina* as a member of the fast evolving HNWD gene family. Curr Genet 55:93–102

Coustou V, Deleu C, Saupe S, Begueret J (1997) The protein product of the *het-s* heterokaryon incompatibility gene of the fungus *Podospora anserina* behaves as a prion analog. Proc Natl Acad Sci USA 94:9773–9778

Coutts MP (1969) The mechanism of pathogenicity of *Sirex noctilio* on *Pinus radiata*. I. Effects of the symbiotic fungus *Amylostereum* sp. (Thelophoraceae). Aust J Biol Sci 22:915–924

Coutts MP, Dolezal JE (1969) Emplacement of fungal spores by the woodwasp, *Sirex noctilio*, during oviposition. For Sci 15:412–416

Espagne E, Balhadère P, Penin M-L, Barreau C, Turcq B (2002) *HET-E* and *HET-D* belong to a new subfamily of WD40 Proteins involved in vegetative incompatibility specificity in the fungus *Podospora anserine* Genetics 161:71–81

Fong LK, Crowden RK (1973) Physiological effects of mucus from the wood wasp, *Sirex noctilio* F., on the foliage of *Pinus radiata* D. Don. Aust J Biol Sci 26:365–378

Gaut IPC (1969) Identity of the fungal symbiont of *Sirex noctilio*. Aust J Biol Sci 22:905–914

Gaut IPC (1970) Studies of siricids and their fungal symbionts. PhD thesis, University of Adelaide, Adelaide, Australia

Glass NL, Grotelueschen J, Metzenberg RL (1990) *Neurospora crassa* A mating type protein. Proc Natl Acad Sci USA 87:4912–16

Hansen EM, Stenlid J, Johansson M (1994) Somatic incompatibility in *Heterobasidion annosum* and Phellinus weirii. In: Johansson M, Stenlid J (eds) Proceedings of the eight IUFRO Root and Butt Rot conference, Swedish University of Agricultural Sciences, Uppsala

Haugen DA (1990) Control procedures for *Sirex noctilio* in the Green Triangle: Review from detection to severe outbreak (1977–1987). Aust For 53:24–32

Herre EA, Knowlton N, Mueller UG, Rehner SA (1999) The evolution of mutualisms: exploring the paths between conflict and cooperation. Trends Ecol Evol 14:49–53

Hibbett DS, Pine EM, Langer E, Langer G, Donoghue MJ (1997) Evolution of gilled mushrooms and puffballs inferred from ribosomal DNA sequences. Proc Natl Acad Sci USA 94:12002–12006

Hoebeke ER, Haugen DA, Haack RA (2005) *Sirex noctilio*: discovery of a Palearctic siricid woodwasp in New York. Newsl Mich Entomol Soc 50:24–25

Högberg N, Stenlid J (1994) Genetic structures of rare and common wood decay fungi. In: Abstracts of Fifth international mycological congress, Vancouver, 14–21 Aug 1994

Hsiau PT-W (1996) The taxonomy and phylogeny of the mycangial fungi from *Dendroctonus brevicomis* and *D. frontalis* (Coleoptera: Scolytidae). D.Phil. thesis, Iowa State University

Hurley BP, Slippers B, Wingfield MJ (2007) A comparison of control results for the alien invasive woodwasp, *Sirex noctilio*, in the Southern Hemisphere. Agric For Entomol 9:159–171

Hurley BP, Slippers B, Croft PK, Hatting HJ, van der Linde M, Morris AR, Dyer C, Wingfield MJ (2008) Factors influencing parasitism of *Sirex noctilio* (Hymenoptera: Siricidae) by the nematode *Deladenus siricidicola* (Nematoda: Neotylenchidae) in summer rainfall areas of South Africa. Biol Control 45:450–459

James TY, Srivilai P, Kües U, Vilgalys R (2006) Evolution of the bipolar mating system of the mushroom *Coprinellus disseminatus* from its tetrapolar ancestors involves loss of mating-type-specific pheromone receptor function. Genetics 172:1877–1891

Kauserud H (2004) Widespread vegetative compatibility groups in the dry-rot fungus *Serpula lacrymans*. Mycologia 96:232–239

Kauserud H, Saetre G-P, Schmidt O, Decock C, Schumacher T (2006) Genetics of self/nonself-recognition in *Serpula lacrymans*. Fungal Genet Biol 43:503–510

Kim SY, Jung HS (2000) Phylogenetic relationships of the Aphyllophorales inferred from sequence analysis of nuclear small subunit ribosomal DNA. J Microbiol 38:122–131

Kothe E (1999) Mating types and pheromone recognition in the homobasidiomycetes *Schizophyllum commune*. Fungal Genet Biol 27:146–152

Kothe E, Gola S, Wendland J (2003) Evolution of multispecific mating-type alleles for pheromone perception in the homobasidiomycetes fungi. Curr Genet 42:268–275

Kronstad JW, Staben C (1997) Mating type in filamentous fungi. Annu Rev Genet 31:245–76

Kües U (2000) Life history and developmental processes in the basidiomycete *Coprinus cinereus*. Microbiol Mol Biol Rev 64:316–353

Lind M, Stenlid J, Olson A (2007) Genetics and QTL mapping of somatic incompatibility and intraspecific interactions in the basidiomycete *Heterobasidion annosum s.l.* Fungal Genet Biol 44:1242–1251

Madden JL (1981) Egg and larval development in the woodwasp, *Sirex noctilio* F. Aust J Zool 29:493–506

Maijala P, Harrington TC, Raudaskoski M (2003) A peroxidase gene family and gene trees in *Heterobasidion* and related genera. Mycologia 95:209–221

May GS, Badrane H, Vekemans X (1999) The signature of balancing selection: Fungal mating compatibility gene evolution. Proc Natl Acad Sci USA 96:9172–9177

Miller SL, Larsson E, Larsson K-H, Verbeken A, Nuytinck J (2006) Perspectives in the new Russulales. Mycologia 98:960–970

Morgan F (1968) Bionomics of Siricidae. Annu Rev Entomol 13:239–256

Nielsen C, Williams DW, Hajek AE (2009) Putative source of the invasive *Sirex noctilio* fungal symbiont, *Amylostereum areolatum*, in the eastern United States and its association with native siricid woodwasps. Mycol Res 113:1242–1253

Paźoutavá S, Šrůrk P, Holuša J, Chudíčkova M, Lokařík M (2010) Diversity of xylariaceous symbionts in Xiphydria woodwasps: role of vector and a host tree. Fungal Ecol 3:392–401

Rayner ADM (1991) The challenge of the individualistic mycelium. Mycologia 83:48–71

Saupe SJ, Glass NL (1997) Allelic specificity at the *het-c* heterokaryon incompatibility locus of *Neurospora crassa* is determined by a highly variable domain. Genetics 146:1299–1309

Saupe SJ, Kuldau GA, Smith ML, Glass NL (1996) The product of the *het-C* incompatibility gene of *Neurospora crassa* has characteristics of a glycine-rich cell wall protein. Genetics 143:1589–1600

Saupe SJ, Turcq B, Bégueret J (1995) A gene responsible for vegetative incompatibility in the fungus *Podospora anserina* encodes a protein with a GTP-binding motif and G homologous domain. Gene 162:135–139

Saupe S, Descamps C, Turcq B, Bégueret J (1994) Inactivation of the *Podospora anserin*a vegetative incompatibility locus *het-c*, whose product resembles a glycolipid transfer protein, drastically impairs ascospore production. Proc Natl Acad Sci USA 91:5927–5931

Schirawski J, Heinze B, Wagenknecht M, Kahmann R (2005) Mating type loci of *Sporisorium reilianum*: novel pattern with three *a* and multiple *b* specificities. Eukaryot Cell 4:1317–1327

Slippers B (1998) The *Amylostereum* symbiont of *Sirex noctilio* in South Africa. M.Sc. thesis. University of the Free State, Bloemfontein

Slippers B, Wingfield MJ, Wingfield BD, Coutinho TA (2000) Relationships among *Amylostereum* species associated with siricid woodwasps inferred from mitochondrial ribosomal DNA sequences. Mycologia 92:955–963

Slippers B, Wingfield MJ, Wingfield BD, Coutinho TA (2001) Population structure and possible origin of *Amylostereum areolatum* in South Africa. Plant Pathol 50:206–210

Slippers B, Wingfield BD, Coutinho TA, Wingfield MJ (2002) DNA sequence and RFLP data reflect relationships between *Amylostereum* species and their associated wood wasp vectors. Mol Ecol 11:1845–1854

Slippers B, Coutinho TA, Wingfield BD, Wingfield MJ (2003) A review of the genus *Amylostereum* and its association with woodwasps. S Afr J Sci 99:70–74

Smith ML, Micalli OC, Hubbard SP, Mir-Rashed N, Jacobson DJ, Glass NL (2009) Vegetative incompatibility in the *het-6* region of *Neurospora crassa* is mediated by two linked genes. Genetics 155:1095–1104

Solheim H (2006) Treveps og assosierte sopper i Norge. Agarica 26:87–95

Spradbery JP, Kirk AA (1978) Aspects of the ecology of siricid woodwasps (Hymenoptera: Siricidae) in Europe, North Africa and Turkey with special reference to the biological control of Sirex noctilio F. in Australia. Bull Entomol Res 68:341–359

Staben C, Yanofsky C (1990) *Neurospora crassa* a mating type region. Proc Natl Acad Sci USA 87:4917–4921

Stenlid J, Vasiliauskas R (1998) Genetic diversity within and among vegetative compatibility groups of *Stereum sanguinolentum* determined by arbitrary primed PCR. Mol Ecol 7:1265–1274

Stenlid J, Karlsson JO, Högberg N (1994) Intraspecific genetic variation in *Heterobasidion annosum* revealed by amplification of minisatelite DNA. Mycol Res 98:57–63

Tabata M, Abe Y (1997) *Amylostereum laevigatum* associated with the Japanese horntail, *Urocerus japonicus*. Mycoscience 38:421–427

Tabata M, Abe Y (1999) *Amylostereum laevigatum* associated with a horntail, *Urocerus antennatus*. Mycoscience 40:535–539

Tabata M, Harrington TC, Chen W, Abe Y (2000) Molecular phylogeny of species in the genera *Amylostereum* and *Echinodontium*. Mycoscience 41:585–593

Talbot PHB (1977) The *Sirex-Amylostereum-Pinus* association. Annu Rev Phytopathol 15:41–54

Thomsen IM (1996) *Amylostereum areolatum* and *Amylostereum chailletii,* symbiotic fungi of woodwasps (*Sirex* sp. and *Urocerus* sp.). Ph.D. thesis, Danish Forest and Landscape Research Institute, Horsholm

Thomsen IM (1998) Fruit body characters and cultural characteristics useful for recognizing *Amylostereum areolatum* and *A. chailletii*. Mycotaxon 69:419–428

Thomsen IM, Koch J (1999) Somatic compatibility in *Amylostereum areolatum* and *A. chailletii* as a consequence of symbiosis with siricid woodwasps. Mycol Res 103:817–823

Turcq B, Deleu C, Denayrolles M, Bégueret J (1991) Two allelic genes responsible for vegetative incompatibility in the fungus *Podospora anserina* are not essential for cell viability. Mol Gen Genet 228:265–269

Van der Nest MA, Slippers B, Stenlid J, Wilken PM, Vasaitis R, Wingfield MJ, Wingfield BD (2008) Characterization of the systems governing sexual and self-recognition in the white rot homobasidiomycete *Amylostereum areolatum*. Curr Genet 53:323–336

Van der Nest MA, Slippers B, Steenkamp ET, de Vos L, van Zyl K, Stenlid J, Wingfield MJ, Wingfield BD (2009) Genetic linkage map for *Amylostereum arolatum* reveals an association between vegetative growth and sexual and self recognition. Fungal Genet Biol 46:632–641

Vasiliauskas R, Stenlid J (1999) Vegetative compatibility groups of *Amylostereum areolatum* and *A. chailletii* from Sweden and Lithuania. Mycol Res 103:824–829

Vasiliauskas R, Stenlid J, Thomsen IM (1998) Clonality and genetic variation in *Amylostereum areolatum* and *A. chailletii* from Northern Europe. New Phytol 139:751–758

Vasiliauskas R, Johannesson H, Stenlid J (1999) Molecular relationships within the genus *Amylostereum* as determined by internal transcribed spacer sequences of the ribosomal DNA. Mycotaxon 71:155–161

Worrall JJ (1997) Somatic compatibility in Basidiomycetes. Mycologia 89:24–36

Chapter 7
Siricid Woodwasps and Their Fungal Symbionts in Asia, Specifically Those Occurring in Japan

Masanobu Tabata, Hiroaki Miyata, and Kaoru Maeto

Abstract In Japan seven species of Siricinae have been recorded in four genera, namely *Sirex, Urocerus, Xeris* and *Xoanon*. The fungal symbionts of woodwasps in the genera of *Sirex*, *Urocerus* and *Xoanon* are all in the genera *Amylostereum*. We found that *A. areolatum* was carried by *Sirex nitobei* and *Xoanon matsumurae*, while *A. laevigatum* was the symbiont of *Urocerus antennatus* and *U. japonicus* in Japan. In this chapter, we present an overview of Siricid woodwasps and their fungal symbionts with a specific focus on those occurring in Japan.

7.1 Introduction

Woodwasps (horntails) are primitive hymenopterans whose larvae feed on the sapwood of various conifer and angiosperm trees (Benson 1951; Bradley 1913; Takeuchi 1962). Most species of woodwasps carry their fungal symbionts in mycangia and inoculate the wood of the plant host with hyphal fragments or arthrospores as they oviposit (Francke-Grosmann 1939; Gaut 1970; Tabata and Abe 1997, 1999; Terashita 1970). The fungus is likely to provide essential nutrients (Madden 1988) and/or produce enzymes that decompose the cellulose or lignin in the wood (Kukor and Martin 1983; Martin 1984).

M. Tabata (✉)
Forestry and Forest Products Research Institute, Ibaraki 305-8687, Japan
e-mail: butter@ffpri.affrc.go.jp

H. Miyata
Kochi Prefectural Forest Technology Center, Kami 782-0078, Japan

K. Maeto
Faculty of Agriculture, Kobe University, Kobe 657-8501, Japan

B. Slippers et al. (eds.), *The Sirex Woodwasp and its Fungal Symbiont: Research and Management of a Worldwide Invasive Pest*, DOI 10.1007/978-94-007-1960-6_7,
© Springer Science+Business Media B.V. 2012

Fig. 7.1 Wood discoloration in a cross section of *Cryptomeria japonica* stem caused by *Urocerus japonicus* and *Amylostereum laevigatum*

In Japan, seven species of Siricinae have been recorded and these occur in the four genera *Sirex, Urocerus, Xeris* and *Xoanon* (Takeuchi 1962). *Sirex nitobei* Matsumura, *Urocerus japonicus* Smith, *U. antennatus* Marlatt and *Xeris spectrum* Linne have been studied (Kanemitsu 1978; Okuda 1989; Sano 1992). *Sirex nitobei* attacks damaged or moribund trees of Japanese red pine, *Pinus densiflora* and Japanese black pine, *P. thunbergii* (Kobayashi et al. 1978, M. Tabata unpublished). *Urocerus japonicus* and *U. antennatus* have been recently regarded as the pests in stands of Japanese cedar, *Cryptomeria japonica* D. Don, and Japanese cypress, *Chamaecyparis obtusa* (Sieb. et Zucc.) Endlicher in Japan, because the discoloration induced by the fungal symbionts reduce the commercial value of wood (Shibata 1984; Yamada and Okuda 1987; Tabata and Abe 1997, 1999). Wood discoloration in *Cr. japonica* (Fig. 7.1) and *Ch. obtusa* by the woodwasps and their symbionts occurs widely from the Tohoku to Kyushu regions of Japan (Hosoda et al. 2005; Miyata et al. 2001).

7.2 Symbiosis Between Woodwasps and Fungi

As far as is known, the fungal symbionts of woodwasps in the genera of *Sirex*, *Urocerus* and *Xoanon* are all in the genus *Amylostereum* (Gaut 1970; Tabata and Abe 1997, 1999; Tabata et al. 2000; Vasiliauskas 1999) (Table 7.1). Tabata and Abe (1997, 1999), Tabata et al. (2000) and Terashita (1970) found that *A. areolatum* was

Table 7.1 Relationships between woodwasps and their symbiotic fungi

Woodwasp	Symbiotic fungus	Tree hosts	Reference
Sirex areolatus	*Amylostereum laevigatum*	Unknown	Tabata et al. (2000)
Sirex californicus	*Amylostereum chailletii*	*Pinus*	Gaut (1970)
Sirex cyaneus	*Amylostereum chailletii*	*Abies, Picea*	Gaut (1970)
Sirex imperialis	*Amylostereum chailletii*	Unknown	Gaut (1970)
Sirex juvencus	*Amylostereum areolatum*	*Picea*	Gaut (1970)
Sirex nitobei[a]	*Amylostereum areolatum*	*Pinus*	Gaut (1970); Terashita (1970)
Sirex noctilio	*Amylostereum areolatum*	*Pinus*	Gaut (1970)
Urocerus antennatus[a]	*Amylostereum laevigatum*	*Chamaecyparis, Cryptomeria*	Tabata and Abe (1999)
Urocerus augur augur	*Amylostereum chailletii*	*Abies*	Gaut (1970)
Urocerus augur sah	*Amylostereum chailletii*	*Abies*	Gaut (1970)
Urocerus gigas	*Amylostereum chailletii*	*Abies, Picea*	Gaut (1970); Vasiliauskas (1999)
Urocerus japonicus[a]	*Amylostereum laevigatum*	*Chamaecyparis, Cryptomeria*	Tabata and Abe (1997)
Xoanon matsumurae[a]	*Amylostereum areolatum*	*Larix*	Tabata et al. (2000)

[a]The species that occur in Japan

carried by *S. nitobei* and *Xo. matsumurae* Rohwer, while *A. laevigatum* was the symbiont of *U. antennatus* and *U. japonicus* in Japan (Table 7.1).

There are four recognized species of *Amylostereum*, namely *A. areolatum* (Fr.: Fr.) Boidin, *A. chailletii* (Pers.: Fr.) Boidin, *A. ferreum* (Berk. & Curt.) Boidin and Lanquetin and *A. laevigatum* (Fr.: Fr.) Boidin (Boidin and Lanquetin 1984). All of these *Amylostereum* species occur on coniferous trees. *Amylostereum areolatum*, *A. chailletii* and *A. laevigatum* are associated with wood decay in the Pinaceae and other conifers in the Northern Hemisphere (Breitenbach and Kranzlin 1986; Chamuris 1988; Eriksson and Ryvarden 1973; Eriksson et al. 1978; Ginns and Lefebvre 1993) and *A. ferreum* decays wood of *Podocarpus* spp. in Latin America (Boidin and Lanquetin 1984). Cultures of *A. areolatum* and *A. chailletii* or *A. laevigatum* can be recognized by the color of mycelial mats on PDA and the presence or absence of arthrospores (oidia) in culture (M. Tabata unpublished; Thomsen 1998, Chap. 5).

Amylostereum chailletii is known as a fungal symbiont of *Sirex* and *Urocerus* species (Gaut 1970). Eriksson and Ryvarden (1973), Breitenbach and Kranzlin (1986), Chamuris (1988) and Thomsen (1998) describe the basidiocarps of *A. chailletii* as resupinate to effuso-reflexed with narrow margin and a dimitic hyphal system (Table 7.2). Basidiospores of *A. chailletii* are reported to be 6–7 × 2.5–3 μm (Eriksson and Ryvarden 1973), 5.5–7 × 2.5–3 μm (Breitenbach and Kranzlin 1986), 5–8 × 2–3.5(–4) μm (Chamuris 1988) and 4.8–8 × 2.4–4 μm (Thomsen 1998) (Table 7.2) (also see Chap. 5). The size of basidiospores from basidiocarps of *A. laevigatum* collected on the felled logs and produced on the stem segments

Table 7.2 Comparison of *Amylostereum chailletii* and *A. laevigatum*

	A. chailletii			
Characters	Eriksson & Ryvarden	Breitenbach & Kranzlin	Chamuris	Thomsen
Shape of basidiocarp	Resupinate to effuso-reflexed	Resupinate to effuso-reflexed	Resupinate to effuso-reflexed	Resupinate to effuso-reflexed
Hyphal system	Dimitic	Dimitic	Dimitic	Dimitic
Size of basidiospores (μm)	6–7.5 × 2.5–3	5.5–7 × 2.5–3	5–8 × 2–3.5(–4)	4.8–8 × 2.4–4
	A. laevigatum			
Characters	Eriksson & Ryvarden	Breitenbach & Kranzlin	Chamuris	Tabata & Abe
Shape of basidiocarp	Resupinate	Resupinate	Resupinate	Resupinate
Hyphal system	Monomitic	Monomitic	Monomitic	Monomitic
Size of basidiospores (μm)	7–12 × 3–4	7–9 × 3–4	7–10(–12) × (2.5–)3–4	(6.5–)7–10 × 3–4

partially overlap with those of *A. chailletii*, but *A. laevigatum* has a monomitic hyphal system and its basidiocarps never become reflexed (Tabata and Abe 1997, 1999, Table 7.2). *Amylostereum laevigatum* has been shown as the symbiotic fungus of *U. antennatus* and *U. japonicus* (Tabata and Abe 1997, 1999). *Amylostereum laevigatum* is distributed in Canada, France, Norway, Sweden, Switzerland and USA (Boidin and Lanquetin 1984; Breitenbach and Kranzlin 1986; Eriksson and Ryvarden 1973; Ginns and Lefebvre 1993). It has never been recorded as a fungal symbiont of *Sirex* or *Urocerus* species in these latter regions, but is known to occur on *Abies, Juniperus, Taxus* and *Thuja* (Boidin and Lanquetin 1984; Breitenbach and Kranzlin 1986; Eriksson and Ryvarden 1973; Ginns and Lefebvre 1993).

Xeris spectrum is widely distributed in Asia, Europe and North America (Morgan 1968; Takeuchi 1962). Some *Xe. spectrum* populations emerge in spring during May and June, while others emerge in summer during August and September (Fukuda and Hijii 1997). This species has a specific life history that utilizes fungal symbionts of other woodwasps such as *S. nitobei* and *U. japonicus*, without carrying any symbiotic fungus itself (Fukuda and Hijii 1997).

7.3 Molecular Phylogeny of Fungal Symbionts

The four species of *Amylostereum* appear to form a monophyletic group based on both ITS and peroxidase A sequence analyses (Tabata et al. 2000). Further, *A. ferreum, A. chailletii* and *A. laevigatum* form a separate group, sister to *A. areolatum* (Tabata et al. 2000). Based on sequences of the small-subunit mt-rDNA and IGS rDNA Slippers et al. (2000, 2002) also found that *A. ferreum* and *A. laevigatum* are more closely related to *A. chailletii* than to *A. areolatum*. Likewise, Vasiliauskas et al. (1999) found that *A. chailletii* and *A. laevigatum* from Europe have similar ITS rDNA sequences. These data correlated with the mating studies done by Boidin and Lanquetin (1984), which showed that *A. ferreum* was partially interfertile with *A. laevigatum* and *A. chailletii*, but was completely intersterile with *A. areolatum*.

Isolates of *A. areolatum*, *A. chailletii* and *A. laevigatum* from woodwasps had similar or identical ITS and peroxidase sequences to linked isolates from basidiomes or decayed wood (Tabata et al. 2000). Our DNA sequence analyses from that study also confirmed that *A. areolatum* is the associate of *S. nitobei*, *S. juvencus* L. and *S. noctilio* Fab. from Japan and Europe.

An isolate from the mycangium of *Xo. matsumurae* in Japan had an ITS sequence identical to that of other *A. areolatum* isolates (Tabata et al. 2000). The isolate from *Xo. matsumurae* had the same characteristics as that from *S. nitobei*. Both isolates produced arthrospores that were 5–21 × 1–4 μm. From the ITS sequences and morphological comparisons, we concluded that the fungal symbiont of *Xo. matsumurae* is *A. areolatum* (Tabata et al. 2000).

Amylostereum chailletii is a common decay fungus in both North America and Europe, and it also has been recorded as a symbiont of *S. areolatus* Cr., *S. californicus* Ash., *S. cyaneus* Fab., *Urocerus augur augur* Klug, *U. augur sah* Mocsary and *U. gigas* L. (Gaut 1970). Isolates from *U. gigas* in Europe had ITS and peroxidase A sequences that were similar to those from decayed wood or basidiomes.

Gaut (1970) concluded from anastomosis, dikaryotization and interfertility tests that the fungus associated with *S. areolatus* was *A. chailletii*. An *Amylostereum* isolate from the mycangium of *S. areolatus* was isolated by Stillwell and was presumed to be *A. chailletii*. However, this isolate has unique ITS, peroxidase A and mtSSU sequences that were similar to those of *A. laevigatum* and distinct from those of *A. chailletii* (Slippers et al. 2000; Tabata et al. 2000). *Sirex areolatus* commonly attacks cedar-like trees such as *Libocedrus*, *Cupressus*, *Juniperus* and *Sequoia* (Furniss and Carolin 1977), similar to the hosts of *U. japonicus* and *U. antennatus*, which have *A. laevigatum* as their symbiont. More isolates need to be examined to clarify the species of *Amylostereum* associated with *S. areolatus*, but it appears that *A. laevigatum* is the primary symbiont of *Sirex* and *Urocerus* woodwasps attacking cedar-like tree species, while *A. chailletii* and *A. areolatum* are the primary symbionts of the woodwasps attacking the Pinaceae.

ITS sequences of *A. laevigatum* isolates from Japan and France were distinct, but the isolates from these two countries had similar peroxidase A sequences (Tabata et al. 2000). It has been speculated that *A. laevigatum* in Europe represents two distinct taxa, one occurring on *Juniperus* (with basidiospores 7–9 μm long) and another on *Taxus* (basidiospores 9–12 μm long) (Eriksson and Ryvarden 1973). However, Boidin and Lanquetin (1984) did not find the same distinction between *Taxus* and *Juniperus* isolates from Sweden and France. Specimens of *A. laevigatum* from Japan and those from *Juniperus* in Sweden were examined and each had the shorter basidiospores typical of the *Juniperus* type (Tabata et al. 2000). Thus, the peroxidase A sequences and morphological data support the earlier identification of the symbiotic fungus associated with *U. antennatus* and *U. japonicus* as

A. laevigatum (Tabata and Abe 1997, 1999). Further studies of greater numbers of specimens of *A. laevigatum* from Europe are needed.

7.4 Life-Cycle and Host Interaction of Woodwasps

Most *S. nitobei* emerge from *Pinus densiflora, P. thunbergii* and *Abies firma* from August to mid-October (Fukuda 1997; Kanemitsu 1978). The life cycle of *S. nitobei* is completed in 1 year (Kanemitsu 1978), and the adults live for about 4 days (Fukuda et al. 1993).

Most *U. japonicus* require 1 year to complete development (Fukuda 1997), while a few require 2 years (Miyata 1999; Okuda 1989). Most adults of this woodwasp emerge from July to September (Miyata 1999; Okuda 1989). Female adults of *U. japonicus* oviposit in the sapwood of *Cr. japonica* and *Ch. obtusa* stems. Larvae feed on wood tissue infected by *A. laevigatum* (Tabata and Abe 1997) and molt 10–11 times before pupation (Okuda 1989). Adults are sexually mature by the time of emergence and they copulate soon after emergence (Okuda 1989). The mean longevity of the species ranged from 3 to 4 days (Fukuda 1997; Sano 1992).

Most adults of *U. antennatus* emerge from June to July (Hosoda et al. 2005; Miyata et al. 2001). Female adults of *U. antennatus* oviposit in the sapwood of *Cr. japonica* and *Ch. obtusa* stems (Miyata et al. 2001). Larvae feed on wood tissue infected by *A. laevigatum* (Tabata and Abe 1999).

Sato and Maeto (2006) found that α-pinene was attractive to females of *U. japonicus*. Adhesive traps with α-pinene also attracted females of *U. antennatus* (M. Tabata unpublished). This is useful information for monitoring population levels, research and planning management strategies. To avoid wood damage due to discoloration by the *U. japonicus* and *U. antennatus*, trees felled for thinning are immediately removed from the stands of *Cr. japonica* and *Ch. obtusa* (Hosoda et al. 2005; Miyata 1999), because the woodwasps reproduce on newly felled trees left in the stands.

Acknowledgements We thank Prof. T.C. Harrington of Iowa State University for providing useful information.

References

Benson RB (1951) Hymenoptera – Symphyta (section a). Handbooks for the identification of British insects. Royal Entomological Society of London, London

Boidin J, Lanquetin P (1984) Le genre *Amylostereum* (Basidiomycetes) intercompatibilites partielle entre especes allopatriques. Bull Soc Mycol France 100:211–236

Bradley JC (1913) The Siricidae of North America. J Ent Zool 5:1–36

Breitenbach J, Kranzlin F (1986) Fungi of Switzerland, vol 2. Non-gilled fungi. Verlag Mykologia, Lecerne

Chamuris G (1988) The non-stipitate stereoid fungi in the northeastern United States and adjacent Canada. Mycol Mem 14:41–45

Eriksson J, Ryvarden L (1973) The Corticiaceae of North Europe, vol 2. Fungiflora, Oslo

Eriksson J, Hjortstam K, Ryvarden L (1978) The Corticiaceae of North Europe, vol 5. Fungiflora, Oslo

Francke-Grosmann H (1939) Uber das Zusammenlebeb von Holzwespen (Siricidae) mit Pilzen. Z Angew Ent 25:647–680

Fukuda H, Kajimura H, Hijii N (1993) Fecundity of the woodwasp, *Sirex nitobei* Matsumara, in relation to its body size. J Jpn For Soc 75:405–408

Fukuda H (1997) Resource utilization and reproductive strategy of three woodwasp species (Hymenoptera: Siricidae). Nogoya Univ For Sci 16:23–73

Fukuda H, Hijii N (1997) Reproductive strategy of a woodwasp with no fungal symbionts, *Xeris spectrum* (Hymenoptera: Siricidae). Oecologia 112:551–556

Furniss RL, Carolin VM (1977) Western forest insects. USDA For Serv Misc Publ 1339: 453–457

Gaut IPC (1970) Studies of siricids and their fungal symbionts. PhD thesis, University of Adelaide, Australia

Ginns J, Lefebvre MNL (1993) Lignicolous corticioid fungi (Basidiomycota) of North America, systematics, distribution, and ecology. Mycol Mem 19:21

Hosoda K, Onagamitsu J, Inada T, Sano A, Kon J, Kato T, Hougen T, Inoue M, Sudo S, Sugimoto H, Takemoto M, Miyata H, Yoshimoto K (2005) Study on control methods against the damage of discoloration caused by horntails. For Pest 54:3–14

Kanemitsu K (1978) Woodwasps and their hymenopterous parasitoids in Japanese conifers. Kontyu Tokyo 46:498–508

Kukor JJ, Martin MM (1983) Acquisition of digestive enzymes by siricid woodwasps from their fungal symbiont. Science 220:1161–1163

Kobayashi T, Sasaki K, Enda N (1978) Correlation between *Sirex nitobei* and *Amylostereum areolatum*, associated with the death of Japanese pine trees during winter season. J Jpn For Soc 60:405–411

Madden JL (1988) *Sirex* in Australia. In: Berrymann AA (ed) Dynamics of forest insect populations: patterns, causes, implications. Plenum Press, New York/London

Martin MM (1984) The role of ingested enzymes in the digestive process of insects. In: Anderson AD, Rayner ADM, Walton DWH (eds) Invertebrate-microbial interactions. Cambridge University Press, London

Miyata H (1999) Damage of wood discoloration caused by *Urocerus japonicus* in Kochi prefecture. Ringyo-to-yakuzai 147:13–18

Miyata H, Kato T, Yoshioka S, Fukuhara N, Hosoda K, Hougen T, Inoue M, Sudo S, Okubo M, Inada T, Onagamitsu J (2001) States of the damage of discoloration caused by horntails and survey of control methods against the damage. For Pest 50:105–113

Morgan FD (1968) Bionomics of Siricidae. Annu Rev Entomol 13:239–256

Okuda M (1989) *Urocerus japonicus*. Ringyo-to-yakuzai 108:1–8

Sano A (1992) *Urocerus japonicus*. Ringyo-to-yakuzai 122:1–8

Sato S, Maeto K (2006) Attraction of female Japanese horntail *Urocerus japonicus* (Hymenoptera: Siricidae) toα-pinene. Appl Entomol Zool 41:317–323

Shibata E (1984) Injury of polished logs of Japanese cedar, *Cryptomeria japonica*, by *Urocerus japonicus*. For Pest 33:12–14

Slippers B, Wingfield MJ, Wingfield BD, Coutinho TA (2000) Relationships among *Amylostereum* species associated with siricid woodwasps inferred from mitochondrial ribosomal DNA sequences. Mycologia 92:955–963

Slippers B, Wingfield BD, Coutinho TA, Wingfield MJ (2002) DNA sequence and RFLP data reflect geographical spread and relationships of *Amylostereum areolatum* and its insect vectors. Mol Ecol 11:1845–1854

Tabata M, Abe Y (1997) *Amylostereum laevigatum* associated with the Japanese horntail, *Urocerus japonicus*. Mycoscience 38:421–427

Tabata M, Abe Y (1999) *Amylostereum laevigatum* associated with a horntail, *Urocerus antennatus*. Mycoscience 40:535–539

Tabata M, Harrington TC, Wei C, Abe Y (2000) Molecular phylogeny of species in the genera of *Amylostereum* and *Echinodontium*. Mycoscience 41:585–593

Takeuchi K (1962) Insecta Japonica, Hymenoptera: Siricidae, ser 2 part 2. Hokuryukan, Tokyo

Terashita T (1970) A Basidiomycete symbiotic to a siricid in Japan. J Jpn For Soc 52:313–316

Thomsen IM (1998) Characters of fruitbodies, basidiospores and cultures useful for recognizing *Amylostereum areolatum* and *A. chailletii*. Mycotaxon 69:419–428

Vasiliauskas R (1999) Spread of *Amylostereum areolatum* and *A. chailletii* decay in living stems of *Picea abies*. Forestry 72:95–102

Vasiliauskas R, Johannesson H, Stenlid J (1999) Molecular relationships within the genus *Amylostereum* as determined by internal transcribed spacer sequences of the ribosomal DNA. Mycotaxon 71:155–161

Yamada T, Okuda K (1987) Wood discoloration of hinoki and sugi living trees inoculated with *Amylostereum* sp. symbiotic to the Japanese horntail (*Urocerus japonicus* Smith). Trans Ann Mtg Jpn For Soc 98:515–516

Chapter 8
Parasitoids in the Management of *Sirex noctilio*: Looking Back and Looking Ahead

E. Alan Cameron

Abstract The control of the woodwasp *Sirex noctilio* can be visualized as a three-legged stool. Silvicultural practices to reduce tree competition and stress and biological control using the nematode *Deladenus siricidicola* are two important legs. The third, and equally important, component of a total control program for this pest is the use of parasitoid wasps. Parasitoids were the first form of control used against *S. noctilio* in the Southern Hemisphere and remain important, especially as a constant repressor at low population levels. However, there is significant variation in the effect and application of the various parasitoid wasps in different regions. There is evidently scope for renewed efforts to better understand and optimize the use of parasitoids to achieve stable, long lasting population repression of invasive *S. noctilio* populations.

8.1 Introduction

Sirex noctilio Fabr. (Hymenoptera: Siricidae: Siricinae) has caused severe problems in Southern Hemisphere plantation forests in at least seven countries, in some for almost a century. It is interesting that only one of the pine hosts attacked in Southern Hemisphere plantations, Maritime pine, *Pinus pinaster* Aiton, occurs in the native range of *S. noctilio*. Further, all of the pines in the Southern Hemisphere countries where problems occur are deliberately introduced species being grown under plantation management. Repeatedly, this exotic insect pest has colonized exotic pine species, especially Monterey pine, *Pinus radiata* D. Don.

E.A. Cameron (✉)
Department of Entomology (Emeritus), Penn State University,
University Park, PA 16802, USA
e-mail: eajabaka@comcast.net

B. Slippers et al. (eds.), *The Sirex Woodwasp and its Fungal Symbiont: Research and Management of a Worldwide Invasive Pest*, DOI 10.1007/978-94-007-1960-6_8,

© Springer Science+Business Media B.V. 2012

Control efforts for *S. noctilio* have had variable outcomes across the Southern Hemisphere (Hurley et al. 2007, Chaps. 14–18). Sirex is no longer considered to be a forest pest in New Zealand, and is even difficult to find in many areas (Chap. 13). In Australia, damage is not nearly as severe or extensive as it was in the past, but occasional outbreaks are still cause for concern. In four South American countries, Uruguay, Argentina, Brazil, and Chile, success has been variable as researchers and foresters attempted to learn how to manage the insect over the last several decades. But in South Africa, following its introduction into Western Cape Province in 1994 and especially into Eastern Cape and KwaZulu-Natal provinces in the early 2000s, this exotic pest has at times become a catastrophic problem in pine plantations and has expanded its range explosively. In many of these regions suitable and effective pest management protocols are in urgent need of development.

In 2004, *S. noctilio* was detected in New York state, United States (Hoebeke et al. 2005). It has since been found in Scots pine, *P. sylvestris* L., and with the use of monitoring traps in at least 25 counties in New York state and at least two counties in neighboring Pennsylvania (Schneeberger 2007). de Groot et al. (2007) reported trapping *S. noctilio* in relatively nearby parts of the province of Ontario, Canada, where it is now widespread even if in fairly low numbers. Dodds et al. (2010) recently described the impact of *S. noctilio* in New York and Ontario in both Scots pine and red pine (*P. resinosa* Ait.). The information strongly suggests that the North American infestation was present much earlier than when it was first found. These reports represent the first documented movement to another northern hemisphere country. In North America, *S. noctilio* has been recorded in abandoned plantations and in the native and diversified forest.

The author was first involved with work to control Sirex more than 45 years ago. Based on experience, it is evident that successful management of Sirex, that is, reduction of this pest to a minor or inconsequential concern, can be visualized by looking at a three-legged stool. The seat of the stool (a healthy and productive forest) is supported by first, and most importantly, the maintenance of healthy conditions in the growing stand in which trees are not under stress. Stress, caused by prolonged and severe drought lasting for a period of years, contributed to the devastating losses in extensive plantations of Monterey pine in New Zealand in the 1950s. Stressed and newly-dead trees, including smog-killed trees (Cameron 1968), are those most commonly attacked by Sirex. While there is little that can be done regarding such stress factors, stress caused by overstocking can be addressed. Good silvicultural practices are thus essential in the light of the importance of stress for Sirex control, and the unpredictability of some stress factors. Thinning during a rotation to reduce the number of stems initially planted is mandatory; lower bole pruning brings added benefits.

The second and third legs of the stool both concern biological control. The second leg is the active promotion of the presence of the nematode, *Deladenus* (=*Beddingia*) *siricidicola* Bedding (Nematoda: Neotylenchidae), which has played a significant role as a biological force against increases in Sirex populations. In some localized areas, infection rates approaching 100% have been recorded (Bedding and Akhurst 1974), while following other introductions less than 5% of the hosts were infected (Hurley et al. 2007). The nematode, however, cannot do the job as the only biological agent working against Sirex, even in stands under ideal silvicultural management. For this reason, the third leg supporting the stool is the presence of various insectan parasitoids of Sirex. If any one of these legs is neglected, the stool collapses. In the opinion of the author, success will come only from a program that integrates all three of these tactics in a sound pest management strategy.

The parasitoids of Sirex, the third leg of the stool described above, are the subject of this chapter. The literature on the numerous parasitoids known to attack the genera *Sirex, Urocerus*, and *Xeris* (Hymenoptera: Siricidae: Siricinae) was summarized during a period of active collection of these agents in North America and Asia for biological control purposes (Cameron 1965a). At the time, no records were found of parasitoids attacking the siricid genera *Xoanon* or *Siricosoma*. Benson (1943) had noted that the Siricinae attack only gymnosperms, while the Tremicinae (*Tremex* spp.) are restricted to angiosperms. No records were found of parasitoids of the Tremicinae, even though they may be in the same genus (e.g., *Megarhyssa*), crossing over to the Siricinae.

8.2 Use of Parasitoids as Biological Control Agents

8.2.1 New Zealand and Australia

Historically, the first attempts to use parasitoids in a biological control program against Sirex were undertaken in New Zealand. *Sirex noctilio* had been introduced accidentally very early in the twentieth century and spread widely through exotic Monterey pine stands. Monterey pine is native to coastal areas of southern California, U.S.A., and grows rapidly under plantation management in a number of countries in the Southern Hemisphere where it has been introduced. Hudson (1919) is the first report in the literature of '*Sirex juvencus*' [which was in fact *S. noctilio*] in New Zealand, although it was first found in 1904 (Anon 1960). It was not until 1926 that the insect was recognized as becoming a serious pest (Clark 1933), which stimulated early attempts to introduce parasitoids for purposes of biological control. In the late 1920s and early 1930s, *Rhyssa persuasoria* L. (Hymenoptera: Ichneumonidae) was introduced from England and adults were field released in 1929, 1931 and 1932. Small numbers of *Ibalia leucospoides* Hochenwald (now *I. leucospoides leucospoides*) (Hymenoptera: Ibaliidae) larvae were also introduced, but did not produce adults for release (Miller and Clark 1935). *Ibalia* was reintroduced in the 1950s, and successfully established. The country suffered a prolonged and severe drought in the 1950s, which caused severe stress over hundreds of thousands of hectares of *P. radiata* plantations. At about this time, rigorous thinning and pruning practices were initiated to improve the health of the overstocked and generally untended and stressed stands. In the 1960s, *Megarhyssa nortoni nortoni* Cresson (Hymenoptera: Ichneumonidae) from California (USA), was introduced, released, and established, along with *Ibalia ensiger* Norton (now *I. leucospoides ensiger*). *Rhyssa persuasoria himalayensis* Wilkinson was introduced from the Indian subcontinent and released, but did not become established. These introductions were made through a cooperative programme between the Department of Scientific and Industrial Research (DSIR) and Australia's Commonwealth Scientific and Industrial Research Organisation (CSIRO). The nematode, *Deladenus (=Beddingia) siricidicola*, was also discovered in New Zealand by Zondag (1962, 1969), and exploited. Healthy stands, along with the biological agents, have reduced populations such that Sirex is no longer considered to be a forest problem in New Zealand and, indeed, is seldom seen (A. Cameron, personal observations and conversations while in New Zealand in 2004, Chap. 13). The three-legged stool stands.

In response to the discovery of *S. noctilio* on mainland Australia about 1960, CSIRO, under the leadership of Dr. D. F. Waterhouse (Chief, Division of Entomology), initiated a major initiative to combat the pest. Since the early 1950s, *S. noctilio* had been present in *P. radiata* plantations on the island state of Tasmania, but discovery of this pest on the mainland posed a much more serious threat to the extensive plantation forestry in southeastern Australia. The services of the Commonwealth Institute of Biological Control (CIBC) were engaged to carry out extensive scouting and collection activities especially in North America through the Californian Station. More modest programs were initiated at the Indian and Pakistan Stations as well as the Japanese sub-Station. These searches resulted in a number of parasitoid species being imported directly into quarantine facilities in Australia for rearing to increase numbers for field release and, through the cooperative program with DSIR, into New Zealand quarantine facilities as well. DSIR, for its part, supplied *I. leucospoides leucospoides* to Australia for colonization and release.

By the mid-1960s, CSIRO established its own Sirex unit based in Great Britain. This laboratory served as a base of foreign exploration operations throughout areas of interest in Eurasia and northern Africa where *S. noctilio* is native, and activities continued for more than a decade. Much-needed biological studies of the host in its native range, available parasitoids, and the nematode were also carried out by personnel associated with this unit.

Beginning in 1962, as an employee of the CIBC, the author began a 3 year program to provide breeding stocks of potentially useful parasitoids from North America. The first summer was devoted to extensive scouting across the western United States and eastern Canada to determine the most productive sites for more intensive collections to be made in subsequent years. Sites in California were chosen for more intensive work in 1963 and 1964, based on relative likelihood of collection success of a variety of species indicated to be of interest by the Australians and New Zealanders, the availability of field station support facilities in or near collection areas, and accessibility of air freight shipment depots relatively near to collection areas. Primarily as a test of shipment methods and routes, two shipments of parasitoids were sent to Rotorua, New Zealand, from California, USA, and one from New Brunswick, Canada, in 1962. Insect survival in transit was minimal, highlighting the urgent need to develop acceptable protocols.

In 1960, two large forest fires burned in conifer forests in midsummer in northern California, USA. One of these, the 'Donner Burn,' covering slightly more than 16,000 ha, was on the east side of the Sierras and burned from mid-August through the end of September through a mixture of ponderosa pine (*Pinus ponderosa* Laws.) and, at higher elevations, Jeffrey pine (*P. jeffreyi* Murr.) mixed with white fir (*Abies concolor*) and red fir (*A. magnifica*). Elevation varied between approximately 1,775 and 2,300 m. At the same time, the 'Volcano (Foresthill) Burn,' also just in excess of 16,000 ha, burned on the west side of the Sierras at elevations of approximately 765–1,225 m. Predominant tree species included *A. concolor*, incense-cedar (*Libocedrus decurrens*), and Douglas-fir (*Pseudotsuga menziesii* (Mirb.)), with some ponderosa pine at the higher elevations (Cameron 1962). While there had already been considerable salvage activity to recover useable timber, especially in the Volcano Burn, extensive areas of standing dead and fire-scorched dying trees remained. In addition, huge slash piles throughout both burns contained unsalvaged logs up to ~9.75 m and ~25 cm in diameter [a '32–10'] at the small end (which were considered to be uneconomical to salvage at the time). These fire-scorched trees and large slash provided ideal host material, and siricid populations exploded.

In 1963, collection efforts were concentrated in the Donner and Volcano burns in California, with shorter excursions to surrounding areas with recently killed conifers (Cameron 1963). Parasitoids were collected as adults in the field, and by caging infested logs in the field as well as transporting Sirex-infested logs to the CIBC laboratory in Fontana, California, to be held for parasitoid emergence. During the season, a total of 118 *I. leucospoides ensiger*, 3 *Ibalia* sp., 1 *Schlettererius cinctipes* Cresson (Hymenoptera: Stephanidae), and 5 *Rhyssa alaskensis* Ashmead adults were shipped to the Tasmanian Regional Laboratory in Hobart, and 43 *I. leucospoides ensiger* and 9 *M. nortoni nortoni* were shipped to the Forest Research Institute, Rotorua, New Zealand. All insects shipped to Hobart arrived alive and in good condition. Only 2 (of 4) females and 1 (of 5) males of *M. nortoni nortoni* arrived alive in Rotorua. These stocks were used to initiate laboratory rearing in both countries, using host material that had been manipulated through cold storage treatment. This approach was followed to retard development of the hosts and to compensate for the necessary shift from northern hemisphere seasons to Southern Hemisphere seasons.

The most common siricid associated with *Ibalia* in California was *Sirex areolatus* (Cresson) which developed in incense-cedar, *L. decurrens,* logs taken from the Volcano burn. *Xeris macgillivrayi* Bradley was also frequently recovered from these same logs that yielded *Ibalia*, but in only about 1/3 the numbers of *S. areolatus*. *Ibalia* was seldom observed in the field; emergence in the laboratory (164 individuals) was slightly skewed toward males (55%). Emergence occurred primarily from late September through early November. In late September 1962, Wickman (1964) had observed a 'mass flight' of *Ibalia* around siricid-infested incense-cedar at a mill handling logs that were salvaged from the Volcano burn. *Sirex areolatus* adults were also common at the time. *Ibalia*, an egg parasitoid, is well-synchronized with the emergence of its host in the fall of the year.

Three females and one male of *S. cinctipes* were collected. One of the females was observed to mate, in the laboratory, and it was sent to Hobart. At the time, virtually nothing was known of the biology of this parasitoid; no records of parasitism of siricids were available, but it had been associated with wood-boring insects. A lab colony was established from this single female specimen, leading, years later, to

Table 8.1 Parasitoids shipped from North America to Australia and New Zealand, 1963–1964[a]

	Australia		New Zealand	
Year	1963	1964	1963	1964
Species				
Ibalia leucospoides ensiger Norton	50♀; 68♂	5[b]	24♀, 19♂	
Ibalia sp.	3[b]			
Megarhyssa nortoni nortoni Cresson		222♀, 83♂	4♀, 5♂	241♀, 158♂
Rhyssa alaskensis Ashmead	4♀, 1♂			
'*Rhyssa*' spp.		22♀, 11♂		12♀, 19♂[c]
Schletterius cinctipes Cresson	1♀			5♀, 2♂[d]

[a]In 1962, trial shipments of *M. nortoni nortoni* and *Rhyssa* spp. were sent overseas; survival was very poor. An improved shipping container was used in 1963 and 1964 (Cameron 1965)
[b]Unknown sex
[c]Mixture of ♂ *R. persuasoria persuasoria*, ♂ and ♀ *R. alaskensis*, and ♂ and ♀ *Pseudorhyssa maculicoxis* Kriechbaumer, a cleptoparasite (Nuttall 1989)
[d]No progeny from lab rearings

successful field release and establishment in Tasmania. Very few *Rhyssa* were collected, and small numbers were sent to Hobart. Other species of *Rhyssa,* especially *R. persuasoria himalayensis* Wilkinson, were being received in both Hobart and in Rotorua from collaborators at the Indian and Pakistan Stations of CIBC. While no *Megarhyssa* were recovered from caged logs, adults were captured in the field from late June through July. From these, as single shipment was made to Rotorua.

With apparently thriving laboratory colonies of *I. leucospoides ensiger* having been established in both Australia and New Zealand, the emphasis for 1964 collections was shifted to obtaining numbers of *M. nortoni nortoni*. A University of California, Berkeley, field station is located at Sagehen Creek, north of Truckee, California, and on the edge of the Donner burn. During the summer of 1964, a number of students taking an entomology field course were excited to be able to earn income on a piece work basis for collection of adult *Megarhyssa*. Students were only paid for females, and the author was easily able to collect enough males to insure mating of virtually all females. Between June 24 and August 3, a total of 704 *M. nortoni nortoni* adults was shipped to Hobart and Rotorua, almost all of which were collected in the field and, therefore, of unknown age. Of these, 676 (96%) arrived alive and in good condition. An economical and efficient shipping container was devised, which was also used to ship smaller numbers of *I. leucospoides ensiger*, *Rhyssa* spp., and even *S. juvencus californicus* Ashmead to various distant locations. Survival in transit for all shipments combined was 94.8% (Cameron 1965b).

During 1963 and 1964, the author shipped a total of 959 adults of potentially useful parasitoids from California, U.S.A., to Australia and New Zealand. Species included were *I. leucospoides ensiger, M. nortoni nortoni, R. alaskensis, 'Rhyssa'* spp., and *S. cinctipes* (Table 8.1).

Hurley et al. (2007) provided a thorough and valuable review of biological control efforts directed against *S. noctilio* in the Southern Hemisphere, complete with a table summarizing the various parasitoid species introduced, liberated, and recovered in the field in New Zealand, Australia (separated by states, namely, Tasmania, Victoria, and New South Wales), and South Africa (separated by Western Cape, and KwaZulu-Natal). In New Zealand, since 1929, *R. persuasoria persuasoria, I. leucospoides leucospoides, I. leucospoides ensiger*, and *M. nortoni nortoni* have been released and recovered; *R persuasoria himalayensis* was released but never recovered. Except for earlier releases of *R. persuasoria persuasoria* and *I. leucospoides leucospoides*, these releases were made during the 1960s and 1970s, with some releases of *Megarhyssa* as late as 1984.

In Australia, *R. persuasoria persuasoria* was first released in Tasmania in 1957 and *I. leucospoides leucospoides* in 1959, with additional releases through the mid-1960s. These parasitoids had been supplied from colonies maintained in New Zealand, with local rearing undertaken to provide numbers adequate for liberation in Tasmania. *Rhyssa persuasoria himalayensis, R. lineolata, I. leucospoides leucospoides, I. leucospoides ensiger, S. cinctipes*, and *M. nortoni nortoni* were all released in the early to mid-1960s, with all but *R. persuasoria himalayensis* subsequently being recovered in the island state. During this time also, the nematode was inoculated into *P. radiata* stands on the island as well. Between the mid-1960s and the mid-1970s, Sirex populations were largely reduced to a level that did not cause continuing concern in Tasmania. *Megarhyssa nortoni nortoni* and *R. persuasoria persuasoria* provided the greatest pressure on the host populations. It was not until the mid-1970s that *S. cinctipes* and *D. siricidicola* were released in Tasmania (Taylor 1978). On the mainland (Victoria and New South Wales states), releases of all of these species except *R. persuasoria himalayensis* and *R. lineolata* were made in the 1970s and 1980s as the insect moved northward; *R. hoferi* was also released. All species were subsequently recovered. Numbers released were in the tens of thousands for the two *Ibalia* sub-species and for *Megarhyssa*; numbers in the thousands of *R. persuasoria persuasoria* were released (Hurley et al. 2007). From the single fertilized female *S. cinctipes* received in Tasmania from California, over 1700 male and female progeny were ultimately released over a period of 13 years. *Ibalia* was the most successful of the parasitoids on the mainland, but by itself the combined parasitism by all of the species was inadequate to reduce host levels to an acceptable level.

Especially on mainland Australia, improved silvicultural practices were undertaken in some of the dense stands to reduce stocking density and thereby increase the vigor of the remaining trees, and the nematode was widely inoculated into trees. Siricid populations were reduced over much of the area where outbreaks were causing serious problems, but economic considerations, largely the cost of labor, precluded as much silvicultural manipulation as might have been desired. The effectiveness of thinning as stands mature was clearly illustrated in a trial in South Australia during the outbreak in the 'Green Triangle' area from 1987 to 1991 (Chap. 14). With both nematode and parasitoid activity also present, no mortality occurred in a thinned compartment, bordered on all four sides by unthinned *P. radiata* in a 13 year old plantation. The unthinned surrounding forest suffered 68% mortality (D. Haugen, personal communication). Once again, the combination of parasitoids, the nematode, and good silvicultural practices reduced the loss of trees. To the present time there remain pockets of Sirex activity in south-eastern Australia even when parasitoids and/or the nematode are present, especially where stands are not adequately thinned as the trees grow and mature (Chaps. 12 and 14).

8.2.2 South America

The history of *S. noctilio* in the four South American countries – Uruguay, Brazil, Argentina, and Chile – is not nearly as long. In 1980, *Sirex* was reported for the first time in Uruguay (Maderni 1998). It is believed to have arrived through accidental introduction in saw timber imported from overseas and in wooden packing materials. Loblolly pine, *P. taeda* L., native to the southeastern United States, is the most heavily attacked species, but Monterey pine and Maritime pine, *P. pinaster*, a Mediterranean area native, are also commonly attacked. Slash pine, *P. elliottii* Engelm., is also attacked, but appears to be less susceptible than the other species. *Ibalia leucospoides leucospoides*, also accidentally introduced presumably along with the woodwasp, was discovered soon after the initial discovery of Sirex. With the addition of the nematode, *D. siricidicola*, but especially with improved silvicultural practices to maintain healthy, vigorous stands, it appears that the insect has been held to relatively minor pest status in recent years (Maderni 1998, Chap. 16).

Argentina first reported established populations of *S. noctilio* in 1985 in the eastern part of the country, and in Patagonia by 1990 from where it spread more widely. Initially there were attempts to eradicate the insect by removing and incinerating infested trees, but these were unsuccessful. *I. leucospoides leucospoides* once again appeared along with the pest, established readily in some areas, and has been reported to have rates of parasitization as high as 20–40% (Klasmer et al. 1998). *Deladenus siricidicola* has also been introduced into siricid-infested trees, and improved silvicultural management practices have been initiated. The insect still poses problems in Argentinian plantations (also see Chap. 15).

A similar pattern has been recorded in southern Brazil. Loblolly pine is the primary plantation host attacked, and slash pine is also recorded as a host; Sirex was first reported in 1988. *Ibalia leucospoides leucospoides* was first detected as an accidental introduction in 1990. This parasitoid has spread, along with its host, with parasitism rates averaging about 25%, but often reaching higher levels. It is present in almost all of the stands attacked by Sirex. *Deladenus siricidicola* was introduced, beginning in 1989, and has been very successful. Improved forest management, especially thinning, was widely implemented in an effort to maintain healthier, and thus more resistant, stands. Both *M. nortoni nortoni* and *R. persuasoria* have been cultured in the lab and released, beginning in 1996 and 1997 (Iede et al. 1998), but neither species has yet become established (Iede 2007, Chap. 16). Breeding colonies of these parasitoids were obtained from colleagues in Australia.

Monterey pine, as in a number of other Southern Hemisphere countries, dominates plantation forestry in Chile. The first report of *S. noctilio* in Chile was in 2001.

An aggressive quarantine and monitoring programme delayed introduction into Chilean *P. radiata* stands, and with suppression measures aimed at eradication in place since the appearance of this pest, extensive infestations have so far been prevented (Cisternas 2007). Beeche et al. (Chap. 17) report that *I. leucospoides leucospoides* arrived in Chile along with another exotic siricid, *Urocerus gigas* L., and has readily adapted to *S. noctilio*. *M. nortoni nortoni*, and the nematode, *D. siricidicola,* were first released in 2006. Improved plantation management, especially in privately owned plantations, has been undertaken (Ahumada and Poisson 2007).

8.2.3 South Africa

In South Africa, *S. noctilio* was first recorded in Cape Province in 1994 in *P. radiata.* The nematode, *D. siricidicola*, was quickly introduced and established; *I. leucospoides leucospoides*, obtained from Uruguay, was released and also later recovered in the field and is considered established. *Megarhyssa nortoni nortoni,* obtained from Australia, was also introduced, but it is not recorded as having become established (Tribe and Cillie 2004). Damage has been relatively minor, probably in large part because most of the pines in the western Cape are grown for saw-timber and stands are thinned during a rotation. In 2002, *S. noctilio* was first recorded in the Eastern Cape and KwaZulu-Natal provinces, and exploded to become a major problem. Most of the plantation pines in KwaZulu-Natal are grown for pulpwood and are not thinned during rotation. *Sirex* has been especially destructive in the widely-grown Mexican weeping pine, *P. patula* Schiede, causing over 35% mortality in some stands in the mid-2000s, but this damage has declined in recent years (Hurley 2010). In 2004, nematode inoculations were initiated; these have been continued annually, with increasing natural spread being recorded. In January 2006, *I. leucospoides leucospoides,* collected from the Western Cape, was released in KwaZulu-Natal and the Eastern Cape; field recoveries were made in late 2007 and further releases have been made. Development of laboratory-reared stocks for continuing releases is underway (Hurley 2010).

8.2.4 North America

To complete the international picture, *S. noctilio* is so recently recognized as being active in North America that no parasitoid introductions have been attempted, which is appropriate in the view of the author. It is probable that some of the parasitoids of native siricids, already present, will also serendipitously attack *S. noctilio*, and in any case the insect is expected to pose little if any threat of consequence in the northeastern part of Canada and the United States.

8.3 Conclusions

Climate matching is a factor that bears investigation when considering the introduction of parasitoids. Carnegie et al. (2006) used the climate-matching programme CLIMEX to develop predictions of the potential distribution of *S. noctilio*. But it has seldom, if ever, been a consideration in parasitoid introduction programmes in the Southern Hemisphere over the years. An after-the-fact analysis of Argentine climate with respect to *I. leucospoides leucospoides* suggests a good match (Villacide and Corley 2003). In broad general terms, it is in those countries or states that have a cooler, more moist, temperate climate, such as parts of New Zealand and the island state of Tasmania, Australia, where deliberately introduced parasitoids have been most easily established and most successful. In both of these areas, *Megarhyssa* has been an important parasitoid. The adults that were introduced to both Australia and New Zealand were collected in the early 1960s from higher elevations (~2,000 m) in the Sierra Nevada Mountains of California. The various species of *Rhyssa*, and of *Ibalia*, on the other hand, seem to have established more easily (and generally originated) in areas with warmer and more arid climates. Most of the collections of these genera in California were at elevations that were lower by about 1,000 m, and in forests growing under more arid conditions. Collections of *Rhyssa* spp. at the same time from the Indian subcontinent were also from more arid areas, as were collections by Kirk (1974, 1975) in the early '70's in the southeastern and southwestern United States. Unfortunately, *Rhyssa* spp., except for *R. persuasoria*, have a generally spotty record of establishment where introductions have been attempted.

These sweeping generalizations certainly are not absolute, and no detailed analyses of climates in both parasitoid source and later establishment areas that would confirm these impressions appear to have been done. From personal experience, I can provide a contradiction in the case of the stephanid, *S. cinctipes*. The one fertilized female sent to Tasmania was from the higher elevation location. While the species was established in Tasmania, it has also been established (and sometimes reaches rates of parasitization of up to 20%) on the Australian mainland. Also, *I. leucospoides leucospoides,* of European (English) origin, has done well in some of the more arid Southern Hemisphere countries into which it has been introduced. But there is enough general experience to justify further pursuit of this idea with at least a few of the more promising species before large sums are devoted to breeding programmes for a species that might not be well suited for release.

It is noteworthy that none of the parasitoids that were shipped by the author to Australia and New Zealand 45 years ago originated from *S. noctilio*, and none of the native conifer hosts from which they were collected included *P. radiata*. *Ibalia leucospoides leucospoides*, sent from England to New Zealand in the late 1920s and early 1930s (Hanson 1939), and *R. persuasoria* from England and Wales, and *Ibalia* again in the 1950s from England (Nuttall 1989), may have included *S. noctilio* as a native host, but *S. cyaneus* F. was much more common and *S. gigas* (now *Urocerus gigas* L. and *S. juvencus* L. were also present where they were collected (Hanson 1939). None of these parasitoids had previous exposure to *P. radiata*. The results of field releases have demonstrated clearly that the various parasitoids do not have a narrow tree host acceptance behavior, nor are they insect host-specific. The most common siricid host for *Megarhyssa* in California was *S. juvencus californicus*; other hosts included *S. longicauda* and *S. areolatus*. *Rhyssa* spp. collected in California were usually found in association with *X. macgillivrayi* and *X. spectrum* L. *Ibalia* was collected mainly from caged logs of incense-cedar, *L. decurrens,* from which *S. areolatus* emerged in numbers and *X. macgillivrayi* and *X. spectrum spectrum* L. also emerged. At least two species of *Sirex* and one of *Urocerus* were also recovered as adults in flight in the area from which the logs were collected, but in substantially smaller numbers than *Xeris*. One unequivocal statement can be made, however, and that is that parasitoids of the subfamily Tremicine, which develop in hardwood hosts, show no crossover to hosts in the subfamily Siricinae, which develop in conifers. Siricinae are most common in *Pinus*, but also attack firs, *Abies* spp., Douglas-fir, *Pseudotsuga menziesii,* incense-cedar, *L. decurrens,* and occasionally larches, *Larix* spp.

A healthy stand of trees is usually a Sirex-resistant stand of trees. Tree health is the first determinant of susceptibility to attack by *Sirex*. Therefore, tending is required *throughout the rotation of the stand.* Top down forces (parasitoids and other biological enemies) are important in pest population regulation, but the bottom up forces provided by the host tree set the stage for attack or for resistance to attack. This is not a new idea. At least as far back as, Rawlings and Wilson (1949) stressed the importance of sound silviculture to maintain healthy individual trees. Indeed, maintenance of healthy, non-stressed trees and stands is the touchstone that appears over and over through the great majority of papers addressing Sirex pest management. This is also very evident in its native range of southern Europe, eastern Asia, and northern Africa, where *S. noctilio* is only a secondary insect and not considered to be a pest; it is a part of the 'clean-up crew' of insects and pathogens involved in recycling nutrients from dead and dying trees (see also Chap. 5).

Reduction of established Sirex populations to an acceptable level of infestation, and, more importantly, maintenance of that acceptable level (i.e. population regulation), depends primarily on rigorous silvicultural management to create and maintain healthy stands. Initial stocking density in pine plantations, depending on species, usually runs from about 800 to 2,000+ trees/ha. Once the canopy begins to close, there is increased competition among trees for sunlight, moisture, and nutrients depending on the individual site. In short, trees become stressed, some fall into the suppressed category, and occasional mortality occurs in unthinned stands. Thinning, often more than once during a rotation, contributes significantly to maintaining stand health; pruning, especially in stands grown to produce sawtimber, adds an additional benefit. At harvest, well-managed and thinned stands have usually been reduced to approximately 400–600 trees/ha. Stands grown for pulpwood in South Africa are not thinned. Initial stocking density at planting is 1,666 trees/ha (on 3×2 m spacing). The desire is to maximize biomass production on every hectare; thinning is an unwanted and, under current market conditions, an uneconomical option. A huge population of Sirex built up in a landscape where there is a banquet of available food in the form of overcrowded, and stressed, maturing trees in monocultures. An extended drought in the early 2000s added to the stresses under which trees were being grown. The consequence was that heavy Sirex infestations severely disrupted management plans by forcing premature cutting of compartments. Dead trees not only reduced biomass production, but also presented post-harvest handling challenges. Perhaps ironically, the extensive areas cut to salvage badly damaged stands, and in the process the removal of very large numbers of resident Sirex populations as immatures within the cut trees, may have helped to slow the explosive increases in woodwasp populations in *P. patula* grown for pulpwood KwaZulu-Natal in recent years (also see Chap. 18).

Contrast the South African situation with New Zealand (Chap. 13), where Sirex today is not a pest of concern even though it caused devastating damage in the 1950s and 1960s. Their extensive *P. radiata* plantations, established in the 1930s, had not been well tended; drought exacerbated the situation as the trees were maturing in the '50's. Standard practice in New Zealand plantation forests now is to prune three times during a rotation (at age 4 years, to 1 ½ to 2 m; at age 6 years, to 4 m; and at age 8–9 years, to 6 m), and to reduce the initial stocking density of only 600–800 trees/ha to about 400 by maturity. Rotations take 22–30 years. The bottom-up forces maintain stands that are not hospitable to Sirex.

Of the biological agents used for management of *S. noctilio* in plantations throughout the Southern Hemisphere countries, clearly the nematode, *D. siricidicola*, has played the most significant role in reducing populations. Research on the interaction between host and parasite continues (see for example Chaps. 9 and 18) to identify the best strain, to improve culturing methods for laboratory replication, and to improve field inoculation techniques so the nematode can establish and flourish, and be spread naturally by Sirex adults through their normal life cycle activities. But *D. siricidicola* as the only biological antagonist has not been able to maintain *S. noctilio* populations at acceptable levels in any of the Southern Hemisphere countries where it has become established.

A number of parasitoids are available, and have been shown to provide the necessary added top-down pressure on Sirex populations to bring them to, and to maintain, levels that are acceptable in plantation management programmes. *Ibalia leucospoides leucospoides* and *I. leucospoides ensiger*, egg parasitoids, and *R. persuasoria persuasoria* and *Megarhyssa nortoni nortoni*, larval parasitoids, are the species that have been used most commonly because they are the most effective in areas where they have been introduced and established. Each one is amenable to laboratory propagation; where multiple species are present, their actions are additive and not competitive. *Rhyssa lineolata, R. amoena, R. crevieri* and *R. persuasoria himalayensis* have also been shipped to quarantine facilities in New Zealand. Of these, only *R. persuasoria himalayensis* was field-released, but it was never recovered (Nuttall 1989). *Ibalia leucospoides ensiger* has been released and recovered in New Zealand and Australia (Taylor 1978; Nuttall 1989). *Schlettererius cinctipes* is well established on the island state of Tasmania, and in Victoria and South Australia, especially in cooler areas. Its value is especially noticeable in years when there are cooler than normal temperatures. Parasitoids, especially *I. leucospoides leucospoides*, contribute to population reduction in some of the South American countries, and this species is established, but not yet a force in South Africa. Once established, parasitoids are a self-maintaining regulatory force, persisting through low host population densities.

The use of introduced parasitoids is an important tactic to use in an integrated pest management strategy designed to manage *S. noctilio*. By themselves, either as a single species or as a complex of species, they have not been shown to provide the pressure necessary to reduce Sirex numbers to a level acceptable to forest managers. But then neither has the nematode when used as the only biological control. The only tactic capable of achieving this goal is sound silvicultural management of the stand, from planting through harvest, as continues to be demonstrated throughout New Zealand. Parasitoids do, however, provide that additive pressure on Sirex that makes the difference between an unacceptably high level of tree attack and an acceptable level. It is for that reason that they should be considered to be an essential tactic to be considered when *S. noctilio* is a problem insect.

The best choice of species will be determined by practical factors such as ease of obtaining stock cultures, ease of rearing, and ease of establishment as shown through experience in countries where introductions have been made previously. The pressure that an individual species may bring to bear on the population is not well understood. Taylor (1978), as a result of studies extending over 11 years, concluded that *R. persuasoria persuasoria* and *M. nortoni nortoni* were responsible for reducing populations of Sirex, the two species acting as one delayed density-dependent factor, whereas *I. leucospoides leucospoides* is apparently density-independent. Most literature reports, however, express results in terms of the percent of hosts killed; data are not presented in a life table form where indispensible mortalities can be calculated. There is much to be learned in this area. With cryptic development of parasitoids in a host that itself is cryptic, and may of which have a life cycle that lasts for a year and may extend over two or even 3 years, the challenges are not ones easily overcome.

When biological control programmes are undertaken, frequently the effort is put into introduction and establishment of an agent. It is not put into the underlying biological, behavioral, population, and community studies that would provide a basis for better decision-making. Time is of the essence; economic losses are occurring, and demands of forest managers and agencies for solutions do not allow the luxury of years of research to improve our understanding of the dynamics of the complex of organisms with which we are working. Unfortunately, once a problem appears to be 'solved,' or at least the damage has been reduced significantly, research support funds tend to be redirected to another problem of the moment. In spite of well coordinated, long term, and large scale efforts that have occurred at different times during the last century, involving both the forest industry and appropriate government agencies and, in best case scenarios, also University researchers, in countries such as New Zealand, Australia, and South Africa, many critical questions remain unanswered. We are left to move ahead based on the best available interpretations of information from similar occurrences in other parts of the world, and piece by piece to add to the foundation of understanding that is being built. What we do know is that parasitoids have an important role in effective integrated management programmes targeting the woodwasp. There is no silver bullet. Doubtless the specific choices of parasitoid(s) to be used will vary with the local situation. Parasitoids should always be included among the tactics employed at some stage of any pest management strategy directed against *S. noctilio*. They are an essential third leg of that three-legged stool that also includes good silvicultural practices and the parasitic nematode.

References

Ahumada RJ, Poisson MA (2007) The control of *Sirex noctilio* in the *Pinus* plantations of the Arauco companies. In: Abstract, International Sirex symposium and workshop, Pretoria & Pietermaritzburg, 9–16 May 2007

Anon (1960) Biological control of Sirex. N Z Timber J 7:1

Bedding RA, Akhurst RJ (1974) Use of the nematode *Deladenus siricidicola* in the biological control of *Sirex noctilio* in Australia. J Aust Entomol Soc 13:129–135

Benson RB (1943) Studies in Siricidae, especially of Europe and Southern Asia (Hymenoptera, Symphyta). Bull Entomol Res 34:27–51

Cameron EA (1962) North American survey for natural enemies of the Siricidae, May–Oct 1962. Internal report, Commonwealth Institute Biological Control, Californian Station, Fontana

Cameron EA (1963) 1963 Field work for the collection of siricid parasites, California, U.S.A. Internal report, Commonwealth Institute Biological Control, Californian Station, Fontana

Cameron EA (1965a) The Siricinae (Hymenoptera: Siricidae) and their parasites. Tech B Commonw Inst Biol Control 5:1–31

Cameron EA (1965b) A method for shipping adult parasites of the Siricidae. Can Entomol 97:945–946

Cameron EA (1968) *Sirex juvencus californicus* in smog-killed trees in southern California (Hymenoptera: Siricidae). Pan-Pacific Entomol 44:168

Carnegie AJ, Matsuki M, Haugen DA, Hurley BP, Ahumada R, Klasmer P, Sun J, Iede ET (2006) Predicting the potential distribution of *Sirex noctilio* (Hymenoptera: Siricidae), a significant exotic pest of *Pinus* plantations. Ann For Sci 63:119–128

Cisternas MB (2007) Official program for detection and control of *Sirex noctilio* (Hymenoptera: Siricidae) in Chile. In: Abstract, International Sirex symposium and workshop, Pretoria & Pietermaritzburg, 9–16 May 2007

Clark AF (1933) The horntail borer and its fungal association. N Z J Sci Tech 15:188–190

de Groot P, Nystrom K, Scarr T (2007) Discovery of *Sirex noctilio* (Hymenoptera: Siricidae) in Ontario, Canada. Gt Lakes Entomol 39:49–53

Dodds KJ, de Groot P, Orwig DA (2010) The impact of *Sirex noctilio* in *Pinus resinosa* and *Pinus sylvestris* stands in New York and Ontario. Can J For Res 40:212–223

Hanson HS (1939) Ecological notes on the Sirex wood wasps and their parasites. Bull Entomol Res 30:27–65

Hoebeke ER, Haugen DA, Haack RA (2005) *Sirex noctilio*: discovery of a palearctic siricid woodwasp in New York. Newsl Mich Entomol Soc 50:24–25

Hudson GV (1919) *Sirex juvencus* in New Zealand. Entomol Mon Mag 55:181

Hurley BP (2010) Factors influencing the control of the Sirex woodwasp in South Africa. Ph.D. dissertation, Faculty of Natural and Agricultural Sciences, University of Pretoria, Pretoria

Hurley BP, Slippers B, Wingfield MJ (2007) A comparison of control results for the alien invasive woodwasp, *Sirex noctilio*, in the southern hemisphere. Agr For Entomol 9:159–171

Iede ET (2007) The woodwasp *Sirex noctilio* in Brazil – monitoring and control. In: Abstract, International Sirex symposium and workshop, Pretoria & Pietermaritzburg, 9–16 May 2007

Iede ET, Penteado SdoRC, Schaitza EG (1998) Sirex noctilio problem in Brazil: detection, evaluation, and control. In: Iede ET, Schaitza E, Penteado S, Reardon RC, Murphy ST (eds)

Proceedings of a conference: training in the control of *Sirex noctilio* by the use of natural enemies, Colombo, PR State, Brazil, Nov 1996. Forest Health Technology Enterprise Team 98–13, USDA, Forest Service

Kirk AA (1974) Siricid woodwasps and their associated parasitoids in the south eastern United States. J Ga Entomol Soc 9:139–144

Kirk AA (1975) Siricid woodwasps and their associated parasitoids in the south western United States. Pan-Pacific Entomol 51:75–81

Klasmer P, Fritz G, Corley J, Botto E (1998) Current status of research on *Sirex noctilio* F. in the Andean-Patagonian region in Argentina. In: Iede ET, Schaitza E, Penteado S, Reardon RC, Murphy ST (eds). Proceedings of a conference: training in the control of Sirex noctilio by the use of natural enemies, Colombo, PR State, Brazil, Nov 1996. Forest Health Technology Enterprise Team 98–13, USDA, Forest Service

Maderni JFP (1998) Sirex noctilio F.: present status in Uruguay. In: Iede ET, Schaitza E, Penteado S, Reardon RC, Murphy ST (eds) Proceedings of a conference: training in the control of Sirex noctilio by the use of natural enemies, Colombo, PR State, Brazil, Nov 1996. Forest Health Technology Enterprise Team 98–13, USDA, Forest Service

Miller D, Clark AF (1935) *Sirex noctilio* (Hym.) and its parasite in New Zealand. Bull Entomol Res 26:149–154

Nuttall MJ (1989) *Sirex noctilio* F., sirex wood wasp (Hymenoptera: Siricidae). In: Cameron PJ, Hill RL, Bain J, Thomas WP (eds) A review of biological control of invertebrate pests and weeds in New Zealand 1874 to 1987, technical communication no. 10, CAB International Institute of Biological Control

Rawlings GB, Wilson NM (1949) *Sirex noctilio* as a beneficial and destructive insect to *Pinus radiata* in New Zealand. N Z J For 6:20–29

Schneeberger NF (2007) Response to the recent find of *Sirex noctilio* in the United States. In: Abstract, International Sirex symposium and workshop, Pretoria & Pietermaritzburg, 9–16 May 2007

Taylor KL (1978) Evaluation of the insect parasitoids of *Sirex noctilio* (Hymenoptera: Siricidae) in Tasmania. Oecologia 32:1–10

Tribe GD, Cillie JJ (2004) The spread of *Sirex noctilio* Fabricius (Hymenoptera: Siricidae) in South African pine plantations and the introduction and establishment of its biological control agents. Afr Entomol 12:9–17

Villacide JM, Corley JC (2003) The potential distribution of the parasitoid *Ibalia leucospoides* (Hymenoptera: Ibaliidae) in Argentina. Rev Cienc For-Quebracho 10:7–13

Wickman BE (1964) Observations on the siricid parasite *Ibalia ensiger* (Hymenoptera: Ibalidae). Pan-Pacific Entomol 40:19–20

Zondag R (1962) A nematode disease of *Sirex noctilio* (F.) Interim Research Release, Forest Research Institute, Rotorua, New Zealand Forest Service

Zondag R (1969) A nematode infection of *Sirex noctilio* (F.) in New Zealand. N Z J Sci 12:732–747

Chapter 9
Factors Affecting the Efficacy of *Deladenus siricidicola* in Biological Control Systems

Bernard Slippers, Brett P. Hurley, X. Osmond Mlonyeni, Peter de Groot, and Michael J. Wingfield

Abstract The nematode, *Deladenus* (=*Beddingia*) *siricidicola*, represents the cornerstone of *Sirex noctilio* biological control programs across the Southern Hemisphere. There is, however, significant variation in its efficacy in different regions. In this review, we consider emerging issues related to the biology and handling of the nematode that might influence its efficacy in biological control systems. Most practical aspects concerning the handling of *D. siricidicola* have been streamlined over the past half-century and these appear to be very efficient. However, large gaps remain in our knowledge about some key aspects of the biology of *D. siricidicola*. For example, very little is known regarding the evolution of virulence in the nematode populations, and the consequent evolution of resistance in *S. noctilio* populations. Furthermore, the levels of diversity in *D. siricidicola* and its ability to adapt to fungal, wasp and environmental variation are poorly understood. In this regard, new collections and storage of native populations of the *Deladenus* spp. are critical for the future research and management of this key biological control agent of *S. noctilio*.

B. Slippers (✉) • B.P. Hurley • X.O. Mlonyeni • P. de Groot • M.J. Wingfield
Department of Genetics, Forestry and Agricultural Biotechnology Institute (FABI), University of Pretoria, Pretoria, South Africa

Department of Zoology and Entomology, Forestry and Agricultural Biotechnology Institute (FABI), University of Pretoria, Pretoria, South Africa

Natural Resources Canada, Canadian Forest Service, Great Lakes Forestry Centre, 1219 Queen Street East, Sault Ste. Marie, Ontario, Canada P6A 2E5
e-mail: bernard.slippers@fabi.up.ac.za

B. Slippers et al. (eds.), *The Sirex Woodwasp and its Fungal Symbiont: Research and Management of a Worldwide Invasive Pest*, DOI 10.1007/978-94-007-1960-6_9,
© Springer Science+Business Media B.V. 2012

9.1 Introduction

Sirex noctilio was accidentally introduced into New Zealand around 1900 (Miller and Clark 1935). The wasp soon became a serious pest in *Pinus radiata* plantations and during the 1940s and 1950s reached epidemic proportions (Rawling 1955). The damage caused by the wasp sparked intensive studies on its biological control, initially focussed on parasitic wasp species (Hanson 1939; Nuttall 1989). During the course of this work, a nematode-infected female *S. noctilio* was discovered on the north island of New Zealand (Zondag 1962). The nematode was thus naturally introduced together with *S. noctilio* into this region. It soon became clear that this nematode held promise as a biological control agent (Zondag 1965, 1967, 1969, 1971, 1979). This stimulated the emergence of a research field that continues today and that has made *D. siricidicola* one of the best-studied entomopathogenic nematodes in any system.

Deladenus siricidicola has a bi-cyclic life cycle (Fig. 9.1), including a mycetophagous or free-living and a parasitic cycle (Bedding 1967). The two morphological forms associated with this unusual life history are so distinct that it might initially have been described in two families, the Neotylenchidae (where it is currently placed) and the Allantonematidae (Bedding 1967, 1974). In the free-living cycle, the nematode feeds exclusively on *Amylostereum* spp., the fungal symbionts of Siricid wasps, and it reproduces oviparously (Bedding 1967, 1972). In the parasitic cycle, female nematodes enter and develop in the haemocael of the Siricid larvae and reproduce ovoviviparously (Bedding 1967, 1972). Nematode larvae produced by parasitic females are released inside the haemocael of Siricid larvae, and migrate towards and then infest the testes and developing eggs.

Deladenus siricidicola sterilizes the female of *Sirex noctilio*. The nematode does not affect oviposition and is consequently spread by the female wasps through infected eggs (Zondag 1969; Bedding 1972). Furthermore, the free-living cycle makes it possible to rear the nematode in large quantities in the laboratory and thus to be able to achieve mass releases in the field (Bedding 1974; Bedding and Iede 2005) (Fig. 9.1).

To date, seven *Deladenus* spp. have been described associated with the Siricid-*Amylostereum* symbiosis (Bedding 1974). Several of these species also infest the parasitoids of Siricids and the beetle *Serropalpus barbatus* (Bedding 1967, 1972, 1974). In addition, *D. siricidicola* is highly specific to Siricids and *A. areolatum*, including not infesting other hymenopteran parasitoids of Siricids (Bedding and Akhurst 1974, 1978). These characters together have made *D. siricidicola* an ideal biological control tool. Initially it was introduced between plantations by moving infested logs , wasps or contents from infested wasps (Zondag 1969, 1971). Subsequently, artificial inoculation has become the preferred method of introduction into plantations, followed by natural spread by female wasps. This followed intensive work on artificial rearing of the nematode on *A. areolatum* cultures (Bedding and Akhurst 1974; Bedding 1979), and subsequent development of the method to inoculate it into cavities punched into tree stems together with a carrier gel solution (Bedding and Iede 2005).

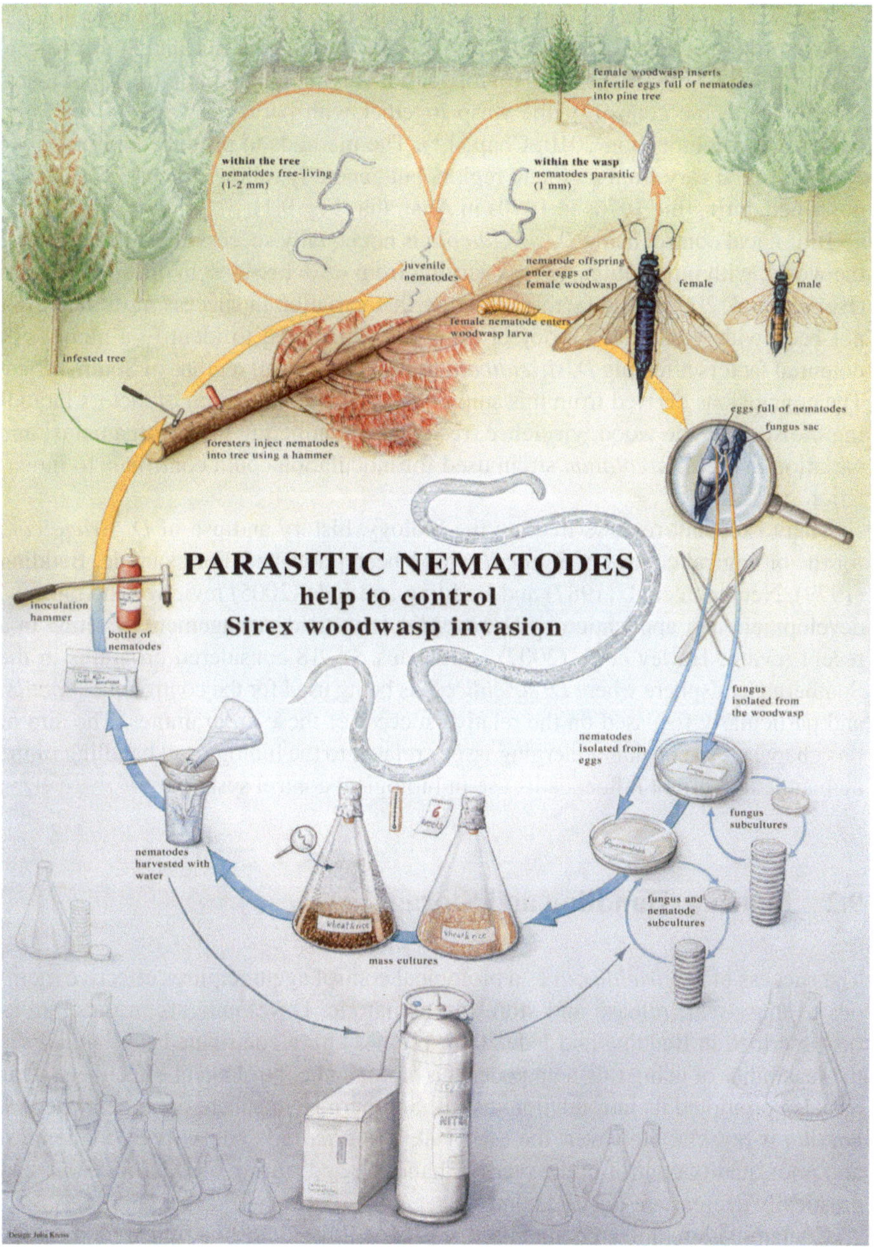

Fig. 9.1 *Deladenus siricidicola* mass rearing and release strategy and bi-cyclic life cycle. The image was produced by Julia Kreiss

Today, *D. siricidicola* is considered to be the cornerstone of ongoing biological control programs against *S. noctilio* in Australia, South America and South Africa. In New Zealand the nematode is no longer actively inoculated into trees, but it still contributes to the control of the wasp together with parasitic wasps (John Bain, personal communication 2010; Chap. 13). The methods to produce and inoculate *D. siricidicola* vary from region to region, but generally follows the basic principles described during the 1970s to 1990s in Australia (Fig. 9.1).

Biological control using *D. siricidicola* is not equally successful in all regions of the world, with inoculation success varying from <5% to >99% in different regions (Hurley et al. 2007). The factors affecting this variation in success have, however, not been widely studied. Hurley et al. (2008) attempted to identify a number of potential factors affecting *D. siricidicola* in summer rainfall regions of South Africa. The conclusions derived from this study were that a combination of factors including moisture in the wood, virulence, resistance, competing micro-organisms, and variation in the *A. areolatum* strain used for inoculation could contribute to inoculation success.

Many excellent reviews treating the biology, history and use of *D. siricidicola* for the biological control of *S. noctilio* have been published. For example, Bedding (1979), Neumann et al. (1987) and Bedding and Iede (2005) reviewed the biology, development and application of *D. siricidicola* in field management systems. In a recent review, Hurley et al. (2007) and Chaps. 14–18 considered programs in the Southern Hemisphere where *D. siricidicola* is being used for the control of *S. noctilio* and particularly focussed on the relative success of these programmes. The aim of this chapter is to consider emerging issues related to the biology and handling of the nematode that might influence its use in biological control systems.

9.2 Rearing, Handling and Storage

The success of *D. siricidicola* as a biological control agent requires effective methods to mass-rear, release and store the nematode. Developments in these areas (as described in Bedding and Iede 2005; Fig. 9.1) have contributed substantially to the feasibility of using this nematode as a biological control agent effectively. This has also promoted its introduction into all Southern Hemisphere countries where *S. noctilio* is present. However, the use of *D. siricidicola* is not without challenges, and poor quality control in the rearing, handling and storing of *D. siricidicola* can drastically decrease *S. noctilio* control.

Contamination of the medium used to grow *A. areolatum*, the fungal food source for *D. siricidicola,* can result in greatly reduced numbers of nematodes produced. This is exacerbated by the fact that the nematode cultures take many weeks to mature. In worst cases, the majority of a rearing population can be lost in a short time. The contaminants, including bacteria and mites, are often transferred by mites and can spread rapidly within a rearing facility. It is thus imperative to ensure that working conditions are as clean as possible. However, even under the sterile conditions, contamination is likely to occur during the mass rearing process and this needs to be considered when planning control programs.

The temperature of the nematodes during their transit to the field is a major factor influencing their survival. Nematodes are initially transported in water in sealed, breathable plastic bags and later suspended in a polyacrylamide gel before being inoculated into the trees, typically under cool conditions (<5°C) (Bedding and Iede 2005). Recent studies examining the survival of nematodes in water have confirmed that nematode survival decreases over time and as temperature increases (BP Hurley, unpublished data). For example, at 5°C and 10°C, over 80% of the nematodes survived after 150 h, whereas at 25°C only 39% survived after 30 h, and at 30°C only 7% survived after 24 h. Similarly, nematode survival in the polyacrylamide gel has also been shown to decrease with an increase in temperature (authors, unpublished data). The temperature of the gel is, however, less of a concern as the nematodes generally remain in the gel for only a short time before being inoculated into trees. In contrast, the nematodes can remain in the water-filled bags for numerous days. This increases the chance for them to be exposed to high temperatures and this can consequently greatly reduce the success of the inoculations.

A specially designed rebound hammer punch is used to inoculate trees with *D. siricidicola* (Bedding and Iede 2005, Chap. 14). The hammers are designed to make clean holes in the wood without bending the tracheids, thus allowing entry of the nematodes into the wood. Nematode numbers introduced into trees are significantly reduced when blunt punches are used (BP Hurley, unpublished data). Furthermore, the level of care with which contractors use the hammer punches to produce inoculation holes can also influence the quality of the inoculation site and thus nematode establishment.

Continuous rearing of the mycetophagous stage of *D. siricidicola* on *A. areolatum* over a number of years can result in a loss of virulence of the nematode (Bedding and Iede 2005). This can have serious negative consequences for biological control efforts, as was observed in the Green Triangle of Australia (Haugen 1990; Haugen and Underdown 1990; Haugen and Underdown 1993). This problem can be largely solved by storing nematodes in liquid nitrogen outside of the inoculation season (Bedding and Iede 2005).

9.3 Evolution of Nematode Virulence and Wasp Resistance

Variation and natural change in resistance of the *S. noctilio* populations to infection by *D. siricidicola*, and equally, changes in the virulence of *D. siricidicola* populations should be expected over time. Such variation and change are common patterns in biological interactions and they are thought to be linked to the evolution and maintenance of sexual reproduction in biological systems. The red queen hypothesis (van Valen 1973) postulates that hosts and their parasites are in a continual "arms race" involving cycles of evolution of resistance (including tolerance) in the host and over-coming resistance (including higher levels of virulence) in the parasite.

The "trade-off" hypothesis between transmission and virulence predicts that parasites will evolve towards lower levels of virulence in situations where there is a restriction on spread linked to high levels of virulence (Alizon et al. 2009). This idea has been most intensively explored in human-pathogen interactions, but not in agricultural or forestry situations. It is especially relevant to some biological control systems, in particular classical biological control, such as the *S. noctilio-D. siricidicola* system, which relies on the natural dispersal of the parasite that is often linked to the dispersal of the host.

Virulence is defined here as the number of adults in a given *S. noctilio* population that are infested by *D. siricidicola* and are sterilized by it. In female wasps this only includes individuals with infested eggs. Usually all eggs in such females will be infested. The nematodes sometimes infest the females, but they do not enter the eggs (Bedding 1972, 1974; Zondag 1975; Yu et al. 2009). For the purposes of this discussion on nematode virulence, these female wasps are not sterile and are thus not included in counts of infected wasps in a population.

It is not known how effectively *D. siricidicola* can spread if it does not infect the eggs. How this condition influences the fitness of the nematode populations is thus not known. This is an important question to answer, especially because the condition appears to be common in some regions. Bedding (1972, 1974) reported this condition in a number of populations. The latter studies also reported this to be the case for *D. imperialis* and *D. rudyi*. Yu et al. (2009) found that none of the 102 nematode-infected *S. noctilio* females collected from various sites in Canada had infected eggs. Zondag (1975) also noticed nematode infected *S. noctilio* females, where the eggs were not also infested. The possibility that the nematodes can spread without infecting the eggs is suggested by the fact that they can be found in the oviducts of the female wasp through which the eggs will pass during oviposition. If there is no selective advantage to *D. siricidicola* spread via eggs and thus sterilizing the host over spreading without infecting the eggs, then it is hard to imagine why all populations would not exclusively spread outside the eggs. Furthermore, if this condition is genetically controlled, then cross-breeding between infective and non-infective strains of *D. siricidicola* should be vigorously avoided where *S. noctilio* control using this nematode is important. There is thus a critical need to better understand the mechanisms that underlie and drive the ability of the nematode to infect Siricid eggs. Futhermore, variation in this condition should be compared in the different regions of the world where *D. siricidicola* is being used for biological control.

Bedding (1974) noted that there is a direct correlation between time of release of juvenile nematodes by infective female wasps, and subsequent infection of the host eggs. He also speculated that the nematode forms that do not enter eggs have evolved in host species that are more solitary. This is in contrast to wasps that often infest the same trees that would allow the evolution of highly virulent, egg-infecting nematode forms, because non-infected hosts are likely to also oviposit in the same tree (Bedding 1972, 1974). Bedding and Iede (2005) report high levels of virulence even under very low levels of plantation infestation (<1%) where attacks are very sparse. There are, however, no other data to make an evaluation of this hypothesis possible.

The infection of wasps by *D. siricidicola* is expected to have an effect on the population of *S. noctilio* even if the nematode does not infect the eggs. A number of researchers (Zondag 1969; Bedding and Iede 2005; Corley and Villacide 2005) have noted that infection by *D. siricidicola* leads to lowered fat reserves in the larvae and adults and that this affects their size and ability to fly. Their fitness would thus be negatively affected, in terms of numbers of eggs produced, dispersal distance and energy to oviposit. The effect of wasp infection without entering the eggs on the fitness of the *S. noctilio* population, however, cannot be quantified at present.

Zondag (1969) noticed that heavily infected wasps sometimes have no fat bodies remaining and they were often observed to die. He stated that 'The most important deviation from the normal pattern of the nematode infection is that the hosts can die when the immature hosts are heavily infected.' Bedding (1972) disagreed with this view and concluded that heavy infestation by the nematode does not lead to the death of the larvae. This is an important question as it underpins a potential driving mechanism for the evolution towards lower virulence in *D. siricidicola*, which would emerge where inordinately high levels of virulence in the nematode increase mortality in the larvae. This could then also be affected by levels of artificial inoculation, where heavy inoculation might lead to high infection levels and high larval mortality. The consequence would be that nematodes that are less virulent and consequently kill less larvae would survive. In this regard, Bedding and Akhurst (1974) noticed that heavily inoculated logs (at 10 cm intervals) produce smaller and fewer numbers of females, seemingly confirming this view.

Mechanisms of natural resistance in *S. noctilio* populations, other than the non-infection of eggs described above, might also exist. For example, Bedding (1972) noticed that nematodes, which infect and sterilize the Australian population of *S. noctilio*, do not do so in the Belgian population of *S. noctilio* where it originated. It is also obvious that any resistance that arises by mutation in a particular population of *S. noctilio* is likely to become fixed rapidly, especially under the strong selective pressure that would emerge where nematode infection sterilizes all non-resistant individuals. The mechanisms that might drive such resistance are not understood and only open to speculation. One possibility might be that resistance to infestation by *D. siricidicola* involves co-ordination of the development of eggs and the nematodes. The nematodes must enter the egg at a very specific stage of development, or they will either stop egg development completely or fail to penetrate the eggs (Bedding 1972). This is clearly an issue that needs urgent consideration.

Another source of variation in virulence in *D. siricidicola* might be by the loss of the ability to develop infective females due to continuous artificial rearing in the free-living cycle in the laboratory (Bedding 1972). The continual selection of those individuals that develop into mycetophagous forms (as opposed to infective females) might lead to such a loss. This is thought to have been the cause of the collapse of the biological control system in Australia during the 1987 to 1990 outbreak in Victoria (Haugen and Underdown 1993). The nematode that had previously consistently resulted in infection levels of above 95% after inoculation, then infected less than 30% of the wasp population after inoculation. To overcome this problem, a new and virulent nematode strain was re-isolated, which again brought parasitism levels after inoculation to >90% (Bedding and Iede 2005). There is, however, no direct evidence that the loss of virulence was caused by this rearing process. Attempts to produce a non-infective strain, for research purposes, by continuously rearing the nematode artificially for the past 6 years at the University of Pretoria, has failed to re-create this effect (Hurley et al. 2008 and subsequent work). The underlying reasons why *D. siricidicola* lost its virulence in Australia and Brazil remain unclear and this could be due to one or more of the factors discussed previously in this review.

9.4 Introduction History and Genetic Diversity

In 1962, more than half a century after *S. noctilio* was first reported in New Zealand, Rudi Zondag noticed size changes in female and male reproductive organs of wasps from the Rotoehu forest (Zondag 1962). Upon closer inspection, this was found to result in nematode infection. Subsequently, nematodes were also described from the parasitoid wasp *Rhyssa* spp. in New Zealand and from various Siricids and their parasitoids from England (Bedding 1967, 1968a, b, 1972; Hocking 1967), including the description of the bi-cyclic life cycle of the species of *Deladenus*.

Subsequent to the discovery of *D. siricidicola* and the realization of the obvious potential it has for biological control, extensive surveys were initiated by the CSIRO (Australia). Ultimately thousands of logs and tens of thousands of wasps from across Europe, North Africa, various sites in North America, Japan, Pakistan and India were collected and screened for nematodes (Bedding 1972; Bedding and Akhurst 1974; Bedding and Iede 2005). These collections represented various species of *Deladenus*, as well as strains of *D. siricidicola* that produced lower levels of infection, or smaller wasps. After extensive screening of strains in the early 1970s, four strains from Corsica, Thasos, Sopron and New Zealand that gave high levels of infestation were selected for final trials. Of these, strain 198 from Sopron in Hungary was finally selected for wide scale application (Bedding and Iede 2005).

The Sopron strain of *D. siricidicola* has been the predominant strain inoculated throughout Australia since the early 1970s. However, it is expected that other strains were dispersed on a limited scale in early years of development of the biological control program in that country. Bedding (1972) reports that '…maintaining strains from many countries and already hundreds of millions of nematodes have been reared and distributed throughout many of the Sirex infested forests of Australia with encouraging results.' The nematode was not distributed in New Zealand and consequently its populations in that country are expected to still reflect original, accidental introductions of another strain(s). The 'Sopron' strain has subsequently also been used widely for inoculations in South America (Iede et al. 1998; Klasmer et al. 1998; Maderni 1998; Bedding and Iede 2005; Chaps. 15–17).

The Sopron strain of *D. siricidicola* has been reported to have lost its virulence in three cases, once in Australia in the late 1980s (Haugen and Underdown 1993; Bedding and Iede 2005), once in Brazil (Bedding and Iede 2005) and once in Argentina (Eskiviski et al. 2003, 2004). This reported to be due to repeated culturing of the nematode for more than 15 years in the free-living form and consequently, its loss of ability to covert to the infective form, at least in culture (Bedding and Iede 2005).

In Australia, the loss of virulence in the Sopron strain was resolved by re-isolating the nematode from the Kamona forest in Tasmania, where original releases of the Sopron strain had been made years before (Bedding and Iede 2005). The culture was selected from a limited number of individual nematodes from a single tree. Subsequent inoculations resulted in high levels of virulence. This strain has subsequently been used extensively in the biological control program of *S. noctilio* in South Africa. In Brazil and Argentina, nematode strains were also isolated from infested wasps to establish new colonies (Eskiviski et al. 2003, 2004; Chap. 16). A strain resulting from these isolations and known as ‘Encruzilhado do Sol’ (Southern Hybrid), is widely used today in South American countries.

One of the consequences of introductions of *D. siricidicola* into Australia, South America and South Africa is that there is a lack of genetic diversity in populations of the nematode. A recent study has shown that the nematode populations from across this region are homozygous for 17 microsatellite regions and 3,291 bp of sequence data (Mlonyeni et al. 2011). This most likely result from a genetic bottleneck in the nematode population created during every round of sub-culturing. A selection of between 100 to 2,000 nematodes is typically transferred between plates and this process is often repeated numerous times. Furthermore, inbreeding levels would also be expected to be high in this system and this would be expected to rapidly reduce heterozygosity. This lack of diversity can be a problem, because the nematode is used in a variety of environments, and in different populations of the wasp and fungus. Its selection for a specific environment is expected to be high due to the strong human selection during isolation, rearing and inoculation.

9.5 Interaction Specificity – *Amylostereum* and *Sirex*

Specificity of *D. siricidicola* to specific wasp populations, and *vice versa*, has been observed. As discussed above Bedding (1972) and Bedding and Akhurst (1974) noticed that a strain of *D. siricidicola* from Japan never infects the eggs of *S. noctilio* females. Similarly, a strain that did not infect the eggs of females in the New Zealand population of *S. noctilio* was reported by Zondag (1975), alongside strains that could sterilize females in this wasp population. This reflects a process that is influenced by the particular strain of the nematode. In contrast, a strain of the nematode from Belgium sterilizes Australian populations, but not Belgian populations of the wasp. In the latter case, it is the wasp population that clearly influences the effect. The drivers behind these apparently strain specific interactions are not clear. Molecular tools now available should make it possible to better understand the relationships between different wasp and/or nematode populations. Irrespective of the driving forces behind host specificity in *D. siricidicola*, this factor clearly needs to be considered in *S. noctilio* management. This is especially because differences in wasp populations or invasions of new populations of the wasp can have far-reaching consequences on the efficacy of biological control programs. This is especially relevant given the lack of diversity in the nematode populations discussed above.

The fungal strain that has been used to rear *D. siricidicola* in Australia (here referred to as the "nematode strain") is thought to have originated from early collections of wasps in Europe, possibly from *S. juvencus*. This strain has been shown to be distinct from the strain in the field in Australasia, South Africa and South America using VCG and molecular markers (Slippers et al. 2001, 2002; Nielsen et al. 2009). The "nematode strain" of the fungus is easily spread during the inoculation of the nematode, as harvesting of the nematodes from fungal cultures also contains many viable propagules of the fungus. This specific fungal strain is widely used across the Southern Hemisphere for mass rearing the nematode.

The difference between the "nematode strain" of the fungus and the strain of *A. areolatum* found in the field across the Southern Hemisphere is potentially important for biological control programs (Hurley et al. 2007). It has been observed that the nematode feeds and develops better on the "nematode strain" of the fungus than on other strains (Authors observations during mass rearing of the nematode). These preliminary observations could thus far not be quantified in experiments (BP Hurley, unpublished data). Nor could the nematode be selectively bred to reproduce more effectively on the field strains collected from *S. noctilio* in South Africa, despite multiple generations over a 2 year period (BP Hurley, unpublished data). Given the importance of potentially lower fitness of *D. siricidicola* on fungal strains other than the "nematode strain", this question needs to be urgently addressed.

9.6 Variable Environmental Factors

For many years it was assumed that *S. noctilio* populations performed best in Mediterranean, winter rainfall regions similar to its most common distribution in Europe (Kirk 1974; Spradbery and Kirk 1978). During the course of the last two decades the wasp has spread and prospered in winter and summer rainfall areas, in particular in Brazil and South Africa (Iede et al. 1998; Hurley et al. 2007). Projections based on current distribution also show that large parts of North and South America, Africa and Australasia would be suitable to future invasion by the wasp (Carnegie et al. 2006). From the South American and South African experiences in particular, it has become evident that the efficacy of the standard biological control programs developed in winter rainfall regions of Australasia, in particular with the nematode, will not be equally effective in all these regions.

Sirex noctilio populations are known to differ substantially in phenology in different climatic zones, which could affect the interaction with the nematode (Neumann and Minko 1981; Carnegie et al. 2005; Hurley et al. 2007). In New Zealand, the Cape region of South Africa and south-eastern pine-growing regions of Australia (Victoria, New South Wales), *S. noctilio* emergence is from December to April (peaking in February), while it occurs between October to January in the summer rainfall regions of South Africa (peaking in November).

One of the major differences between summer and winter rainfall regions that affect *D. siricidicola* is the rate of change in wood moisture after infestation. In the winter rainfall regions the majority of the time that the nematodes are in the trees is during the wet season, while in the summer rainfall regions the nematodes are in the trees mainly during the dry season. In the summer rainfall area of South Africa, moisture content of the trees, especially in the upper sections of the trunks, often drops below 20% (Hurley et al. 2007, 2008). It is not known what the threshold of moisture content is, below which the nematodes and/or the symbiotic fungus, *A. areolatum*, will survive, but lower moisture content has been linked to lower nematode parasitism (Hurley et al. 2008).

Recent studies have shown that some sap stain fungi compete strongly with *A. areolatum* for resources (authors, unpublished data). In particular, *Diplodia pinea*, a commonly occurring and important latent pathogen of pine in South Africa (Swart and Wingfield 1991), grows faster than *A. areolatum*, especially in environments of lower moisture availability. Although *D. pinea* has not been found to overgrow *A. areolatum*, its faster growth enables it to capture more wood resources and thus to limit the growth of *A. areolatum*. Where the growth of *A. areolatum* is severely limited, this will influence the survival and reproduction of *D. siricidicola*. Thus, the composition of sap stain fungi, and conditions that favour the establishment and growth of these fungi in an area, is likely to influence the successful establishment of *D. siricidicola*, and consequently effective biological control.

Various other factors vary between regions and could possibly influence the establishment of *D. siricidicola*, but their effects have not been studied. For example, many different *Pinus* species are planted in the regions where *D. siricidicola* is applied for biological control. These include *Pinus radiata*, *P. patula*, *P. taeda*, *P. carribea*, *P. pondersae*, *P. elliottii*, *P. contorta* var. *latifolia*, and hybrids of some of these species and some could be more suitable as hosts for *S. noctilio*, *A. areolatum* or *D. siricidicola*. Besides these factors that might influence variation, the effects of interactions between the above mentioned factors are also unknown. It is, for example likely that factors such as moisture content in the trees and the effect of sap-stain fungi would be correlated. Many smaller effects working in an additive fashion, as opposed to a single dominant effect, could also result in major differences in parasitism in control programs in different regions.

9.7 Conclusions

The discovery and description of the *D. siricidicola* in New Zealand and Australia represented a significant and exciting scientific breakthrough. The work that followed in this area during the subsequent decades led to the development of a biological control application that has saved potentially billions of dollars of damage to pine plantation industries across the Southern Hemisphere. Despite this success, there are regions where the nematode has not been as successful or has failed completely. As *S. noctilio* continues to spread throughout previously unaffected areas, understanding the cause of this variation and improving the efficacy of biological control will be important. Information and interpretations provided in this review will hopefully clearly illustrate that much work remains to understand the causes of variation in efficacy of *D. siricidicola* in various parts of the world. Such understanding should make it possible to predict problems that will result in ineffective control and to ultimately also avoid these effects.

This review has treated many factors that could potentially affect the usefulness of *D. siricidicola* as a biological control agent. These include factors such as handling of the nematode, environmental factors, variation in the wasp and nematode populations, competing fungi and others. It seems unlikely from current evidence that a single factor is responsible for the variation in nematode success as a biological control agent. Rather, an interaction between several of these factors, or a number of factors in concert will more likely combine to cause the overall effect of dramatically different parasitism results.

It is unfortunate that variation, both molecular and phenotypic, in the populations of *D. siricidicola* has not been considered previously. The potential variation in factors such as the ability of the nematode to adapt to variable environments and populations of the wasp are all of great relevance to control strategies. Furthermore, the mechanisms driving evolution of resistance and virulence in the *D. siricidicola* and *S. noctilio* populations should be high on the research agenda for the future. Recently developed molecular markers should assist in this process. The potential to use rapidly developing novel approaches to DNA sequencing and thus to be able to analyse genetic and genomic factors underlying aspects such as virulence in *D. siricidicola* is great. Studies arising from the application of these technologies and others that have yet to emerge will surely improve our understanding of the biology of *D. siricidicola* and this will enable improved control.

Perhaps the greatest factor hampering research on *D. siricidicola* currently is a lack of availability of natural variation in available *D. siricidicola* strains. During the height of the collection and research programmes on *D. siricidicola* supported by the Australian and UK governments in the 1970s, numerous strains of this nematode (and its relatives) were collected from across the Northern Hemisphere. Today none of those strains from Europe remain available for study and it is only the Kamona strain, and recently strains collected in Canada, that can be investigated. Without appropriate investment in this aspect of the research on *D. siricidicola*, it should be expected that failures in biological control programmes will occur in the future and especially as a result of resistance emerging in populations of *S. noctilio*.

References

Alizon S, Hurford A, Mideo N, Baalen V (2009) Virulence evolution and the trade-off hypothesis: history, current state of affairs and the future. J Evol Biol 22:245–259

Bedding RA (1967) Parasitic and free-living cycles in entomogenous nematodes of the genus *Deladenus*. Nature 214:174–175

Bedding RA (1968a) *Deladenus wilsoni* n.sp. and *D. siricidicola* n.sp. (Neotylenchidae), entomophagous-mycetophagous nematodes parasitic in siricid woodwasps. Nematologica 14:515–525

Bedding RA (1968b) Nematode parasitism of Siricidae and their hymenopterous parasites by *Deladenus* spp. (Neotylenchidae). In: Proceedings of the XIIIth international congress of entomology, Moscow

Bedding RA (1972) Biology of *Deladenus siricidicola* (Neotylenchidae) an entomophagous-mycetophagous nematode parasitic in siricid woodwasps. Nematologica 18:482–493

Bedding RA (1974) Five new species of *Deladenus* (Neotylenchidae), entomophagous-mycetophagous nematodes parasitic in siricid woodwasps. Nematologica 20:204–225

Bedding RA (1979) Manipulating the entomophagous-mycetophagous nematode, *Deladenus siricidicola*, for biological control of the woodwasp *Sirex noctilio* in Australia. Current topics in forest entomology, USDA-Forest Service General Technical Report, WO-8, Hobart

Bedding RA, Akhurst RJ (1974) Use of the nematode *Deladenus siricidicola* in the biological control of *Sirex noctilio* in Australia. J Aust Entomol Soc 13:129–135

Bedding RA, Akhurst RJ (1978) Geographical distribution and host preferences of *Deladenus* species (Nematoda: Neotylenchidae) parasitic in siricid woodwasps and associated hymenopterous parasitoids. Nematologica 24:286–294

Bedding RA, Iede E (2005) Application of *Beddingia siricidicola* for Sirex woodwasp control. In: Grewal PS, Ehlers R-U, Shapiro-Ilan DI (eds) Nematodes as biocontrol agents. CAB International, London

Carnegie AJ, Eldridge RH, Waterson DG (2005) History and management of Sirex wood wasp in pine plantations in New South Wales, Australia. N Z J For Sci 35:3–24

Carnegie AJ, Matsuki M, Haugen DA, Hurley BP, Ahumada R, Klasmer P, Sun J, Iede ET (2006) Predicting the potential distribution of *Sirex noctilio* (Hymenoptera: Siricidae), a significant exotic pest of *Pinus* plantations. Ann For Sci 63:119–128

Collett NG, Elms S (2009) The control of sirex wood wasp using biological control agents in Victoria, Australia. Agric For Entomol 11:283–294

Corley JC, Villacide J (2005) Una visión dinámica del manejo de la plaga forestal *Sirex noctilio*. idia XXI:132–134

Eskiviski E, de Coll O, Faraldo GS, Bennesch MA (2003) Efficiency of new isolations of *Deladenus siricidicola* B. in the biological control of *Sirex noctilio* F. X Jornadas Tecnicas Forestales y Ambientales-FCF. UNaM – EEA Monetecarlo, INTA

Eskiviski E, Faraldo G, de Coll O (2004) Effects of the nematode *Beddingia* (=*Deladenus*) *siricidicola* parasitism in adult stage of wood wasp *Sirex noctilio*. 11as Jornadas Técnicas Forestales y Ambientales-FCF. UNaM – EEA Monetecarlo, INTA

Hanson HS (1939) Ecological notes on the *Sirex* wood wasp and their parasites. Bull Entomol Res 30:27–65

Haugen DA (1990) Control procedures for *Sirex noctilio* in the green triangle: review from detection to severe outbreak (1977–1987). Aust For 53:24–32

Haugen D, Underdown MG (1990) *Sirex noctilio* control program in response to the 1987 green triangle outbreak. Aust For 53:33–40

Haugen D, Underdown MG (1993) Reduced parasitism of *Sirex noctilio* in radiata pines inoculated with the nematode *Beddingia siricidicola* during 1974–1989. Aust For 56:45–48

Hocking H (1967) A nematode (*Deladenus* sp.: Neotylenchidae) associated with *Rhyssa* spp. (Hymenoptera: Ichneumonidae), parasites of Siricid woodwasps. J Aust Entomol Soc 6:52–56

Hurley BP, Slippers B, Wingfield MJ (2007) A critical review of control of the alien invasive woodwasp, *Sirex noctilio*, in the southern hemisphere, with a special emphasis on biological control. Agric For Entomol 9:159–171

Hurley B, Slippers B, Croft PK, Hatting HJ, van der Linde M, Morris AR, Dyer C, Wingfield MJ (2008) Factors influencing parasitism of *Sirex noctilio* (Hymenoptera: Siricidae) by the nematode *Deladenus siricidicola* (Nematoda: Neotylenchidae) in summer rainfall areas of South Africa. Biol Control 45:450–459

Iede ET, Penteado S, Schaitza EG (1998) *Sirex noctilio* problem in Brazil – detection, evaluation and control. In: Iede E, Schaitza E, Penteado S, Reardon R, Murphy T (eds) Proceedings of a conference: training in the control of *Sirex noctilio* by use of natural enemies, USDA Forest Service (FHTET 98–13), Morgantown, MV

Kirk AA (1974) Bioclimates of Australian *Pinus radiata* areas and *Sirex noctilio* localities in the northern hemisphere. Aust For 37:126–131

Klasmer P, Fritz G, Corley J, Botto E (1998) Current status of research on *Sirex noctilio* F. in the Andean – Patagonian region in Argentina. In: Iede E, Shaitza E, Penteado S, Reardon R, Murphy T (eds) Proceedings of a conference: training in the control of *Sirex noctilio* by use of natural enemies, USDA Forest Services (FHTET 98–13), Morgantown, MV

Maderni JFP (1998) *Sirex noctilio* present status in Uruguay. In: Iede E, Shaitza E, Penteado S, Reardon R, Murphy T (eds) Proceedings of a conference: training in the control of *Sirex noctilio* by use of natural enemies, USDA Forest Services (FHTET 98–13), Morgantown, MV

Miller D, Clark AF (1935) *Sirex noctilio* (Hym.) and its parasites in New Zealand. Bull Entomol Res 26:149–154

Mlonyeni XO, Wingfield BD, Wingfield MJ, Ahumada R, Klasmer P, Leal I, de Groot P, Slippers B (2011) Extreme homozygosity in Southern Hemisphere populations of *Deladenus siricidicola*, a biological control agent of *Sirex noctilio*. Biol Control (in press) doi:10.1016/j.biocontrol.2011.09.009

Neumann FG, Minko G (1981) The sirex wood wasp in Australia pine plantations. Aust For 44:46–63

Neumann FG, Morey JL, McKimm RJ (1987) The Sirex wasp in Victoria. Bulletin no 29. Victoria. Department of Conservation, Forests and Lands, Melbourne

Nielsen C, Williams DW, Hajek AE (2009) Putative source of the invasive *Sirex noctilio* fungal symbiont, *Amylostereum areolatum*, in the eastern United States and its association with native siricid woodwasps. Mycol Res 113:1242–1253

Nuttall MJ (1989) *Sirex noctilio* F., sirex wood wasp (Hymenoptera: Siricidae). In: Cameron PJ, Hill RL, Bain J, Thomas WP (eds) A review of biological control of pests and weeds in New Zealand 1974–1987, Technical communication no 10. C.A.B. International & DSIR, Wallingford

Rawling GB (1955) Epidemics in *Pinus radiata* forests in New Zealand. J For 7:53–55

Slippers B, Wingfield MJ, Coutinho T (2001) Population structure and possible origin of *Amylostereum areolatum* in South Africa. Plant Pathol 50:206–210

Slippers B, Wingfield MJ, Coutinho T (2002) DNA sequence and RFLP data reflect geographical spread and relationships of *Amylostereum areolatum* and its insect vectors. Mol Ecol 11:1845–1854

Spradbery JP, Kirk AA (1978) Aspects of the ecology of siricid woodwasps (Hymenoptera: Siricidae) in Europe, North Africa and Turkey with special reference to the biological control of Sirex noctilio F. in Australia. Bull Entomol Res 68:341–359

Swart WJ, Wingfield MJ (1991) Biology and control of *Sphaeropsis sapinea* on *Pinus* species in South Africa. Plant Dis 75:761–766

van Valen L (1973) A new evolutionary law. Evol Theory 1:1–30

Yu Q, de Groot P, Leal I, Davis C, Ye W, Foord B (2009) Characterization of *Deladenus siricidicola* (Tylenchida: Neotylenchidae) associated with *Sirex noctilio* (Hymenoptera: Siricidae) in Canada. Int J Nematol 19:23–32

Zondag R (1962) A nematode disease of *Sirex noctilio*. Interim Research Release, New Zealand Forest Service, Rotorua

Zondag R (1965) Biological control of *Sirex noctilio*. Forestry Research Institute Report 1964, Rotorua

Zondag R (1967) Biological control of *Sirex noctilio* F. Forestry Research Institute Report 1966, Rotorua

Zondag R (1969) A nematode infection of *Sirex noctilio* (F.) in New Zealand. N Z J Sci 12:732–747

Zondag R (1971) Control of *Sirex noctilio* (F.) with *Deladenus siricidicola* Bedding, part I. 1967 field trial. N Z J Sci 1:5–14

Zondag R (1975) A non-sterilising strain of *Deladenus siricidicola*. Forestry Research Institute Report 1974, Rotorua

Zondag R (1979) Control of *Sirex noctilio* (F.) with *Deladenus siricidicola* Bedding, part II. Introduction and establishments in the South Island 1968–1975. N Z J For Sci 9:68–76

Chapter 10
Ecological Considerations in Using *Deladenus (=Beddingia) siricidicola* for the Biological Control of *Sirex noctilio* in North America

David W. Williams, Kelley E. Zylstra, and Victor C. Mastro

Abstract *Sirex noctilio* was discovered in North America in New York, USA in 2004. The woodwasp has the potential to damage pine forest resources if it is not controlled. *Sirex* has been managed with varying degrees of success in pine plantations throughout the Southern Hemisphere using biological control. The most successful control agent is the "Kamona strain" of the nematode, *Deladenus (=Beddingia) siricidicola*. The technology for mass rearing and applying the nematode has been developed through many years of research in Australia. Thus, implementing a biological control program in North America seemed at first to be a simple matter of technology transfer. This has proved not to be the case because of environmental factors that both affect and are affected by use of the nematode. The first set of factors includes climate and the discovery of an indigenous strain of *D. siricidicola*. Climate may affect the overwinter survival and population increase of the nematode, as well as the timing of program operations, whereas the presence of another strain complicates nematode identification and, hence, the evaluation of effectiveness of the Kamona strain. In addition, strain hybridization may occur, potentially decreasing the efficacy of the Kamona strain. The second set of factors includes possible effects of the nematode on non-target North American siricids. As many as five species of native pine-feeding Siricinae potentially may be exposed to infection by the nematode when it is released.

D.W. Williams (✉) • K.E. Zylstra • V. C. Mastro
Otis Laboratory, United States Department of Agriculture, Animal and Plant Health Inspection Service, 1398 West Truck Road, Buzzards Bay, MA 02542, USA
e-mail: david.w.williams@aphis.usda.gov

B. Slippers et al. (eds.), *The Sirex Woodwasp and its Fungal Symbiont: Research and Management of a Worldwide Invasive Pest*, DOI 10.1007/978-94-007-1960-6_10,
© US Government 2012

10.1 Introduction

Sirex noctilio F. was identified in bark beetle trap collections in Oswego County, New York, U.S.A., in the autumn of 2004 (Hoebeke et al. 2005). The arrival time of *S. noctilio* in North America is unknown. It was apparently introduced into New York and Ontario, Canada in recent years but it may have entered North America on multiple occasions starting as early as the nineteenth century (Benson 1943). It seems to have been accompanied in its invasion by both its symbiotic fungus, *Amylostereum areolatum* (Fries) Boidin, and its parasitic nematode, *Deladenus (=Beddingia) siricidicola* Bedding (Yu et al. 2009). The exact origins of the invading woodwasp, fungus, and nematode are uncertain, but all are likely from Europe (Bedding and Akhurst 1978; Ciesla 2003; Nielsen et al. 2009).

Since its discovery in New York, *S. noctilio* seemed an ideal candidate for a classical biological control program because it has been the subject of 30 years of research and development, primarily in Australia. The entomopathogenic nematode, *D. siricidicola*, has been used as a biological control agent in management programs with varying degrees of success in Australia, New Zealand, South America, and South Africa (Bedding and Akhurst 1974; Bedding 1993; Bedding and Iede 2005; Hurley et al. 2007, 2008; Collett and Elms 2009). The current biological control program in Australia uses the Kamona strain of *D. siricidicola*, which was chosen originally from a natural population in Hungary for its high infection rate and minimal effect on host size (Bedding and Iede 2005). Transfer of this technology to North America seemed a simple matter; cultures of the Kamona strain were readily available from Ecogrow Ltd, the licensed vendor in Australia, and the procedures for culture and mass rearing were well documented (Calder and Bedding 2002). Moreover, the delivery system for inoculating trees, including a special punch hammer and calibration of the dosage of nematodes in a polyacrylamide gel carrier, was well developed and tested (Bedding and Iede 2005).

At the outset in 2006, it seemed that developing a biological control program in the United States Department of Agriculture (USDA) using *D. siricidicola* would involve simply fine-tuning operational activities and events, especially the timing of releases, to compensate for climatic differences between North America and Australia. As it turned out, this view was naïve on a number of counts. For example, pine management in Australia is relatively simple. *Pinus* species are not native, and they are grown in large monocultures of a few species imported from the Northern Hemisphere. Through strict quarantines, the Australians have managed to exclude most major pests of pine, with the notable exception of *S. noctilio*, and they have no native borer species that may be harmed by the biological control program or compete for resources with *S. noctilio*. The Kamona strain was chosen for its high infectivity in the Sirex strain that invaded Australia.

By contrast with Australia, North America has many native pine species growing in both natural forests and plantations, and those species host a well developed mycoflora and fauna that may compete with *S. noctilio* and its associated fungus and nematode species. For example, cerambycid larvae, most notably *Monochamus* and *Tetropium* spp. (Williams, unpublished data), may compete directly with Sirex larvae inside wood, and bark beetles may colonize host trees early in the season rendering them unavailable for use by *S. noctilio*. Large areas of the heartwood may be colonized by blue stain fungi mainly in the genera *Ophiostoma* and *Ceratocystis*, and a native white rot fungus, *Amylostereum chailletii* (Pers. ex Fries) Boidin, reducing the wood volume available for the growth of *A. areolatum.* This could consequently reduce the food supply for *S. noctilio* as well as the nematode, *D. siricidicola*. The potential for competition of *A. areolatum* with fungal associates of southern pine beetle has been investigated in the southeastern United States where the woodwasp may invade in time (Klepzig et al. 2009). In addition, the nematode may have a direct competitor in *Deladenus wilsoni* Bedding, a North American species that feeds on both *Amylostereum* species and infects ichneumonid species that parasitize siricids (Bedding 1968).

Discovery of the invasive species associated with Sirex woodwasp, *A. areolatum* and *D. siricidicola*, as well as increased knowledge of the ecology of native North American siricids pose challenges for the development of a classical biological control program that were not anticipated 5 years ago. Specifically, the presence of a "native strain" of *D. siricidicola* may jeopardize the effectiveness of biological control releases against *S. noctilio*. Moreover, possible impacts of the Kamona strain on non-target native siricid species are a real concern. In the following section, we discuss recent findings on the Sirex system in North America and their possible ramifications for the developing biological control program, including factors that may affect and be affected by its implementation.

10.2 Ecological Factors in North America Affecting the Biological Control Program

10.2.1 Climate

Climatic patterns have long been recognized to affect the distribution and abundance of animal populations (Andrewartha and Birch 1954). In particular, temperature determines the rates of survival, growth, and reproduction of poikilothermic animals (Uvarov 1931). Survival and development of *D. siricidicola* follows patterns typical of other animals, with growth rate decreasing at lower temperatures and an upper lethal temperature threshold. Akhurst (1975) reported that egg hatch in *D. siricidicola* occurred after 3 days at 25°C and 13 days at 10°C, whereas hatch was minimal at 5°C. Mortality of all nematode life stages transferred from 24°C to 30°C was 100%.

Regional climatic regimes undoubtedly affect the population dynamics of *D. siricidicola.* Figure 10.1 shows the long term average seasonal temperature

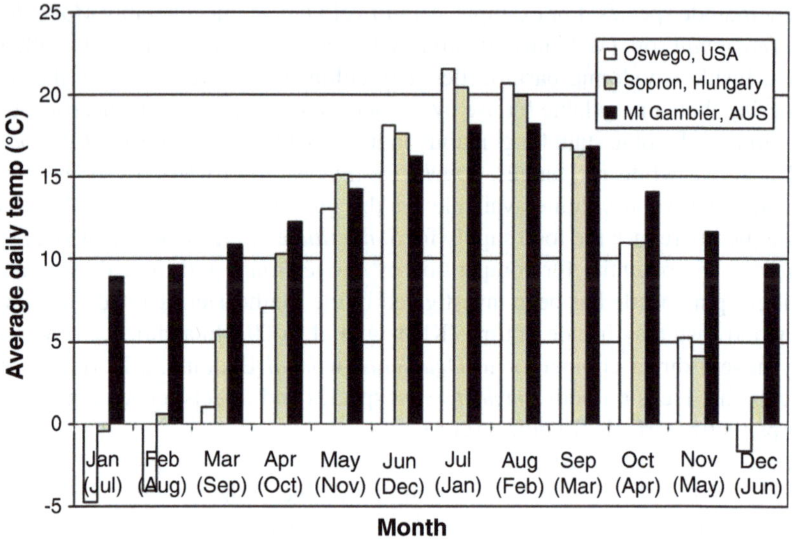

Fig. 10.1 Monthly averages of mean daily temperatures from historical weather data collected in Oswego, New York, USA; Sopron, Hungary; and Mt. Gambier, South Australia, Australia. Months in parenthesis are seasonal equivalents in the Southern Hemisphere

patterns from three locations relevant to nematode biological control in North America: Mt. Gambier, South Australia, Australia, where *S. noctilio* has been controlled successfully with the Kamona strain; Sopron, Hungary, where the progenitors of the Kamona strain were collected; and Oswego, New York, U.S.A., an area where the Kamona strain may be applied for biological control. The temperatures in Mt. Gambier are relatively mild and without extremes, dipping slightly below 10°C only in mid-winter. By contrast, the temperatures in Oswego exhibit extremes in both summer and winter, with winter temperatures near or below 5°C for 5 months of the fall and winter. It is reassuring that Sopron, where the Kamona strain originated, has temperatures that are slightly less extreme but similar overall to those in Oswego. Given the amount of time with temperatures below 5°C, including 3 months below 0°C, the nematode must not produce many generations during winter in Oswego. Assuming that it survives the winter, the major opportunity for population growth occurs in the spring and early summer.

The timing of release in the northeastern United States is critical: nematodes must be applied late enough that they can find mature Sirex larvae to infect but early enough that they can establish and reproduce before the onset of temperatures below their thermal threshold. Guidelines established by the Australian Sirex National Coordination Committee (Haugen et al. 1990) suggest that nematodes be applied in May through July, which translates to November through January in the Northern Hemisphere. This range obviously was fully encompassed by winter in New York, so we have shifted our controlled releases (see below) to late September. Even at this earlier time, temperatures average over 10°C only during October, so we may need to consider releases as early as late August. Release times will also need to be adjusted as *S. noctilio* inevitably invades the southern and western regions of the United States. In addition, research studies to date have used naturally struck trees, but as the biological control program moves into an operational mode, trap trees will be used. Their creation will need to be carefully timed and calibrated to optimize the nematode infection rate (Zylstra et al. 2010).

10.2.2 The North American Nematode

A nematode was discovered infecting *S. noctilio* in 2006 in New York and Ontario (Shields 2009; Williams et al. 2009; Yu et al. 2009). Specimens were examined morphologically and identified as a strain of *D. siricidicola* (Bedding 2009, personal communication; Yu et al. 2009). Molecular analysis of the Ontario strain has also confirmed that it is *D. siricidicola* (Yu et al. 2009). The nematode is presumed to have entered North America with an invasion of *S. noctilio*. It is well established in central New York State with variable populations across more than 100 study sites within 80 km of Syracuse (Table 10.1). The sites were even-aged, about 30–40 years old, unmanaged, and contained Scots, red, or white pines. The adult trapping mainly used Lindgren funnel traps that were usually placed on girdled pines or, less frequently, baited with experimental lures. The traps were deployed from mid-June through the end of September and checked weekly. With over 5,100 *S. noctilio* dissected, percentage infection by nematodes was variable among studies, but generally low, ranging from about 1% to 19%. However, note that the infection rates in the trapping studies increased from 2007 through 2009. By contrast, Canadian researchers working at six sites in Ontario reported average rates of nematode infection around 50% and a rate of 95% at one location (Yu et al. 2009).

The presence of a preexisting strain of *D. siricidicola*, hereafter referred to as the "North American strain", complicates the development of a nematode biological control program. A primary problem is that the North American and Kamona strains are distinguishable only by molecular techniques. Thus, expensive

Table 10.1 Infection of *Sirex noctilio* by *Deladenus siricidicola* at various sites in central New York, USA, 2006–2009 (K. Zylstra, unpublished data)

Studies	Number of sites	Total *S. noctilio* dissected	% infected by nematodes
2006 larval dissections	13	120	2.5
2007 larval dissections	3	1,293	0.7
2007 lab reared adults	17	2,152	4.6
2007 adult trapping	47	541	8.7
2008 adult trapping	39	427	14.1
2009 adult trapping	26	645	18.9

and time-consuming analysis will be needed to discriminate the two strains and evaluate the effectiveness of the Kamona strain in field studies. Fortunately, the strains can be distinguished by DNA sequences and microsatellite markers (Yu et al. 2009, Chap. 9; E. Braswell 2010, personal communication), and efforts are currently underway to develop operational tests for samples collected in our field experiments.

Another potential problem is that the two strains may hybridize in the field. Hybrids will not be recognizable by the molecular techniques used to discriminate pure strains. In addition, hybridization may dilute the high infectivity rate of the Kamona strain, rendering it less useful for Sirex control. Tests are currently underway at the Otis Laboratory to determine whether or not hybridization occurs. Strain hybridization seems very likely given the numerous compatible conspecific crossbreedings reported by Akhurst (1975).

In addition to *D. siricidicola*, four *Deladenus* species are indigenous to North America: *D. wilsoni* Bedding, *D. canii* Bedding, *D. proximus* Bedding, and *D. nevexii* Bedding (Bedding and Akhurst 1978). The possibility of hybridization between *D. siricidicola* and the other species is very unlikely. In an extensive study, Akhurst (1975) reported that crosses between different *Deladenus* species very rarely produced more than a few F_2 larvae. The one exception did not involve North American species.

A very useful characteristic of the Kamona strain in Australia is that juvenile nematodes enter the eggs of the Sirex female shortly after pupation (Bedding 1972). Doing so has two advantages from the standpoint of biological control: it sterilizes the eggs, controlling Sirex reproduction, and it provides an efficient mechanism for dispersing nematodes to other trees so as to spread the infection (Bedding and Iede 2005). However, some nematode strains do not sterilize the host female (Bedding 1972; Nuttall 1980). Juveniles remain in the body cavity or along the egg sheaths but do not enter the host egg. This is thought to result from an inherent asynchrony between the timing of release of juveniles by the mother nematode and the timing of egg shell hardening. That is, the juveniles arrive too late to enter an egg (R. Bedding 2008, personal communication). This phenomenon of "non-sterilization" has been observed in the North American strain in both the United States and Canada (Yu et al. 2009; Williams, unpublished data). Obviously, it is not a desirable trait from the standpoint of developing a biological control program because the Sirex egg is not killed and the transmission of a few juveniles stuck to the outer surface of an egg does not provide an efficient method of dispersal to new trees. However from an evolutionary standpoint, this trait may be advantageous to the long term survival of nematode populations in nature when Sirex population density is low (Nuttall 1980, Chap. 9). It is not known whether the Kamona strain will also be non-sterilizing with the North American Sirex strain, but it is a possibility. Bedding (1972) reported that a strain of *D. siricidicola* isolated from *S. noctilio* in Belgium was highly infective in Australian Sirex but did not enter the eggs of Belgian Sirex. He suggested that this reflected a strain difference in *S. noctilio* populations. DNA analysis of samples from our controlled release studies, when it is available, will reveal whether or not the Kamona strain sterilizes the strain of *S. noctilio* present in North America.

10.3 Possible Effects of the Biological Control Program on Non-target Pine Insects in North America

Since the late 1980s, ecologists in the United States have become concerned over unintended consequences of classical biological control introductions (Carruthers and Onsager 1993; Lockwood 1993). At the center of the controversy are effects on non-target indigenous organisms, with population extinction in rare and endangered species as the most dire scenario (Howarth 1991; Follett and Duan 1999). Changes in populations of non-target organisms as an unintended consequence of biological control clearly may alter the biodiversity and trophic interactions of the native communities of which they are a part (Lockwood 1993).

As a result of this concern, applications to the USDA Animal and Plant Health Inspection Service (APHIS) for importation, movement, and release permits of biological control organisms have come under closer scrutiny in recent years. Permits for environmental release currently require a review by the North American Plant Protection Organization and the development of detailed environmental and biological assessments that evaluate all foreseeable impacts of releasing an exotic species. Because of concerns about possible negative effects of the Kamona strain on non-target siricid species as well as speculation that *S. noctilio* might not prove to be a primary pest in North American forests, the Kamona strain has not been approved for environmental release. Instead, nematode releases between 2006 and 2009 have been "controlled". That is, infested pine trees are inoculated with nematodes in the fall, sample billets are removed for rearing in late winter, and what remains of the trees is chipped before insect emergence in the spring.

Besides *S. noctilio* and another siricid species, *Xeris spectrum* (L.), only one other insect species has been cited in the literature as attacked by *D. siricidicola* in North America. Bedding and Akhurst (1978) reported that "both *D. wilsoni* and *D. siricidicola* have been found producing juveniles in *Serropalpus barbatus* (Schaller), a beetle commonly associated with siricids." This melandryid beetle is distributed widely across Europe, Siberia, and northeastern North America (Baker 1972; Kolk and Starzyk 1996). Its hosts are common fir and Norway spruce, and although common, *S. barbatus* is not considered a serious timber pest. Bedding (1972) and Bedding and Akhurst (1978) reported finding juveniles of *D. siricidicola* in the hemocoel of *S. barbatus*, but not in the reproductive organs, so this occasional host cannot disperse the nematode and is unlikely to be affected much by it.

10.3.1 North American Siricids as Potential Hosts

A biological control program against *Sirex noctilio* in North America could have an impact on the twenty native species and subspecies of Siricinae, which comprise three genera: *Sirex*, *Urocerus*, and *Xeris* (Table 10.2). All species attack conifers, and all *Sirex* species, as well as four *Urocerus* species, attack *Pinus* species. Each species

Table 10.2 Species of Siricidae, Subfamily Siricinae, reported from North America with their host groups and fungus associations

Species	Region of N. America	Attacks *Pinus*?	Fungal symbiont	Reference
Sirex areolatus Cresson	East, West	Yes	*A. chailletii*	(4)
Sirex behrensii Cresson	West	Yes	*A. chailletii*	(8)
Sirex californicus Ashmead	West	Yes	*A. chailletii*	(4)
Sirex cyaneus F.	East, West	Yes	*A. chailletii*	(5)
Sirex edwardsii Brulle	East	Yes	*A. areolatum*	(7)
Sirex longicauda Middlekauff	West	Yes	*A. chailletii*	(5)
Sirex nigricornis F.	East	Yes	*A. chailletii*	(7)
Sirex nitidus Harris	East, West	Yes	*A. areolatum*	(7)
Sirex noctilio F.	East	Yes	*A. areolatum*	(3)
Sirex varipes Walker	West	Yes	Unknown	–
Urocerus albicornis F.	East, West	Yes	*A. chailletii*	(1)
Urocerus californicus Norton	West	Yes	*A. chailletii*	(5)
Urocerus cressoni Norton	East	Yes	*A. chailletii*	(8)
Urocerus gigas flavicornis F.	East, West	Yes	*A. chailletii*	(1)
Urocerus taxodii Ashmead	East	No	*A. chailletii*	(6)
Xeris morrisoni indecisus MacGillivray	West	Yes	None	(2)
Xeris morrisoni morrisoni Cresson	West	No	None	(2)
Xeris spectrum spectrum L.	East, West	Yes	None	(1)
Xeris spectrum townesi Maa	West	Yes	None	(2)
Xeris tarsalis Cresson	West	No	None	(2)

All species have coniferous hosts. *Sirex* species names follow the recent revision by Henri Goulet (Chap. 1) with one exception. Whereas Goulet has sunk *S. edwardsii* as a color morph of *S. nigricornis*, we have kept the two as separate species

References: (1) Stillwell (1966), (2) Morgan (1968), (3) Gaut (1969), (4) Gaut (1970), (5) Smith and Schiff (2002), (6) Wilson and Schiff (2007), (7) Nielsen et al. (2009), (8) Williams unpublished data

has several recorded hosts, and no species is especially rare. Unlike *S. noctilio*, which can attack living trees (Spradbery and Kirk 1978), the North American siricids use trees that are dead or dying from natural or anthropogenic causes. They apparently are not considered serious forest pests (Furniss and Carolin 1977; Madden 1988) and can be considered beneficial insects in the natural forest environment to the extent that they and their fungal symbionts hasten wood decomposition. Evidence of the relative prevalence of native siricids and natural infection by nematodes is presented in Table 10.3. Five indigenous siricid species were trapped commonly during 2007–2009 in study sites within 80 km of Syracuse, New York. Adult trapping sites and methods of capture were the same as those reported in Table 10.1 above. Note that trap catches of the four North American species were much lower than those for *S. noctilio*. Neither of the native *Sirex* species was infected with nematodes, whereas the *Urocerus* species had infection rates of 15.4% and 7.7% in 2008 and 2009, respectively. One specimen of *Tremex columba* (L.), which feeds in hardwood, was infected. None of the nematodes were identified.

Table 10.3 Numbers of North American siricid females caught in trapping studies within 80 km of Syracuse, New York: 37 sites in 2007, 39 sites in 2008, and 26 sites in 2009

Siricid species	2007 totals	2007 % infected	2008 totals	2008 % infected	2009 totals	2009 % infected
Sirex nigricornis	16	0.0	4	0.0	25	0.0
Sirex edwardsii	4	0.0	0	–	4	0.0
Urocerus spp.	26	0.0	13	15.4	52	7.7
Tremex columba	28	0.0	39	0.0	16	6.3
Sirex noctilio	541	8.7	427	14.1	645	18.9

Nematode infection of *S. noctilio* was likely by the North American strain of *D. siricidicola*. Nematodes associated with other species were not identified. *Urocerus* spp. included *U. cressoni* and *U. albicornis* (Zylstra et al. 2010)

The primary factor affecting exposure of a siricid species to *D. siricidicola* is its fungal symbiont. The two most common siricid symbionts worldwide are *A. areolatum*, which apparently originated in Eurasia, and *A. chailletii*, which apparently originated in North America (Bedding and Akhurst 1978; Nielsen et al. 2009). *Deladenus siricidicola* feeds and reproduces primarily on *A. areolatum* (Bedding and Akhurst 1978) and does not feed or reproduce to any appreciable extent on *A. chailletii* (Wilson et al. 2007; Williams, unpublished data). Thus, siricid species that feed on *A. chailletii*, including most of those in North America (Table 10.2), effectively have a refuge from *D. siricidicola*.

Although the fungal associate of a siricid species appears to be the key to its susceptibility to *D. siricidicola*, the faithfulness of the association between a woodwasp and a fungus is still a matter of debate. Francke-Grosmann (1939, quoted in Talbot 1977) stated that woodwasp species are not constantly associated with the same fungus species. Alternatively, Talbot (1977) stated that the relationship between individual woodwasp species and fungus species is fixed without regard to their geographical distributions. A recent study by Nielsen et al. (2009) provides some additional information. Contrary to expectations, they found that the North American species, *Sirex edwardsii* (a synonym of *S. nigricornis* according to H. Goulet; Chap. 1), emerged from the same tree as *S. noctilio*, carrying the same strain of *A. areolatum* as that species, although it should be cautioned that this result was based on just two specimens of *S. edwardsii*. Moreover, they reported that another North American species, *Sirex nigricornis*, emerged from the same part of the same tree but carrying *A. chailletii*. Thus it appears that different *Amylostereum* and *Sirex* species can coexist in the same tree. Nielsen et al. (2009) also reported that the native species, *Sirex nitidus* (also known in the literature as *S. juvencus juvencus*), emerged from a spruce tree (*Picea* species) in Maine carrying *A. areolatum* and noted that this occurrence was outside the currently known range of *S. noctilio* and had a unique gene sequence. They speculated that this isolate of *A. areolatum* may be native to North America. Nielsen et al. (2009) concluded that the associations of North American siricids and their fungal symbionts are less clear cut than they appeared previously.

Only a few other North American siricid species potentially may feed on wood colonized by *A. areolatum*. In particular, *Xeris* species are not symbiotic with a fungus species, but instead, oviposit in trees already attacked by another siricid species and inoculated with its fungus (Stillwell 1966; Fukuda and Hijii 1997). In North America, they can feed on either *A. areolatum* or *A. chailletii*. The subspecies, *X. morrisoni indecisus*, *X. spectrum spectrum*, which is also distributed widely in Eurasia, and *X. spectrum townesi*, will all be exposed to *D. siricidicola* when it is released and *A. areolatum* disperses across North America with *S. noctilio*. However, the exposure will occur only if and when those woodwasps infest trees already infested with *S. noctilio*. Alternatively, the North American *Xeris* species did not evolve with *A. areolatum* as did *X. spectrum* populations in Eurasia. Thus, it is by no means certain that any of the American *Xeris* populations will be able to use *A. areolatum* or will be attracted to it in the way that they are to *A. chailletii*.

In summary, according the best available information, three *Xeris* subspecies and two native *Sirex* species that are associated with *A. areolatum* may potentially be exposed to *D. siricidicola* if or when it is released as a biological control agent.

10.3.2 Questions on Non-target Effects on North American Siricids

There are two main questions related to non-target effects of a biological control program using *D. siricidicola*. First, what siricid species will be exposed to attack by the nematode? And second, what are the likely physiological impacts on North American woodwasp populations that are capable of being infected?

The first question has been addressed above. Where they occur in the same area as *D. siricidicola*, all *Xeris* species potentially will be exposed to infection if they develop in pine trees containing *A. areolatum*. In addition, *S. edwardsii* and *S. nitidus* may be susceptible to the nematode if they have a fixed association with *A. areolatum*.

In general, it is important to remember that populations of the North American *Sirex*, *Urocerus*, and *Xeris* species feeding on pine currently develop in trees invaded by *A. chailletii*. That fungus represents a refuge from the nematode; siricid larvae feeding on it are not exposed to nematode infection. It is still unknown if North American siricid species are capable of shifting fungal symbionts. However, selective pressures imposed by the nematode on any woodwasp populations shifting to *A. areolatum* should serve to maintain a strong association of North American pine-feeding siricids with *A. chailletii*.

The answer to the second question is less certain. From a physiological standpoint, it is known that the effects of nematode infection vary widely with host species and strain. Bedding (1972) reported that infected *S. noctilio* from Australia exhibited suppressed ovarian development, with fewer and smaller eggs produced, and those eggs were generally packed with juvenile nematodes. In addition, nematode infection has been reported to reduce the size of *S. noctilio* females and lower their dispersal rate as a result (Bedding 2008; Villacide and Corley 2008). European

Sirex juvencus females exhibited normal ovarian development when infected, and their eggs were also packed with nematodes (Bedding 1972). In *S. nitobei* from Japan and *S. noctilio* from Belgium, *D. siricidicola* juveniles did not enter the eggs, but instead remained in the ovaries and oviducts and were transmitted outside the egg at oviposition; that is, they were non-sterilizing. By contrast with the *Sirex* species, infected *X. spectrum* had normal ovarian development, and only 10–30% of their eggs contained nematodes (Bedding 1972). Thus, physiological impacts of nematode infection are difficult to predict among woodwasp species and are not necessarily deleterious.

From a population standpoint, it is difficult to predict the impact of infection on the North American *Sirex* and *Xeris* species except that those populations developing in trees containing *A. chailletii* exclusively will not be affected. The cosmopolitan subspecies, *X. spectrum spectrum*, is infected by *D. siricidicola* in Eurasia, yet it apparently survives there. However, the natural control situation in Eurasia and a classical biological control program in North America may not be entirely comparable. The Kamona strain was chosen for the Australian biological control program for its high infectivity in *S. noctilio.* Its impact on non-target siricid species may be greater than that of naturally occurring *D. siricidicola* strains in the Palaearctic. Based on the results of Bedding (1972), it seems unlikely that impacts on *Sirex* and *Xeris* populations will be great. Ultimately, any possible negative impacts of the biological control program on North American siricids must be weighed against the costs to forest resources if *S. noctilio* is not managed.

10.4 Conclusions and Future Directions

Clearly, the implementation of a biological control program in North America using the *Sirex* nematode has turned out to be more challenging than it seemed at the outset. Further work is needed to find the optimum timing for releases in North America to assure that nematodes have a sufficient period of warm temperatures for establishment and growth in the trees before the onset of winter. The success of the program hinges on our ability to identify strains of *D. siricidicola,* so top priority is being given to the development of DNA probes. Possible hybridization of the native and Kamona strains must be investigated to anticipate any negative effects from interbreeding. After the Kamona strain can be identified by molecular techniques, we must determine whether or not it is non-sterilizing. If the Kamona strain is non-sterilizing, it must not be released in the field, and foreign exploration should be undertaken, initially in New Zealand, for sterilizing strains. The fungal associations of all the North American species of Siricinae should be determined, and any propensity for siricid species' shifting symbionts must be explored. Continued controlled releases will answer questions critical to the program until DNA tools are available and the status of the Kamona strain is known. After these knowledge gaps are filled, full environmental releases can be undertaken if the Kamona strain is demonstrated to be more effective in controlling *S. noctilio* than the North American strain. Although we hope that the ecology of native Siricinae will be better known by the time of full release, the decision to proceed may ultimately be based on a cost-benefit analysis of possible impacts to the woodwasp fauna versus the health of North American pine forests.

Acknowledgements We thank Robin Bedding, Peter de Groot, Ann Hajek, Andrew Liebhold, Nathan Schiff, Bernard Slippers, and Michael Stefan for their helpful comments on the manuscript.

References

Akhurst RJ (1975) Cross-breeding to facilitate the identification of *Deladenus* spp., nematode parasites of woodwasps. Nematologica 21:267–272

Andrewartha HG, Birch LC (1954) The distribution and abundance of animals. University of Chicago Press, Chicago

Baker WL (1972) Eastern forest insects. USDA Forest Service Misc Publ No 1175, Washington, DC

Bedding RA (1968) *Deladenus wilsoni* n. sp. and *D. siricidicola* n. sp. entomophagous-mycetophagous nematodes parasitic in siricid woodwasps. Nematologica 14:515–525

Bedding RA (1972) Biology of *Deladenus siricidicola* an entomophagous-mycetophagous nematode parasitic in siricid woodwasps. Nematologica 18:482–493

Bedding RA (1993) Biological control of *Sirex noctilio* using the nematode *Deladenus siricidicola*. In: Bedding RA, Akhurst RJ, Kaya HK (eds) Nematodes and the biological control of insect pests. CSIRO Publications, Melbourne

Bedding RA (2008) Controlling the pine-killing woodwasp, *Sirex noctilio*, with nematodes. In: Hajek AE et al (eds) Use of microbes for control and eradication of invasive arthropods. Springer, Dordrecht/London

Bedding RA, Akhurst RJ (1974) Use of the nematode *Deladenus siricidicola* in the biological control of *Sirex noctilio* in Australia. J Austr Entomol Soc 13:129–135

Bedding RA, Akhurst RJ (1978) Geographical distribution and host preferences of *Deladenus* species parasitic in siricid woodwasps and associated hymenopterous parasitoids. Nematologica 24:286–294

Bedding RA, Iede ET (2005) Application of *Beddingia siricidicola* for *Sirex* woodwasp control. In: Grewal PS, Ehlers R-U, Shapiro-Ilan DI (eds) Nematodes as biocontrol agents. CABI International, London

Benson RB (1943) Studies in Siricidae, especially of Europe and Southern Asia. Bull Entomol Res 34:27–51

Calder J, Bedding RA (2002) Standard operating procedures (SOPs) for production, experimentation and quality assurance testing of the *Sirex* nematode, *Beddingia siricidicola*. CSIRO, Clayton

Carruthers RI, Onsager JA (1993) Perspective on the use of exotic natural enemies for biological control of pest grasshoppers. Environ Entomol 22:885–903

Ciesla WM (2003) European woodwasp. A potential threat to North America's conifer forests. J For 101:18–23

Collett NG, Elms S (2009) The control of sirex wood wasp using biological control agents in Victoria, Australia. Agric For Entomol 11:283–294

Follett PA, Duan JJ (1999) Nontarget effects of biological control. Kluwer Academic Publishers, Norwell

Francke-Grosmann H (1939) Über das zusammenleben von holzwespen (Siricinae) mit pilzen. Z Angew Entomol 25:647–680

Fukuda H, Hijii N (1997) Reproductive strategy of a woodwasp with no fungal symbionts, *Xeris spectrum*. Oecologia 112:551–556

Furniss RL, Carolin VM (1977) Western forest insects. USDA Forest Service Misc Publ No 1339, Washington, DC

Gaut IPC (1969) Identity of the fungal symbiont of *Sirex noctilio*. Aust J Biol Sci 22:905–914

Gaut IPC (1970) Studies of siricids and their fungal symbionts. PhD thesis, University of Adelaide, Adelaide

Haugen DA, Bedding RA, Underdown MG, Neumann FG (1990) National strategy for control of *Sirex noctilio* in Australia. Austral For Grower 13(2):1–8

Hoebeke ER, Haugen DA, Haack RA (2005) *Sirex noctilio*: discovery of a palearctic siricid woodwasp in New York. Newsl Mich Entomol Soc 50:24–25

Howarth FG (1991) Environmental impacts of classical biological control. Ann Rev Entomol 36:485–509

Hurley BP, Slippers B, Croft PK, Hatting HJ, van der Linde M, Morris AR, Dyer C, Wingfield MJ (2008) Factors influencing parasitism of *Sirex noctilio* by the nematode *Deladenus siricidicola* in summer rainfall areas of South Africa. Biol Control 45:450–459

Hurley BP, Slippers B, Wingfield MJ (2007) A comparison of control results for the alien invasive woodwasp, *Sirex noctilio*, in the southern hemisphere. Agric For Entomol 9:159–171

Klepzig KD, Wingfield MJ, Slippers B (2009) Competition between fungi associated with *Sirex* woodwasp and southern pine beetle. In: Proceedings 19th U.S. Department of Agriculture interagency research forum on invasive species. Gen Tech Rep NRS-P-36. USDA Forest Service, Newtown Square

Kolk A, Starzyk JR (1996) Atlas of forest insect pests. Multico Warszawa, Warsaw

Lockwood JA (1993) Environmental issues involved in biological control of rangeland grasshoppers (Orthoptera: Acrididae) with exotic agents. Environ Entomol 22:503–518

Madden JL (1988) Sirex in Australasia. In: Berryman AA (ed) Dynamics of forest insect populations: patterns, causes, implications. Plenum Press, New York

Morgan FD (1968) Bionomics of Siricidae. Annu Rev Entomol 13:239–256

Nielsen C, Williams DW, Hajek AE (2009) Putative source of the invasive *Sirex noctilio* fungal symbiont, *Amylostereum areolatum*, in the eastern United States and its association with native siricid woodwasps. Mycol Res 113:1242–1253

Nuttall MJ (1980) *Deladenus siricidicola* Bedding. Nematode parasite of Sirex. Forest and Timber Insects in New Zealand no. 48, NZ Forest Service. Reprinted and updated in http://www.nzffa.org.nz

Shields L (2009) Update on *Sirex noctilio* in Canada. In: Proceedings. 19th US Department of Agriculture interagency research forum on invasive species. Gen Tech Rep NRS-P-36. USDA Forest Service, Newtown Square

Smith DR, Schiff NM (2002) A review of the siricid woodwasps and their ibaliid parasitoids in the eastern United States with emphasis on the mid-Atlantic region. Proc Entomol Soc Wash 104:174–194

Spradbery JP, Kirk AA (1978) Aspects of the ecology of siricid woodwasps in Europe North Africa and Turkey with special reference to the biological control of *Sirex noctilio* F. in Australia. Bull Entomol Res 68:341–359

Stillwell MA (1966) Woodwasps in conifers and the associated fungus *Stereum chailletii* in Eastern Canada. For Sci 12:121–128

Talbot PHB (1977) The *Sirex-Amylostereum-Pinus* association. Annu Rev Phytopathol 15:41–54

Uvarov BP (1931) Insects and climate. Trans Entomol Soc Lond 79:1–247

Villacide JM, Corley JC (2008) Parasitism and dispersal potential of *Sirex noctilio*: implications for biological control. Agric For Entomol 10:341–345

Williams DW, Zylstra KE, Mastro V (2009) Biological control of *Sirex noctilio* in North America by *Beddingia siricidicola*: 2008 update. In: Proceedings 20th US Department of Agriculture interagency research forum on invasive species. Gen Tech Rep NRS-P-51 USDA Forest Service, Newtown Square

Wilson AD, Schiff NM (2007) Identification of the symbiotic wood decay fungus of *Urocerus taxodii*. Phytopathol 97(7):123

Wilson AD, Schiff NM, Oberle CS (2007) Feeding and reproductive specificity if *Deladenus siricidicola* on *Amylostereum* species and closely related wood decay fungi. Phytopathol 97(7):123

Yu Q, de Groot P, Leal I, Davis C, Ye W, Foord B (2009) Characterization of *Deladenus siricidicola* (Tylenchida: Neotylenchidae) associated with *Sirex noctilio* (Hymenoptera: Siricidae) in Canada. Intl J Nematol 19:23–32

Zylstra KE, Dodds K J, Francese JA, Mastro VC (2010) *Sirex noctilio* in North America: the effect of stem-injection timing on the attractiveness and suitability of trap trees. Agric For Entomol 12:243–250

Chapter 11
The Chemical Ecology of *Sirex noctilio*

Damon J. Crook, Katalin Böröczky, Kelley E. Zylstra, Victor C. Mastro, and James H. Tumlinson

Abstract The recent detection of an established *Sirex noctilio* F. (Hymenoptera: Siricidae) population in New York, USA highlighted an urgent need to develop a trapping system so that the wasp's distribution and density can be monitored effectively. This review is of research that has been done on the insect's biology and chemical ecology towards this goal. This includes detailed work on the volatile composition of pine trees and the antennal responses they elicit in *S. noctilio.* Despite the testing of several promising volatile blends, there is still no effective lure that is as attractive as a funnel trap placed on trees stressed by injection with a herbicide. The possibility of a pheromone attractant between the sexes has also been studied at both the morphological and chemical level. This work has led to the identification of three sex pheromone components, (Z)-7-heptacosene, (Z)-7-nonacosene and (Z)-9-nonacosene to which males respond at short range. There is currently no evidence that *S. noctilio* has a long-range sex pheromone. Behavioral cues for host and mate selection (olfactory, visual and/or auditory) clearly need to be investigated further if we are to develop a monitoring trap that is more effective than a chemically stressed 'trap tree'.

D.J. Crook (✉) • V.C. Mastro
USDA, APHIS, PPQ, CPHST, Otis Laboratory, Bldg. 1398, W. Truck Rd, Buzzards Bay, MA 02542, USA
e-mail: damon.j.crook@aphis.usda.gov

K. Böröczky
Department of Entomology and W. M. Keck Center for Behavioral Biology, North Carolina State University, Raleigh, NC 27695, USA

K.E. Zylstra
USDA, APHIS, PPQ, CPHST, 374 Northern Lights Drive, North Syracuse, NY 13212, USA

J.H. Tumlinson
Center for Chemical Ecology, Department of Entomology, Pennsylvania State University, University Park, PA 16802, USA

B. Slippers et al. (eds.), *The Sirex Woodwasp and its Fungal Symbiont: Research and Management of a Worldwide Invasive Pest*, DOI 10.1007/978-94-007-1960-6_11,
© US Government 2012

11.1 Introduction

Sirex noctilio (F.) has resulted in one of the most damaging biological invasions of pine forestry in the Southern Hemisphere and now poses a serious threat to pine (*Pinus* spp., L.) monocultures in the United States. Although the risk to pine resources in natural northern forests (mixed coniferous and deciduous) is hard to estimate, the risk to the commercial pine monocultures in the southern United States is thought to be considerable (Chap. 19). Since the first *S. noctilio* detection in Oswego, New York, 2004 (Hoebeke et al. 2005) there has been an urgent need within the United States to develop a lure and trap system so that efficient monitoring and management controls can be implemented. This has led to an increased interest in learning more about *S. noctilio* behavior and its chemical ecology.

11.2 *Sirex noctilio* and Stressed Trees

Sirex noctilio primarily attacks stressed pine trees (Madden 1968a, b; Neumann et al. 1982, 1987, Chap. 3). Madden (1968b) found that trees that were damaged had impaired translocation, but enhanced transpiration and phloem respiration. This resulted in changes in tree permeability along with increased monoterpene and water loss through the bark. These changes resulted in attraction of *S. noctilio* to damaged trees, as well as increased oviposition due to a lowered osmotic status within the phloem tissue (Madden 1977). Host tree selection, therefore, involves (a) the attraction of female wasps to a potential breeding location and (b) subsequent assessment of the location as an oviposition substrate (Madden 1968a). During attack the female wasp also inoculates the tree with a muco-secretion that reinforces the tree stress condition and thus favors the recruitment of more females and greater tree debilitation (Madden 1968a).

The duration of attractiveness has been shown to differ between felled and girdled trees. The attraction of felled material is limited because of rapid drying, whereas girdled trees may remain attractive into a second season. Felled trees elicited a peak response by wild *S. noctilio* after 5–7 days and were still attractive for up to 3 weeks post felling (Madden 1971). When trees were ring-girdled they became attractive to wild *S. noctilio* after 9–12 days with attractiveness lasting 3–4 months post treatment. This early work therefore suggested that mechanically girdled trees could be used as "trap" trees to monitor the presence of *S. noctilio* in the field (Madden and Irvine 1971; Morgan and Stewart 1972). Attacks were, however, mainly confined to the region below the mechanical girdle (Madden 1971).

An unexpected side effect of *P. radiata* basal stem injected herbicide trials in north eastern Victoria (Australia) showed the occurrence of *S. noctilio* in many of the treated trees despite there being low population levels of the woodwasp in the area (Minko 1981). Since then, particularly in the Southern Hemisphere, trap trees have mainly been used to detect low-level *S. noctilio* populations (Ciesla 2003, Chaps. 12, 16, 17). Attractive trap trees are usually created by injecting trees with the herbicide Dicamba (3,6-dichloro-2-methoxybenzoic acid) instead of physically girdling them (Neumann et al. 1982; Ciesla 2003). Recent research in the northeastern USA has shown that chemically girdling a tree within 4 weeks of insect flight yields the most *S. noctilio* trap captures and emergence (Zylstra et al. 2010). Concurrently, research conducted by K. Böröczky (unpublished data 2006–2008) has also shown that the host trees have a peak in volatile emission approximately 4 weeks after chemical stem injection.

The wasp is initially attracted to physiologically stressed trees by an attractant that is released from the stem when the supply of nutrients is limiting. Sirex females have shown immediate attraction to felled trees and freshly disturbed phloem suggesting an attractant is present in that tissue or the cambial sap (Madden 1968a; Morgan et al. 1971). This was supported by solvent extracts of phloem being attractive to wasps when placed on healthy trees in field assays (Madden 1968a).

11.3 Pine Tree Volatile Composition of Healthy and Girdled Trees

The volatile composition of Monterey pine (*Pinus radiata* D. Don) and Scots pine (*Pinus sylvestris* L.) is well studied (Simpson 1976; Simpson and McQuilkin 1976a, b; Judžentienė et al. 2006). Scots pine is rich in monoterpene hydrocarbons such as α and β-pinene, δ-3-carene, limonene, α and γ-terpinene, (Z)-β-ocimene, myrcene, camphene, sabinene and terpinolene (see Lawrence 1991 and 2007 for reviews).

A recent study (K Böröczky, unpublished data collected in 2006–2007) used a teflon-film chamber to collect the volatile profiles of chemically girdled and healthy Red (*Pinus resinosa*) and Scots (*Pinus sylvestris*) pine trees. Monoterpenes were the most abundant compounds in both Red and Scots pine samples; α-pinene (40–80%), (S)-(−)-β-pinene (1–20%) and (S)-(+)-3-carene (1–50%). When compared to healthy trees they found higher concentrations of the above volatiles in girdled samples. (S)-(+)-3-carene occurred in higher concentrations in herbicide treated samples and was thought to be a potential indicator of damaged trees. However, this is complicated by the fact that *P. sylvestris* trees can be divided into two chemotype groups; those that have high and low concentrations of 3-carene in their needles and trunk xylem (Sjődin et al. 2000). It is interesting to note that 3-carene has been seen to increase overall woodborer trap catches (including native siricids) when used with a 'standard woodborer' ethanol and α-pinene lure (Costello et al. 2008).

Based on results from volatile collections from both the stem and the needles of Scots pine, Böröczky (unpublished data 2007–2008) suggested two terpene blends for field testing. The stem blend consisted of 60% α-pinene (1:3 ratio of (−):(+) enantiomers), 2% camphene, 3% (S)-(−)-β-pinene, 2% β-myrcene, 30% (S)-(+)-3-carene, 2% limonene and 1% β-phellandrene. The needle blend consisted of 27% α-pinene (3:1 ratio of (−):(+) enantiomers), 1% camphene, 27% (S)-(−)-β-pinene, 15% β-myrcene, 27% (S)-(+)-3-carene, 1.5% limonene and 1.5% β-phellandrene. Despite the formulation and field testing of these promising terpene blends, as well as other treatments (70/30 α/β-pinene, 30/70 α/β-pinene, 33/33/33 α/β-pinene/3-carene, 100% α-pinene, 100% β-pinene, 100% 3-carene), there is unfortunately still no effective bait that is as attractive as a funnel trap placed on a herbicide-injected trap tree (Zylstra et al. 2010).

11.4 Electroantennogram (EAG) Responses of *S. noctilio*

Electrophysiological techniques such as the electroantennogram (EAG) or coupled Gas Chromatography-Electroantennographic Detection (GC-EAD) are proven methods for identifying and elucidating the behavioral role of semiochemicals for insects (Bjostad 1998). The EAG technique allows for rapid evaluation of GC fractions by passing compounds of interest over insect antennae that are connected to an electrode/voltage amplifier system.

The major problem with developing a host-based lure for *S. noctilio* is that the insect is antennally active to almost every compound (approximately 20 or more) in pine aeration samples (DJ Crook, unpublished data 2006–2007). Identifying these host volatiles and finding out which of them are the main behavioral attractants is a difficult and time consuming task that may not be a useful and/or economic option for development of a *Sirex* lure. An additional problem is that the proportion of enantiomers of simple terpenes such as α-pinene can vary depending on the origin of the trees (Kamińska 2007). There is thus the possibility that *Sirex* may respond differently to kairomone-based lures from region to region (Costello et al. 2008).

The first identification of volatiles from Monterey Pine (*Pinus radiata* D. Don) along with the electroantennograms they elicited from *S. noctilio* was done by Simpson and McQuilkin (1976a). They showed that attraction of *S. noctilio* to stressed trees was due to one or more of the following factors: (1) increased permeability of bark tissues to host volatiles; (2) metabolic change causing greater production of some components; (3) metabolic change causing production of components not present under normal conditions.

To better understand the attractiveness of *P. radiata* to *S. noctilio,* Simpson and McQuilkin (1976a) examined the change in composition with time of the volatiles emanating from logs. They then tested the activity of these volatile components over the antennae of *S. noctilio* using electroantennogram (EAG) methods. Monoterpene hydrocarbons were seen to be the major component of all aeration samples constituting over 95% of the total volatiles collected. Monoterpene components identified included α-pinene, camphene, β-pinene, myrcene, 3-carene, limonene, β-phellandrene, p-cymene and terpinolene. The composition of these monoterpene hydrocarbons was seen to be fairly constant on aerations from 1, 23 and 33 day old logs. When the EAG response of *S. noctilio* was measured using aerations from logs that were 3, 7, 11, 15 and 19 days old, an increase in response was observed with increasing log age. A monoterpene hydrocarbon fraction (isolated from the day 19 volatile sample) gave a significantly lower EAG response than the total volatile sample.

Oxygenated components were seen in aerations in only trace amounts immediately after felling but increased to approximately 1% of the total volatile composition by day 23 post felling (a 20–30 fold increase) (Simpson and McQuilkin 1976a). They identified the major oxygenated components as camphor, pinocamphone, isopinocamphone, trans-pinocarveol and fenchone. Fractions of these compounds were seen to give high EAG responses suggesting that these oxygenated components were largely responsible for the increase in response with time after felling.

Simpson (1976) went on to study *S. noctilio* EAG responses in relation to the chemical structure of volatile components in pine oil. This study was based on EAG work by Roelofs et al. (1971) that had shown that compounds corresponding to the natural pheromone produce the highest amplitude EAG signals of all compounds in the same chemical class and that the more distant homologue in a series had a smaller amplitude of response. For the monoterpene hydrocarbons, Simpson (1976) found that compounds that had a bicyclic skeleton and unsaturation, showed the greatest EAG activity, in particular, ß-Pinene (mean response ratio of 1.22), α-Pinene (0.94) and 3-carene (1.15). Moderate activity was shown by ß-phellandrene (0.81, p-menthane skeleton, two double bonds) and myrcene (0.82, acyclic, three double bonds). This is interesting because α and ß-pinene, ß-phellandrene, myrcene and 3-carene were identified as being the major components of the volatiles emanating from felled *P. radiata* (Simpson and McQuilkin 1976a). Low antennal activity in response to the monoterpene hydrocarbons was shown for three ring systems such as tricyclene (0.63) and aromatic structures such as p-cymene (0.55) (Simpson 1976). For the alcohols, the highest antennal activity was shown in response to compounds with a pinane skeleton and one double bond such as cis-verbenol (2.65), trans-verbenol (2.37) and myrtenol (1.55). Of the aldehydes myrtenal (2.6, pinane skeleton, one double bond) showed high EAG activity. Of the acetates the highest response was shown by trans-verbenyl acetate (1.98, pinane skeleton, one double bond). Bicyclic unsaturated acetates produced relatively low EAG responses. Phenols and ethers generally gave weak EAG responses with only 1,8-cineole (1.62, two-ring system) having moderately strong EAG activity. Simpson (1976) reported that ketones exhibited the highest response values of any compounds examined. Pinane skeleton structures such as pinocarvone (3.61) and pinocamphone (3.33) had the highest EAG activity of all compounds tested. Verbenone (2.3) which has the keto-group in the 3-rather than the 2-position had slightly less activity. These higher EAG responsive compounds are interesting because they were identified in the volatiles from 3 to 4 week old *P. radiata* logs that were attractive to *S. noctilio* (Simpson and McQuilkin 1976a).

More recent EAG work has been done on *Sirex edwardsii* and *Sirex nigricornis* by B.T. Sullivan and W.P. Shepherd (USDA FS, personal communication 2006–2007) who found that female adults of both species gave 3–5 times larger responses to oxygenated monoterpenes (in particular fenchone, terpinen-4-ol, camphor, verbenone (also a scolytid pheromone), isopinocamphone and α-terpineol) when compared to monoterpenes such as α-pinene. D.J. Crook (unpublished data 2009) tested female *S. noctilio* in a similar study and found oxygenated monoterpenes (verbenone, fenchone, isocamphone, α-terpineol and terpinen-4-ol) gave much higher antennal responses than α-pinene (+), α-pinene (−) or ß-pinene (−). GC-EAD work showed that mated female *S. noctilio* were particularly sensitive to fenchone in Scots pine bark aeration samples (DJ Crook, unpublished data 2009).

11.5 Recent Lure Trials Based on GC and GC-EAD Results

In 2009, we conducted a field test to see if 70/30 α/ß-pinene 1 g/day lures could be enhanced using fenchone 3–5 mg/day, fenchone 50 mg/day and verbenone 50 mg/day. A six component mix consisting of a 70/30 α/ß-pinene 1 g/day lure + fenchone 3–5 mg/day + α-terpineol 3–5 mg/day + terpinen-4-ol 3–5 mg/day + isocamphone 3–5 mg/day was also tested. Lures were placed on black multi-funnel traps hung at 6 m in infested pine stands (New York State) and checked throughout the July-September *Sirex* flight season. A marginal increase in trap catch was seen for the 6 component lure, but it was not significant overall (Fig. 11.1). Mean trap catch was low and consistent with previous trapping efforts during 2006–2008 (DJ Crook, K Böröczky, unpublished data).

These latest results follow a general trend observed in North American and South African field tests, in that multiple component lures catch marginally more adults than unbaited traps, but are not significant overall. Trap catches in the U.S. are usually very low (mainly ovipositing females, averaging one adult caught on a trap per week). This low female adult trap catch may simply be due to the insect's behavior. Males tend to emerge first and have been observed to form swarms over the tree canopy (Morgan and Stewart 1966; Madden 1988; Taylor 1981). When females emerge they exhibit a strong photopositive response and fly up into these male swarms. After mating the females then search for suitable oviposition sites and can be found on the lower parts of the tree bole. Attracting large numbers of *S. noctilio* down to funnel or cross vane traps in the darker understory may therefore be difficult to achieve. There is also the issue that host volatile based lures may not be a strong

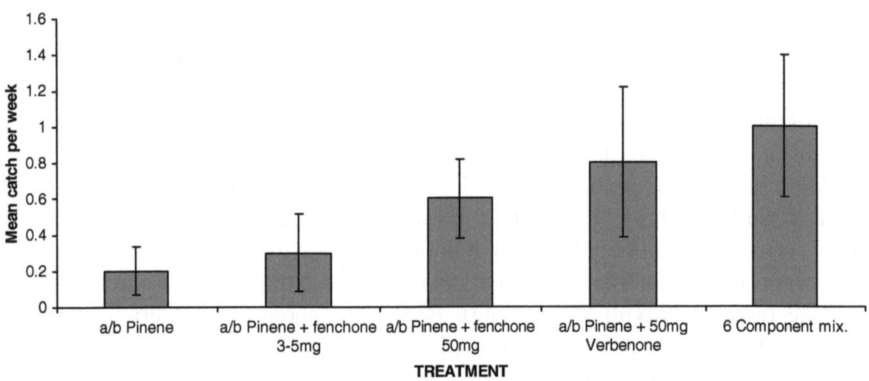

Fig. 11.1 Mean trap catch of *S. noctilio* on funnel traps baited with lures (New York State 2009). n = 12

enough 'point source' to attract adults into a trap and that the lures are outcompeted by surrounding trees.

11.6 Antennal Physiology

A detailed scanning and electron microscopy study by Crook et al. (2008) examined the antennal receptors of *S. noctilio*. Male and female antennae share five sensillum types: Böhm bristles, sensilla chaetica, sensilla basiconica, sensilla ampullacea and uniporous contact chemoreceptors. The uniporous contact chemoreceptors made up approximately 85% and 78% of the total sensilla on male and female antennae, respectively. These contact chemoreceptors increased in number toward the distal tip and covered two thirds of the circumference of each flagellomere. Males were seen to have significantly more contact chemoreceptors on the proximal half of the antennae, thus strongly supporting the evidence that a female produced contact pheromone is used during mate identification.

11.7 Pheromones

Observing the mating behavior of *S. noctilio* in the wild is difficult due to it occurring in the upper canopy of trees. Morgan and Stewart (1966) reported that the female initiates courtship by walking close to a male in the upper canopy branches. The male then follows the female and repeatedly taps her body with his forelegs and antennae, strongly suggesting the presence of a contact cue. The sex ratio of emerging wasps tends to favor males and can be as high as ten to one (Morgan 1968; Tribe and Cillié 2004, Chap. 18) in the Southern Hemisphere, though current data in the north-eastern U.S. describes a ratio closer to 1:1 (Zylstra et al. 2010). These male aggregations of wasps typically stay close to the site of emergence thus negating the need for a long range pheromone to locate mates. Females simply emerge and fly up to the canopy to find a mate before dispersing to other trees. There is currently no evidence that *S. noctilio* has a long range sex pheromone.

In a preliminary laboratory study, *S. noctilio* males gave a strong abdomen bending response when presented with freeze-killed females (DJ Crook, unpublished data 2008, Figs. 11.2 and 11.3). This male response was absent when the cuticular coating of the dead females was removed by dipping the female in hexane. These preliminary results indicated the relevance of female-produced cuticular compounds in *S. noctilio* mating behavior. Further analysis of 'female dipped' hexane extracts by Böröczky et al. (2009) revealed saturated and unsaturated hydrocarbons as the major components. Fractionation of the body wash followed by further male copulatory response tests led to the identification of three sex pheromone components, (Z)-7-heptacosene, (Z)-7-nonacosene and (Z)-9-nonacosene. This is the first description of cuticular hydrocarbons serving as contact sex pheromone components in the hymenopteran sub order Symphyta, in which the family Siricidae belongs (Ayasse et al. 2001; Keeling et al. 2004).

Fig. 11.2 Photograph of male *S. noctilio* antennating the surface of a frozen killed female

Fig. 11.3 Photograph of male *S. noctilio* 'abdomen bending' response that occurs immediately after antennation of the female body surface

11.8 Conclusions

The complicated host-wasp biology and behavior of *S. noctilio* makes it a difficult insect to study, especially with regards to identifying attractants for monitoring purposes. In lab bioassays the wasp does not respond well in glass aeration containers or Y-tube olfactometers and desiccates quickly under the slightest airflow

(DJ Crook, personal observation 2006). Field testing of lures is complicated by the fact that pine stands in the North East US are small, fragmented areas which are often mixed in with other hardwoods (Zylstra et al. 2010). Because of the logistical issues that can arise from using trap trees (e.g., access to private property, legal issues) monitoring for *S. noctilio* in North American pine stands has mainly involved the use of semiochemical baited panel or funnel traps. Behavioral cues for host and mate selection (olfactory, visual and/or auditory) clearly need to be investigated further if we are to develop a monitoring trap that is more effective than a chemically girdled 'trap tree'.

Acknowledgments We thank Dave Williams, Dave Lance, Joe Francese and Miriam Cooperband (USDA, APHIS, PPQ, CPHST, Otis, MA) for reviewing earlier drafts of this manuscript.

References

Ayasse M, Paxton RJ, Tengö J (2001) Mating behavior and chemical communication in the order of Hymenoptera. Annu Rev Entomol 46:31–78
Bjostad LB (1998) Electrophysiological methods. In: Millar JG, Haynes KF (eds) Methods in chemical ecology, vol 1, chemical methods. Kluwer Academic Publishers, Norwell
Böröczky K, Crook DJ, Jones TH, Kenny JC, Zylstra KE, Mastro VC, Tumlinson JH (2009) Monoalkenes as contact sex pheromone components of the woodwasp *Sirex noctilio*. J Chem Ecol 35:1202–1211
Ciesla WM (2003) European woodwasp: a potential threat to North America's conifer forests. J For 101:18–23
Costello SL, Negron JF, Jacobi WR (2008) Traps and attractants for wood-boring insects in Ponderosa pine stands in the Black Hills, South Dakota. J Econ Entomol 101:409–420
Crook DJ, Kerr LM, Mastro VC (2008) Sensilla on the antennal flagellum of Sirex noctilio (Hymenoptera: Siricidae). Ann Entomol Soc Am 101:1094–1102
Hoebeke ER, Haugen DA, Haack RA (2005) *Sirex noctilio*: discovery of a palearctic siricid woodwasp in New York. Newsl Mich Entomol Soc 50:24–25
Judžentienė A, Šližytė J, Stiklienė A, Kupčinskienė E (2006) Characteristics of essential oil composition in the needles of young Scots pine (*Pinus sylvestris* L.) stands growing along an aerial ammonia gradient. Chemija 17:67–73
Kamińska J (2007) Essential oil composition and plant-insect relations in Scots pine (*Pinus sylvestris* L.). Scientific Bulletin of the Technical University of Lodz. Food Chem Biotechnol 71:71–95
Keeling ChI, Plettner E, Slessor KN (2004) Hymenopteran semiochemicals. Top Curr Chem 239:133–177
Lawrence BM (1991) Perfumer Flavorist 16:59
Lawrence BM (2007) Progress in essential oils. Perfumer Flavorist 32(32–40):42–45
Madden JL (1968a) Physiological aspects of host tree favourability for the woodwasp *Sirex noctilio* F. Proc Ecol Soc Aust 3:147–149
Madden JL (1968b) Physiological aspects of host tree favourability for the woodwasp, *Sirex noctilio* F. Proc Entomol Soc Aust 3:147–149
Madden JL (1971) Some treatments which render Monterey pine (*Pinus radiata*) attractive to the wood wasp *Sirex noctilio* F. Bull Entomol Res 60:467–472
Madden JL (1977) Physiological reactions of Pinus radiate to attack by woodwasp, *Sirex noctilio* F. (Hymenoptera: Siricidae). Bull Entomol Res 67:405–426

Madden JL (1988) Sirex in Australasia. In: Berryman AA (ed) Dynamics of forest insect populations. Plenum, New York

Madden JL, Irvine CJ (1971) The use of lure trees for the detection of *Sirex noctilio* in the field. Aust For 35:164–166

Minko G (1981) Chemicals for non-commercial thinning of *Pinus radiata* by basal stem injection. Aust Weeds 1:5–6

Morgan FD (1968) Bionomics of Siricidae. Annu Rev Entomol 13:239–256

Morgan FD, Stewart NC (1966) The biology and behavior of the woodwasp *Sirex noctilio* (F.) in New Zealand. Trans R Soc N Z 7:195–204

Morgan FD, Stewart NC, Horwood DW (1971) Translocation of insecticides in *radiata* pine and effectiveness on *Sirex noctilio*. Austral For 35:133–142

Morgan FD, Stewart NC (1972) Developing and testing a lure-trap for the woodwasp *Sirex noctilio* F. Austral For 36:38–46

Neumann FG, Harris JA, Kassaby FY, Minko G (1982) An improved technique for early detection and control of the *Sirex* wood wasp in *radiata* pine plantations. Austral For 45:117–124

Neumann FG, Morey JL, McKimm RJ (1987) The *Sirex* wasp in Victoria, bulletin 29. Victoria. Department of Conservation, Forest and Lands, Melbourne, 41 p

Roelofs WL, Comeau A, Hill A, Milicevic G (1971) Sex attractant of the codling moth: characterization with electroantennogram technique. Science 174:297–299

Simpson RF (1976) Bioassay of pine oil components as attractants for *Sirex noctilio* (Hymenoptera: Siricidae) using electroantennogram techniques. Entomol Exp Appl 19:11–18

Simpson RF, McQuilkin RM (1976a) Identification of volatiles from felled *Pinus radiata* and the electroantennograms they elicit from *Sirex noctilio*. Entomol Exp Appl 19:205–213

Simpson RF, McQuilkin RM (1976b) Terpenes of the bark oil of *Pinus radiata*. Phytochemistry 15:328–329

Sjődin K, Persson M, Faldt J, Ekberg I, Borg-Karlson A (2000) Occurrence and correlations of monoterpene hydrocarbon enantiomers in *Pinus sylvestris* and *Picea abies*. J Chem Ecol 26:1701–1720

Taylor KL (1981) The Sirex woodwasp: ecology and control of an introduced forest insect. In: Kitching RL, Jones RE (eds) The ecology of pests, some Australian case histories. Commonwealth Scientific and Industrial Research Organization, Hobart, pp 231–248

Tribe GD, Cillié JJ (2004) The spread of *Sirex noctilio* Fabricius (Hymenoptera: Siricidae) in South African plantations and the introduction and establishment of its biological control agents. Afr Entomol 12:9–17

Zylstra KE, Dodds KJ, Francese JA, Mastro V (2010) *Sirex noctilio* (Hymenoptera: Siricidae) in North America: the effect of stem-injection timing on the attractiveness and suitability of trap trees. Agric For Entomol 12:243–250

Chapter 12
The Use of Kairomone Lures for the Detection of *Sirex noctilio* in Susceptible *Pinus radiata* Plantations in Australia

Richard Bashford and John L. Madden

Abstract Exotic woodwasps can cause high levels of mortality in *Pinus* species plantations. One species, *Sirex noctilio* (Sirex), was first detected in Tasmania in 1952 and has subsequently spread into softwood plantations throughout southeastern Australia. Plantations aged 7 years or older are susceptible to Sirex attack. For several decades the method of detecting developing populations of Sirex has been through aerial and ground surveillance and the establishment of trap tree plots. Trap tree plots are sets of ten trees treated with a measured dose of herbicide sufficient to stress the tree and so render it attractive to Sirex. In Australia hundreds of trap tree plots are prepared each year in Sirex-infested plantations and in plantations entering the susceptible age bracket. Approximately half of all prepared plots are attacked by Sirex. The development of a novel method to detect Sirex in young plantations using static traps baited with pinene lures began in 2003 in Tasmania. The use of these traps for the Sirex detection is now a routine operation in Tasmania.

12.1 Identification and Testing of *Pinus radiata* Stem Volatiles

In the mid-1970s several research projects were conducted to determine the volatile oils released by herbicide-treated *Pinus radiata* trees and the 'attractiveness' of those volatile oils, based on electro-antennogram techniques, to Sirex females.

R. Bashford (✉)
Forestry Tasmania, GPO Box 207, Hobart, TAS 7001, Australia
e-mail: dick.bashford@forestrytas.com.au

J.L. Madden
Division of Entomology, Department of Agricultural Science, CSIRO,
University of Tasmania, Hobart, TAS 7005, Australia

B. Slippers et al. (eds.), *The Sirex Woodwasp and its Fungal Symbiont: Research and Management of a Worldwide Invasive Pest*, DOI 10.1007/978-94-007-1960-6_12,
© Springer Science+Business Media B.V. 2012

Forty components of bark oil were identified and a further 20 were noted in varying trace concentrations. Simpson (1976) and Simpson and McQuilkan (1976), working on the chemistry of radiata pine oil, evaluated receptor activity of the wasp. They employed the electro-antennogram (EAG) technique and found that compounds based on the pinane skeleton gave the greatest EAG response and that the monoterpenes accounted for the bulk (>95%) of the volatiles released from felled trees. The presence of such compounds was chiefly responsible for the increased EAG response of volatiles collected from logs 3 weeks after felling. The olfactory response of live wasps was evaluated in a flight mill olfactometer. Borden and Bennett (1969) found that the flight speed of tethered scolytid beetles was initially slightly increased and then suppressed and arrested during a 30 second exposure to pheromones emanating from frass. Flight began again following the removal of the pheromone source. These authors concluded that the cessation of flight was tangible evidence of the arresting effect of pheromones on flying insects (see also Chap. 11). This theory and a similar flight mill design were used to test the effects of a range of pine volatiles against female *S. noctilio*.

The flight arm was fitted above a photoelectric cell. Signals arising from the breaking of the light beam when the arm revolved were amplified, rectified and fed into a DC recording milliameter (Evershed and Vignoles Limited, London). Arm speed and deviations from average velocity were reflected in changes in the frequency and amplitude of the recorded signal. Volatiles were released from the base of the mill below the flight path of the insect and drawn up to an exhaust fan placed 10 cm above the insect flight path. Changes in flight velocity following presentation of a test solution were reflected in changes in the recorded signal. Accelerated flights resulted in a slight reduction of amplitude and an increased frequency of signals per unit time whereas retardation gave signals of increased amplitude and decreased frequency of signals. Accelerated flight, which often predicted retardation, suggested that this response was associated with perception of stimuli. The variations within and between the signal characteristics of retarded flights following presentation of different test materials to different females were taken as a measure of the flight-arresting activity of the respective materials because complete cessation of flight often occurred. Flight cessation was accompanied in the majority of runs by partial lowering of the ovipositor and reciprocation of the valvifers.

The test materials included solvent extracts and distillate fractions of *P. radiata* bark, log volatiles, monoterpenes and ketones. Electro-antennograms were obtained from isolated antennae. The micro-electrodes inserted in the antennae were connected to a Beckman Dynograph DC Recorder, which provided the trace of the antennal response. The flight mill and EAG response were compared to different monoterpenes and age and flight exercise of test females. The results showed that both alpha- and beta- pinene caused strong behavioural responses to increasing volatile release volumes followed to a lesser extent by p-cymene and α-phellandrene then tricylene and D-limonene, myrcene and α-limonene were least active. Both α-pinene and β-pinene gave the most significant response in terms of all parameters measured, with approximately one in three females tested ceasing flight when presented with these materials. The greatest response was to the most dilute solutions (0.5%).

The response to crude steam-distilled oil was markedly lower than the response obtained when the oil was fractionated. The fraction containing the bulk of the monoterpenes resulted in velocity changes comparable to those obtained in response to pure alpha- and beta-pinene solutions. High responses to alcohols, ketones and acetates were recorded but subsequently shown to have a negative reactive response by live insects.

A positive correlation between antennal sensitivity and flight mill response to monoterpene solutions was found in velocity change ($r^2=0.903$), duration of change and flight inhibition ($r^2=0.922$). These results indicate that, for the monoterpenes used, receptor activity and whole insect flight response were directly related.

In a separate trial females were flown in the mill for different periods of time to see if weight loss resulted in increased oviposition response. A 19% weight loss caused by the depletion of the fat reserves of newly emerged females by flying resulted in enhanced oviposition responses. It may be that small females, having fewer eggs and a reduced longevity (Madden 1974), would be expected to become host positive after a short flight period i.e., local dispersal, whereas large females may disperse over long distances before responding to host trees. The post emergence behaviour of *S. noctilio* may be similar to that of the scolytid *Dendroctonus pseudotsugae* in which flight exercise and weight (lipid) loss play an essential role in dispersal and host tree acceptance (Atkins 1966).

Field trials were conducted initially by attaching saturated cheesecloth pads wrapped in aluminium foil to trees. Initially bark extract (1:3 w/v) and 5% pinene solutions were tested. The trees were examined every 2 days for signs of Sirex oviposition. In subsequent field trials the pads were changed to 250 ml glass bottles with cotton wicks. Attacked trees were detected by the presence of drill holes usually located above and below the volatile source. Over 90% of attacks recorded on test trees were on those trees carrying alpha-pinene lures. The frequency of natural attack by Sirex was ascertained each year by scoring six transects consisting of 200 trees. Collectively the incidence of attack on treated trees was 0.16 compared to an expected frequency of 0.04 on untreated trees. This result indicates that trees were preferentially attacked i.e., the presence of the test solutions increased the probability of attack by Sirex. The surface area for the natural release of volatiles through the bark surface of an 18 cm diameter tree at 128 cm above ground level is approximately 1,600 cm^2, compared to the release rate of the lures at 25 cm^2. Lure release rates, increased to 1,000 mg/24 h, resulted in a tenfold (10.2:1) increase of attack on trees releasing alpha-pinene compared to those releasing beta-pinene.

12.2 Development of a Kairomone Baited Trap System in Tasmania

The first capture of *S. noctilio* in a static trap occurred in static traps baited with alpha-pinene in an urban monitoring program in northern Tasmania. In 2004, three Intercept^R panel traps loaded with commercial alpha-pinene lures were specifically set up in each of three *P. radiata* plantations within the Retreat (northern Tasmania) and Pittwater Blocks (southern Tasmania), where a history of Sirex presence had been previously recorded. Small numbers of Sirex females (10 at Retreat and 16 at Pittwater) were captured at all sites. Later in the flight season, three traps were also placed in each of two young northern plantations where Sirex had not previously been recorded (Wurra Wurra and Castra Blocks). Eleven Sirex females were captured in the Wurra Wurra Block.

In the 2005 flight season, panel traps were specifically placed in five compartments, aged 8–12 years, where Sirex had not been previously recorded. The compartments were checked during Forest Health aerial and roadside surveys and no attacked trees were recorded. With advice from Advanced Pheromone Technologies (Portland, Oregon), a series of host volatile lures were formulated for use in Intercept panel traps. The combinations were two volatile release rates of alpha-pinene (300 and 600 mg/day); two of beta-pinene (300 and 800 mg/day) and a combination lure of alpha- and beta-pinene (800 mg/day).

Lures were placed in panel traps with two sets of four traps in each compartment. Each set of four traps was placed at least 300 m apart and each trap within a set was separated from its nearest neighbour by 50 m. Traps ran for 4 weeks in February. In three of the five compartments Sirex females were captured in the traps. Individual traps captured between one and four females. The combination lure captured the highest number of Sirex females. In a follow-up ground survey, several Sirex-killed suppressed trees were located in one of those compartments.

In a separate trial within the Retreat Block (northern Tasmania) a number of lures in combinations of four ratios of alpha- and beta-pinene were field tested in panel traps. (70:30, 60:40, 40:60, and 30:70) against lures containing either alpha- or beta-pinene. Beta-pinene by itself did not attract any females. The attractiveness of the mixed alpha- and beta-pinene lures increased with increasing proportions of alpha-pinene relative to beta-pinene. The combination of 70:30% alpha-/beta-pinene was selected as the most attractive to Sirex females (Bashford 2008).

In the 2006 flight season, panel traps were set up in five northern 'Sirex free' compartments with three traps in each compartment. A lure of 70% alpha-pinene and 30% beta-pinene, release rate 800 mg/day, was used. Traps were run for a period of 8 weeks during peak flight season (January to March) and serviced every 2 weeks. A total of 23 female Sirex and 27 female *Ibalia leucospoides* (an introduced egg parasitoid) were captured in four of the five compartments.

At the same time, a trial using Intercept panel traps and Lindgren multiple funnel traps was established at Pittwater Block in southern Tasmania. Five pairs of traps were used, all baited with the 70% alpha/30% beta-pinene lure. The Intercept panel traps captured approximately 40% more female Sirex than the Lindgren traps (19 females and 12 females respectively for the panel and Lindgren traps respectively). The difference in trap design may influence the capture of fast-flying Sirex females. The panel traps have a larger flat surface area than the funnel traps and also have a larger entry funnel into the collection container (Fig. 12.1).

Fig. 12.1 Intercept R trap used for *Sirex* detection, showing pinene lure (*yellow*) and Ethylene glycol preservative (*green*)

12.3 Integration of a Static Trap Module into the Current Australian NSCC Working Procedures for the Control of *Sirex noctilio*

In 1990, the Australian National Sirex Co-ordination Committee (NSCC; also see Chap. 14) prepared worksheets and a manual to ensure that detection and control methodologies were standardised throughout Australia (Haugen et al. 1990). Establishment of nematode control relied on regular monitoring of plantations to detect struck trees. Detection by ground and aerial surveys and trap tree establishment is both time consuming and expensive. Field trials, as described, demonstrate the potential for a trap/lure combination to detect the presence of Sirex at low population levels in young plantations more effectively than current monitoring techniques. Plantations aged from 7 years of age are now known to be susceptible to Sirex attack.

By placing a small number of traps in young 'Sirex free' plantations the presence or absence of Sirex females can be determined and a decision can be made to establish trap tree plots. If absent, trap tree plots would not be set up in that compartment.

If Sirex is present, then a survey to determine the number, if any, of Sirex struck trees is made within a 100 m radius of a trap. If two or less struck trees are located then no action is taken but recorded for trap tree establishment in the subsequent year. If three or more struck trees are located then trap trees are established at one ten-tree plot per 25 ha of susceptible plantation for nematode *Deladenus siricidicola* Bedding establishment. Each block of susceptible compartments is monitored using traps on a 3 year rotation. In compartments where Sirex is known to be present the routine establishment of trap trees is undertaken.

The modifications indicated in the flow chart (Fig. 12.2) show the additional three steps needed to establish and monitor the results of static trapping. The blue blocks indicate positive actions from trap establishment to testing infectivity of emerging Sirex. In the old strategy, trap trees were placed in plantations aged 10–25 years for the introduction of the nematode into the Sirex population. Trap trees were established usually in plantations where Sirex-struck trees had been previously detected by routine forest health monitoring. This usually meant that Sirex populations were already well established at levels >0.5%.The great advantage of using static traps in younger plantations aged between 7 and 10 years of age is that they detect the very early movement of Sirex into the plantation. That early detection means trap tree establishment and the introduction of nematodes into the population can be undertaken several years earlier than under the old strategy.

Considerable cost savings can be made by reducing the number of trap tree plots routinely established in areas of very low Sirex activity. This saving can be considerable, especially in areas ahead of the spread of Sirex such as northern South Africa, and the Australian State of Western Australia, where Sirex is not yet present.

The introduction of static trapping with a reduced trap tree program in Tasmania has resulted in a 25–30% cost reduction in the Sirex control program. In States such as Queensland where Sirex is not yet present, and Western Australia, where Sirex became established in 2009, the adoption of the static trapping technique in place of detection trap tree plots has saved 38% in monitoring costs.

12.4 Conclusions

The placement of volatile baited traps within Sirex infested plantations is not effective. Stressed trees emit a considerable volume of volatiles with the whole plantation acting as a single tree. The migrating female Sirex detect the mass plume emitted by a plantation and use this olfactory plume as a guide to detecting an area occupied by host trees. It may be that beta-pinene is a defining parameter in attracting Sirex females into a susceptible stressed plantation and then a specific low alpha-pinene concentration leads the wasps to a specific stressed tree.

Once inside a plantation with a wide range of tree health situations, it is almost impossible for a single female to locate the pinene volatiles from a trap as opposed to the large plumes emanating from individual trees. Within a healthy plantation, which may contain only one or two suppressed attacked trees, there is a much greater

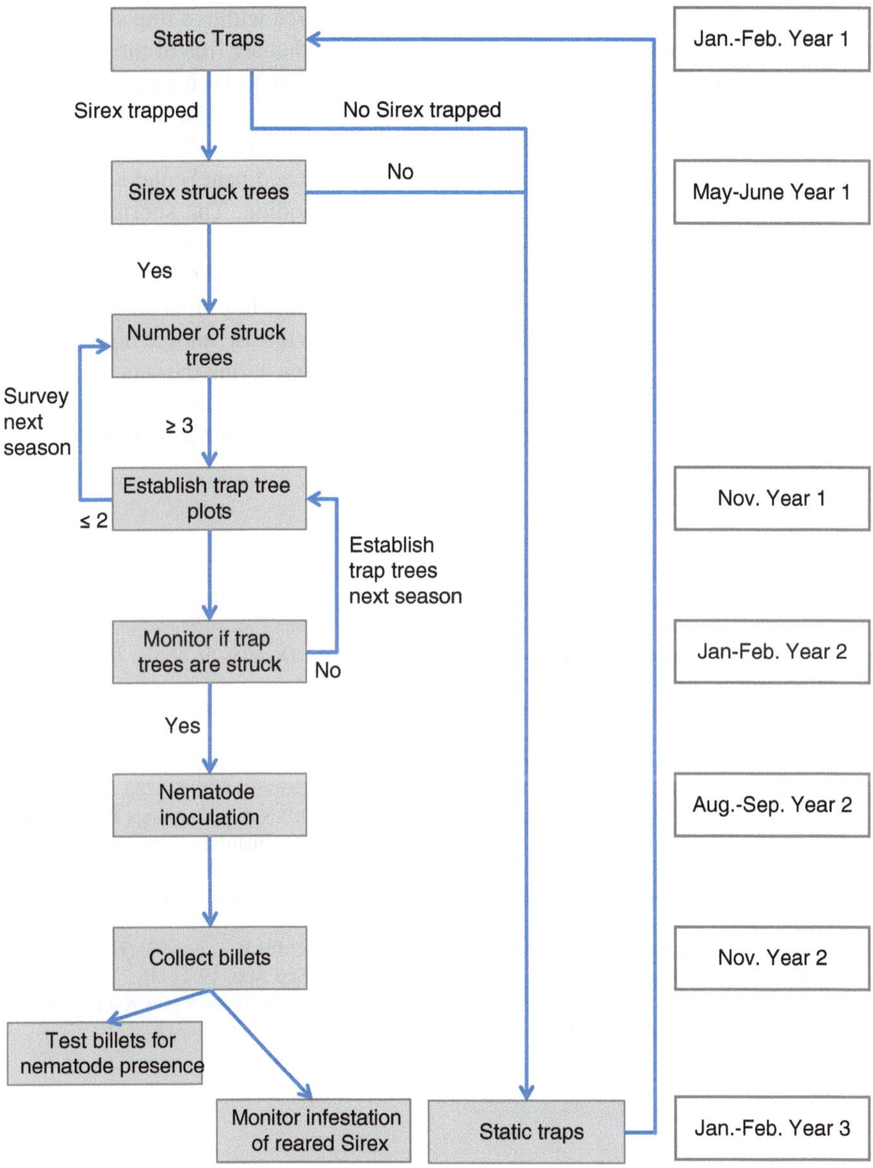

Fig. 12.2 Flowchart illustrating changes to the Australian National Sirex Strategy for the detection and control of *Sirex noctilio* using static traps

chance of a wasp locating the volatile source from a trap. Placement of traps on the perimeter of a plantation, even a 100 m or more away from the plantation boundary increases the chance of Sirex females being able to isolate the volatile source of an 'attractive' tree, i.e., trap. Further lure development in terms of alpha-/beta-pinene ratios is required to enhance the detection of lure source within a trap so that traps can be placed within the perimeter of stressed stands that are free of Sirex.

Trap design also needs to be reviewed. A problem with both panel and funnel traps is that in high rainfall areas, flooding of the traps can occur during the flight season, which has a negative effect on the preservative in the collection containers. A modification of the standard panel trap with shortened panels and a wide brim fixed to the roof of the trap has greatly reduced flooding. The shorter trap also increases the life of the lures by keeping them shaded at all times.

Lure desiccation is a current problem. It is important to ensure that the lures are shaded at all times because direct sunlight can dry out the lures in a couple of days. Lure pouch designs need to be able to regulate volatile release rates at high (+30°C) temperatures, which are common for long periods of time during the summer months in Southern Hemisphere plantations. The continued development of trap design and lure development is recommended to provide a valuable tool for both operational and research use.

References

Atkins MD (1966) Laboratory studies on the behaviour of the Douglas-fir beetle, *Dendroctonus pseudotsugae* Hopkins. Can Entomol 98:953–991

Bashford R (2008) The development of static trapping systems to monitor for wood-boring insects in forestry plantations. Aust For 71:236–241

Borden JH, Bennett RB (1969) A continuously recording flight mill for investigating the effect of volatile substances on the flight of tethered insects. J Econ Entomol 62:782–785

Haugen DA, Bedding RA, Underdown MG, Neumann FG (1990) National strategy for control of *Sirex noctilio* in Australia. Australian national Sirex co-ordination committee. Insert to Australian Forest Grower 13(2) Winter. pp 1–8

Madden JL (1974) Oviposition behaviour of the woodwasp, *Sirex noctilio* F. Aust J Zool 22:341–351

Simpson RF (1976) Bioassay of pine oil components as attractants for *Sirex noctilio* (Hymenoptera: Siricidae) using electro-antennogram techniques. Entomol Exp Appl 19:11–18

Simpson RF, McQuilkan RM (1976) Identification of volatiles from felled *Pinus radiata* and the antennograms they elicit from *Sirex noctilio*. Entomol Exp Appl 19:205–213

Chapter 13
The Sirex Woodwasp in New Zealand: History and Current Status

John Bain, Stephanie L. Sopow, and Lindsay S. Bulman

Abstract The European woodwasp, *Sirex noctilio,* was first found in New Zealand in 1900, but was not recognized as a pest until the late 1920s after deaths of pines in overcrowded plantations were attributed to it. Despite the successful introduction of a parasitoid from Europe as a control measure, damage became more prevalent between 1946 and 1951 when extreme drought conditions confounded the problem. This led to investigation of other parasitoids, some of which were also successfully introduced. In addition, a nematode causing female sterility, discovered to have been introduced along with *S. noctilio*, was effectively propagated and disseminated. To date, both biological control agents and silvicultural methods have caused a significant decrease in *S. noctilio* populations so that it is now rare in plantations and considered to be a minor pest.

13.1 Introduction

New Zealand was the first country to record *Sirex noctilio* outside of its natural range and the first country to initiate biological control measures against it. The role of silviculture with respect to tree mortality was recognized early on, as was the fact that in some stands the mortality of (usually) suppressed trees was in fact beneficial. The sterilizing nematode, *Deladenus* (=*Beddingia*) *siricidicola*, was also first found in New Zealand (Chap. 9).

This chapter outlines the history of *S. noctilio* in New Zealand and summarizes the biological control efforts here using insect parasitoids and the nematode. The current and past pest status of the woodwasp is also discussed.

J. Bain (✉) • S.L. Sopow • L.S. Bulman
Scion (New Zealand Forest Research Institute Ltd), Private Bag 3020, Rotorua, New Zealand
e-mail: john.bain@scionresearch.com

B. Slippers et al. (eds.), *The Sirex Woodwasp and its Fungal Symbiont: Research and Management of a Worldwide Invasive Pest*, DOI 10.1007/978-94-007-1960-6_13,
© Springer Science+Business Media B.V. 2012

13.2 History of Sirex in New Zealand

Sirex noctilio is one of the few insects in New Zealand that is capable of attacking live *Pinus radiata*. The first published record of *Sirex noctilio* in New Zealand (as *Sirex juvencus*) is Miller (1919) but this publication gives no details of where and when. This information is given in Miller and Clark (1935) who state that the first specimen was collected in Wairarapa in 1900. *Sirex noctilio* did not attain "pest status" until the late 1920s when deaths of pines in overcrowded plantations were attributed to it (Clark 1927; Miller and Clark 1935; Nuttall 1989). Between 1946 and 1951, serious outbreaks were associated with extreme drought conditions and overstocked stands (Jackson 1955; Rawlings and Wilson 1949). In the central North Island 33% of trees were killed over an area of about 120,000 ha (Morgan and Stewart 1966; Rawlings and Wilson 1949). These stands were unthinned, overstocked and had suffered a series of unusually dry summers. Since the early 1950s scattered outbreaks have occurred on both islands mostly where trees were under stress from overcrowding, drought, severe pruning or damage caused by logging activities.

Many of those overstocked stands had about 1,500 stems/ha (Elliott 1976). Today initial stocking rates do not exceed 1,000 stems/ha and can be as low as 600 stems/ha with final stocking rates of about 200–350 stems/ha (Maclaren and Knowles 2005). The Ministry of Agriculture and Forestry's national exotic forest description data (NEFD 1998 and NEFD 2010; http://www.maf.govt.nz/mafnet/publications/nefd.html) state the area of *P. radiata* in New Zealand at 1.5 and 1.6 million ha in 1997 and 2009, respectively. However, the percentage of forest pruned decreased from 67% to 60% over that period. The decrease in area of pruned forest is mainly due to growers not receiving sufficient premium on log price to justify the added expense of pruning.

Sirex noctilio is now not considered to be a major pest in New Zealand (Bain 2005). Control has been achieved by a combination of forest management practices and the presence of natural enemies. The latter include the purposely introduced parasitoids *Rhyssa persuasoria persuasoria* (Linnaeus), *Megarhyssa nortoni nortoni* (Cresson) and *Ibalia leucospoides leucospoides* (Hochenwarth), as well as the nematode *Deladenus siricidicola* Bedding (Neotylenchidae), which was introduced along with *S. noctilio*.

13.3 Use of Biological Control Agents

In 1928/1929 and 1931, *R. p. persuasoria* was deliberately introduced from Europe as a control measure (Table 13.1). It was first recovered from the field in 1936 and is well established in nearly all exotic pine plantations (Nuttall 1989). In 1956, *Rhyssa lineolata* (Kirby), a Nearctic species, was found in New Zealand and probably entered the country in timber cut from *Sirex* infected trees; the host siricid could not have been *S. noctilio* because this species does not occur in the Nearctic. *Rhyssa lineolata* is now found in several exotic forests (Nuttall 1974). Between 1963 and 1965 *Rhyssa persuasoria himalayensis* Wilkinson was imported

Table 13.1 Insect parasitoids imported for the control of *Sirex noctilio*

Year	Agent	Origin	Released	Established
1928 1929 1931	*Rhyssa persuasoria persuasoria*	Europe	Yes	Yes
1950–1951	*Ibalia leucospoides leucospoides*	England	Yes	Yes
1966–1968	*Ibalia leucospoides leucospoides*	Australia (original source California)	Yes	Yes
1962–1964	*Megarhyssa nortoni nortoni*	USA	Yes	Yes
1963–1965	*Rhyssa persuasoria himalayensis*	Pakistan and India	Yes	No
1963 1966–1968	*Ibalia leucospoides ensiger*	California and Tasmania (original source USA)	Yes	Yes
1964	*Schlettererius cinctipes*	California	No	
1964	*R. persuasoria persuasoria* *R. alaskensis* *Pseudorhyssa maculicoxis*	California	No	
1965	*Megarhyssa emarginatoria*	Europe	No	
1967–1968	*Rhyssa amoena*	Europe	No	
1966	*Megarhyssa praecellens*	Japan	No	
1968	*Rhyssa crevieri*	Tasmania (original source Canada)	No	

from Pakistan and India. Four field releases were made but this species has never been recovered from the field (Nuttall 1989).

In 1964, two consignments of '*Rhyssa*' spp. were received from California. These were later identified as a mixture of male *R. persuasoria persuasoria* and female *R. alaskensis* Ashmead and male and female *Pseudorhyssa maculicoxis* (Kreichbaumer). This latter genus is cleptoparasitic on sircids by first destroying the rhyssine larva already present on the host (Spradbery 1969) and would have been a unwelcome introduction (Nuttall 1989). No oviposition was seen in the insectary and no progeny emerged. Between 1967 and 1968 three consignments of *Rhyssa amoena* Gravenhorst were received from Europe. Only a few progeny were reared and no liberations were made (Nuttall 1989). *Rhyssa crevieri* (Provancher) were received from Tasmania in 1968. These were the progeny of an importation from Canada. Only a few progeny were reared and no liberations were made (Nuttall 1989).

Ibalia l. leucospoides was introduced in 1950 and 1951 from England (Table 13.1) and again in 1966–1968 from Australia (the original source was California). It was first recovered in 1957 and is well established throughout the country (Nuttall 1989). In 1963, *Ibalia leucospoides ensiger* Norton was introduced from California. Oviposition in the insectary was not seen and no progeny were obtained (Nuttall 1989). Between 1966 and 1968 more *I. l. ensiger* were received from Tasmania; these were from consignments sent there from the USA in 1963. Recoveries were made from 6 of 14 forests where they were released. Because of interbreeding with *I. l. leucospoides* it has not been recognised in collections since 1976 (Nuttall 1989).

Megarhyssa n. nortoni was introduced from the USA between 1962 and 1964 (Table 13.1) and was first recovered from the field in 1968; it is now widespread. This complex of introduced parasitoids may kill over 70% of *Sirex* larvae in particular forest areas and in some localities *M. n. nortoni* has achieved parasitism rates of about 90% (Nuttall 1989). *Megarhyssa emarginatoria* (Thunberg) was imported from Europe in 1965 but oviposition was only observed once in the insectary and no progeny were obtained (Nuttall 1989). In 1966, *Megarhyssa praecellens* (Tosquinet) were received from Japan but very few progeny were reared and no liberations were made (Nuttall 1989). The failure in rearing *M. emarginatoria, M. praecellens, R. amoena* and *R. crevieri* may have been because the natural hosts of these rhyssines are present in genera other than *Pinus* (Taylor 1967a).

In 1964, *Schlettererius cinctipes* (Cresson) (Stephanidae) was imported from California (Table 13.1) but no insectary oviposition was seen and no progeny obtained (Nuttall 1989). This species oviposits satisfactorily on *S. noctilio* (Taylor 1967b) and its failure to do so here is unexplained.

In 1962, *Deladenus siricidicola* Bedding, a nematode associated with Sirex woodwasp, was first found in New Zealand (Zondag 1969). It has subsequently been found in Europe and must have come from there in *Sirex* infested timber. The nematodes feed on the *Amylostereum* fungus in Sirex-infested wood and when they come close to a Sirex larva they change into an "infective form" and penetrate the integument of the larva. When an infected female larva pupates, the nematodes migrate to the ovaries and penetrate the eggs. This renders the female sterile. The nematodes also enter the testes of the male Sirex but this infection does not result in sterility.

The initial discovery of the nematode was in adult female Sirex woodwasps from Rotoehu State Forest in the North Island. Subsequent surveys revealed its presence in nearly all North Island pine forests but it was not found in the South Island. In February-March 1967, the nematode was introduced into three South Island pine plantations with approximately 20 trees inoculated in each forest. In November 1967 sample logs were taken from the inoculated trees and the woodwasps that emerged were dissected to determine the levels of nematode infection. This ranged from 0% to 97%, with an average of 47% (Zondag 1971).

Between 1968 and 1974, the nematode was introduced into another 15 South Island pine plantations and nematode infection rates ranged from 0% to 100% (average 45%). Low infection rates are nearly always in forests with very low Sirex woodwasp populations (Zondag 1979), and in many forests infection rates of up to 90% are common (Nuttall 1980). The nematode has also been recovered from forests where it was not deliberately introduced (Zondag 1975, 1979).

13.4 Pest Status of *Sirex noctilio* in New Zealand 1980–2010

Sirex noctilio is now a rare insect in New Zealand's *Pinus radiata* plantations. Improved stand management and biological control agents have not only prevented the outbreaks seen in the late 1940s and early 1950s, but have reduced populations

Table 13.2 Criteria used to classify *Sirex noctilio* damage

Classification	Criteria
Debris	*S. noctilio* adults, larvae, or emergence holes seen in debris on the forest floor, windrows, thinnings, slash and logging waste, driftwood, etc.
Secondary	Attack secondary and a result of other factors such as over pruning, root rot infection, drought, suppression, nutrient deficiency and stress
Negligible	*S. noctilio* adults seen flying, parasitoids found, light and old damage seen
Light	Abortive or unsuccessful attack
Medium	Single or rare and isolated mortality in a forest stand
High-severe	Multiple mortality in a forest stand

to such a degree that Sirex woodwasp is now considered a minor pest in the forest. This observation is supported by data collected during the forest health surveillance scheme that is supported by all major forest owners in New Zealand. The scheme has two primary objectives. These are firstly to detect introduced pests sufficiently early for eradication to be feasible and secondly to assess and quantify existing forest health problems.

In order to achieve those objectives, expert forest health inspectors conduct aerial surveys, roadside surveys from a slow moving vehicle, and inspections on foot to follow up suspicious symptoms seen during the aerial or roadside surveys. Inspectors also carry out a number of random inspections on foot where wooden debris and living trees are examined for newly established insect pests or fungal pathogens. While carrying out these pest detection surveys, inspectors collect data on forest health disorders and pest status. Details such as the location, inspection date, host, disorder, causal agent, and the incidence and severity of the disorder are recorded and entered into a database. This database contains records from over 100,000 forest health inspections in *P. radiata* plantations, with the bulk of the records made from the 1980s onwards.

Observations of *S. noctilio* damage in New Zealand's plantation *P. radiata* forests were compiled by extracting records of *Sirex* or *Sirex*-related damage from 1980 to 2010. Each record was categorised based on the type and extent of the damage recorded. The criteria used are shown in Table 13.2.

Over the entire 1980–2010 period, just over 98,000 forest health inspections in *P. radiata* forests were made. Records of *S. noctilio* were extremely rare with only 974 records found. One-half of one percent of all *P. radiata* forest health inspections from 2000 to 2009 make mention of the insect. The total number of records has declined over the past three decades, and the number of reports of high to severe damage has declined markedly. In the 1980s there were 248 records of high to severe damage, from 2000 to 2009 the observations of such damage dwindled to 34. Since 2000, well over half the records made were of secondary attack, insects found in forest debris such as thinning slash or stumps, or of negligible damage compared with one third of the records made in the 1980s (Table 13.3).

There were 144 *S. noctilio* records from 2000 to 2009 where age of the forest stand was recorded (records from debris were excluded). *Sirex noctilio* damage

Table 13.3 Number of *Sirex noctilio* records categorised by type of damage recorded

Significance of record	1980–1989	1990–1999	2000–2009	Grand Total
Found in forest debris	61	56	38	155
Secondary attack	78	85	81	244
Negligible	10	13	11	34
Light damage	19	15	28	62
Medium damage	30	17	19	66
High-severe damage	248	131	34	413
Grand total	446	317	211	974

Table 13.4 Number of *Sirex noctilio* records by stand age and damage type for the period 2000–2009

	Age-class				
Significance of record	1–10	11–19	20–24	25+	Grand total
Secondary attack	40	29	2	–	71
Negligible	8	2	–	–	10
Light damage	7	10	5	2	24
Medium damage	6	2	–	2	10
High-severe damage	13	14	2	–	29
Grand total	74	57	9	4	144

declined as tree age increased. Only 9 (6%) were made from trees near harvest age and 4 (3%) were from trees at harvest age 25 years or older. In New Zealand trees are harvested at about 30 years. Of the 29 records of high to severe damage where age was recorded, only 2 (7%) were near harvest and none were at harvest age (Table 13.4).

13.5 Conclusions

Sirex noctilio was first found in New Zealand in 1900. Mortality in pine plantations was first noted in the 1920s and this led to the introduction of the first of several insect parasitoids. A nematode that sterilizes *S. noctilio* females was discovered in 1962 and this was cultured and introduced into plantations where it did not occur. A combination of the parasitoids, the nematode and better silviculture lead to excellent control of *S. noctilio* and today it is regarded as a minor pest. There is no reason to believe that this situation will change.

Acknowledgments Many people have made important contributions to the successful control of the Sirex woodwasp in New Zealand but special mention must be made of the late Rudy Zondag and late Mike Nuttall. They conducted the lion's share of the work on the sterilizing nematode and the insect parasitoids.

References

Bain J (2005) *Sirex noctilio* (Hymenoptera: Siricidae) – the New Zealand experience. In: Gottschalk KW (ed) Proceedings, 16th US Department of Agriculture interagency research forum on gypsy moth and other invasive species 2005, 18–21 Jan 2005; Annapolis, MD. General Technical Report NE-337. US Department of Agriculture, Forest Service, Northeastern Research Station, Newtown Square, PA

Clark AF (1927) The infestation of *Sirex juvencus* in Canterbury. Te Kura Ngahere 2:10–16

Elliott DA (1976) The influence of disease and insect problems on management practice in Kaingaroa forest. New Zealand J For Sci 6:188–192

Jackson DS (1955) The *Pinus radiata/Sirex noctilio* relationship at Rotoehu forest. New Zealand J For 7:26–40

Maclaren JP, Knowles RL (2005) Silviculture of radiata pine. In: Colley M (ed) Forestry handbook. New Zealand Institute of Forestry Inc, New Zealand

Miller D (1919) The economic bearing of hymenopterous insects. New Zealand J Agric 19:201–208

Miller D, Clark AF (1935) *Sirex noctilio* (Hym.) and its parasite in New Zealand. Bull Entomol Res 26:149–154

Ministry of Agriculture and Forestry (1998) National Exotic Forest Description as at 1 April 1997. Ministry of Agriculture and Forestry, Wellington, New Zealand

Ministry of Agriculture and Forestry (2010) National Exotic Forest Description as at 1 April 2009. Ministry of Agriculture and Forestry, Wellington, New Zealand

Morgan FD, Stewart NC (1966) The biology and behaviour of the woodwasp *Sirex noctilio* F. in New Zealand. T Roy Soc New Zealand 7:195–204

Nuttall MJ (1974) *Rhyssa lineolata* (Hymenoptera: Ichneumonidae), as a parasite of *Sirex noctilio* in New Zealand. New Zealand J For Sci 4:487–494

Nuttall MJ (1980) *Deladenus siricidicola* Bedding (Nematoda: Neotylenchidae). Nematode Parasite of Sirex. Forest and timber insects in New Zealand 48. New Zealand Forest Service, Wellington. 8 pp

Nuttall MJ (1989) *Sirex noctilio* F., Sirex wood wasp (Hymenoptera: Siricidae). In: Cameron PJ, Hill RL, Bain J, Thomas WP (eds) Review of biological control of invertebrate pests and weeds in New Zealand 1874–1986. CAB International Institute of Biological Control, Ascot, Technical Communication No. 10

Rawlings GB, Wilson NM (1949) *Sirex noctilio* as a beneficial and destructive insect to *Pinus radiata* in New Zealand. New Zealand J For 6:2–11

Spradbery JP (1969) The biology of *Pseudorhyssa sternata* Merrill (Hym. Ichneumonidae), a cleptoparasite of siricid woodwasps. Bull Entomol Res 59:291–297

Taylor KL (1967a) The introduction, culture, liberation, and recovery of the parasites of *Sirex noctilio* in Tasmania, 1962-67. Division of Entomology Technical Paper 8. Commonwealth Scientific and Industrial Research Organisation, Australia. 19 pp

Taylor KL (1967b) Parasitism of *Sirex noctilio* F. by *Schlettererius cinctipes* (Cresson) (Hymenoptera: Stephanidae). J Australian Ent Soc 6:13–19

Zondag R (1969) A nematode infection of *Sirex noctilio* in New Zealand. New Zealand J Sci 12:732–747

Zondag R (1971) Control of *Sirex noctilio* F. with *Deladenus siricidicola* Bedding. Part I – 1967 field trial. New Zealand J For Sci 1:5–14

Zondag R (1975) Controlling Sirex with a nematode. Proceedings 28th New Zealand weed and pest control conference Hastings, pp 196–199

Zondag R (1979) Control of *Sirex noctilio* F. with *Deladenus siricidicola.* Bedding. Part II. Introductions and establishments in the South Island 1968–1975. New Zealand J For Sci 9:68–76

References

Chapter 14
Sirex Woodwasp in Australia: Current Management Strategies, Research and Emerging Issues

Angus J. Carnegie and Richard Bashford

Abstract It has been almost 60 years since *Sirex noctilio* (Sirex) was first detected in Australia, and it has now spread to all pine-growing states except Western Australia. Major outbreaks occurred soon after detection in both Tasmania and Victoria. However, in the Green Triangle region of western Victoria/south-eastern South Australia, only low numbers of Sirex were initially found. This led to reduced surveillance, which ultimately resulted in the largest recorded outbreak of Sirex in Australia, killing over five million trees between 1986 and 1989. Based on this outbreak, it was predicted that Sirex damage could result in losses of up to AU$60 million per annum in the absence of control. This prediction resulted in the development of the National Strategy for Control of *Sirex noctilio* in Australia. There have been only a few, localised outbreaks of Sirex since, although several potential problems with the Sirex control strategy have recently been identified. In this Chapter we provide a synopsis of the Sirex Control Strategy in light of new developments and procedures that compliment the original strategy. We also describe current research activities and emerging issues, including reduced parasitism of nematode-inoculated trap trees; problems associated with *Ips grandicollis* attacking Sirex trap tree plots; determining whether temperature is limiting the northern spread of *S. noctilio* and *Amylostereum areolatum,* and its effect on the nematode, *Deladenus siricidicola*; and variation in nematode infectivity across south-eastern Australia.

A.J. Carnegie (✉)
Forest Science Centre, NSW Department of Primary Industries, P.O. Box 100, Beecroft, NSW 2119, Australia
e-mail: angus.carnegie@industry.nsw.gov.au

R. Bashford
Forestry Tasmania, GPO BOX 207, Hobart, TAS 7001, Australia

B. Slippers et al. (eds.), *The Sirex Woodwasp and its Fungal Symbiont: Research and Management of a Worldwide Invasive Pest*, DOI 10.1007/978-94-007-1960-6_14,

© Springer Science+Business Media B.V. 2012

14.1 Introduction

Sirex noctilio (Sirex) has been in Australia for six decades. It was first detected in 1952 in Tasmania (Gilbert and Miller 1952), although it may have been introduced several years earlier in softwood packing cases from New Zealand (Madden 1975). In 1961, it was detected in Victoria, in a farm woodlot outside of Melbourne (Irvine 1962), but it is likely to have been introduced several years earlier via imported pine timber (Eldridge and Simpson 1987; Eldridge and Taylor 1989). *Sirex noctilio* moved slowly north, east and west, reaching all pine growing regions in Victoria in less than two decades (Neumann et al. 1987; Collett and Elms 2009). It reached South Australia and southern New South Wales (NSW) in 1980 (Eldridge and Taylor 1989). Sirex is now found throughout South Australia and most of the pine growing regions in NSW, except coastal subtropical areas (Carnegie et al. 2005). It was detected for the first time in Queensland in February 2009 (National Sirex Coordination Committee, unpublished), in *Pinus radiata* plantations just over the border from known areas of infestation in NSW. It has not yet reached Western Australia.

The discovery of an established population of *S. noctilio* in Victoria resulted in the establishment of the National Sirex Trust Fund in 1962, funded on a 1:1 basis by Commonwealth and State Governments, with State contributions on a proportional basis according to area of *Pinus* plantations (Eldridge and Simpson 1987). The Fund provided resources for survey and eradication of Sirex, and later for research on control and management strategies (Haugen et al. 1990).

The National Sirex Trust Fund, and Commonwealth funding, concluded in 1977–1978, but culture and supply of the biological control agents to all States was continued via the then Australian Forestry Council (Haugen et al. 1990), which was funded by State Government agencies and private growers. The Green Triangle outbreak in the late 1980s, where over five million trees were killed (Haugen et al. 1990), highlighted the need for a national Sirex control strategy and led to the establishment of the National Sirex Coordination Committee (NSCC) to oversee and scrutinize Sirex management at a state and national level (Haugen et al. 1990). The *National Strategy for Control of Sirex noctilio in Australia* (Haugen et al. 1990) was developed to provide forest managers and forest health practitioners with a framework within which to construct a control program for their organisations, and included techniques for detecting and monitoring Sirex populations, introducing the nematode *Deladenus siricidicola* into the Sirex population, and monitoring the efficacy of biological control agents.

Several recent reviews have covered the history of Sirex in south-eastern Australia in detail, including spread, impact and biological control (Bedding 2009; Carnegie et al. 2005; Collett and Elms 2009). This chapter provides an overview of current Sirex management in Australia highlighting advances in knowledge and describing new procedures that now supplement the original strategy. It also details the history of the national biological control program, and summarizes current research and emerging issues affecting Sirex control in Australia.

14.2 Control and Management

14.2.1 Australian Sirex Biological Control Unit

In response to the discovery of an established population of Sirex in Victoria, the Sirex Biological Control Unit (Fig. 14.1) was established in 1963 at Silwood Park, United Kingdom, by the Commonwealth Scientific and Industrial Research Organisation (CSIRO), Division of Entomology. The primary aim of the Unit was to collect and identify Sirex parasitoids of potential use in Australia. Two CSIRO officers, Frank Wilson and Ted Woolcock, were sent from Australia to establish the station, which was initially housed at the Australian Scientific Liaison Office in Holborn, London, before being moved to a purpose-built laboratory on the grounds of the Imperial College Field Station (University of London) at Silwood Park. A complex of insect rearing cages was built nearby where Sirex-infested logs (collected throughout Europe) would be stored, under quarantine, to monitor for emergence. Phillip Spradbery was appointed as Research Scientist to collect Sirex

Fig. 14.1 Sirex biological control unit staff – 1966. L to R: Bill Bowyer, Ted Woolcock, Robin Bedding, Secretary, Phillip Spradbery, Dick Bashford, Doug Waterhouse, Frank Wilson

material from the coniferous forests of most of the countries of Europe and North Africa and to research the biology of Sirex parasitoids. Robin Bedding was appointed in 1965 to lead studies on newly discovered nematode parasites of siricids, including *Deladenus siricidicola*.

In 1962, a laboratory was constructed by CSIRO in Hobart, Tasmania, Australia, and staffed by Keith Taylor and later, in 1966, by John Madden. This lab served as a quarantine station to receive parasitoids from Silwood Park. The parasitoids were mass-reared and field-tested in Tasmania to find suitable species for general Australian release. Other officers involved in the scientific program included A. A. Kirk (1967) at Silwood Park, with J. Varley, R. F. Simpson, H. Hocking, R. Ackhurst and D. Bashford in Hobart. Bedding relocated to Hobart in 1969 to conduct nematode releases; following the closure of Silwood Park in 1971, Spradbery also relocated to Hobart to continue research into Sirex. Researchers G. A. Kile, M. P. Coults, J. E. Dolezal and J. F. Titze also conducted research on Sirex/tree physiological relationships at the Forestry and Timber Bureau in Tasmania. Mass rearing of siricid parasitoids, sent from Hobart for release on the mainland, was undertaken at the Waite Agricultural Research Institute, Adelaide, South Australia. The Hobart laboratory closed in 1980 following the retirement of Keith Taylor.

The research output during the first decade of the CSIRO program resulted in the publication of 36 papers and research reports (CSIRO 1965–1972). Seminal papers such as Spradbery (1970a, b), Bedding (1967, 1972), Madden (1971), and Taylor (1967) provided the basis for subsequent research programs. In the following two decades these researchers and colleagues produced a further 44 papers and reports relating to Sirex in Australia. The three decades of research and development of the biological control program culminated in the production of the *National Strategy for Control of Sirex noctilio in Australia* in 1990 (Haugen et al. 1990).

14.2.2 Current Operational Management

Sirex management in Australia since 1990 has closely followed the *National Strategy for Control of Sirex noctilio in Australia* with its 12 recommendations (Haugen et al. 1990). Below is a synopsis of these recommendations, including current operational procedures.

14.2.2.1 National Sirex Coordination

The National Sirex Trust Fund (now the National Sirex Fund, NSF) was administered by the National Sirex Fund Committee to coordinate research activities and to organize an eradication programme ('search and destroy') in Victoria. A Research sub-committee of the NSF visited New Zealand in 1962 on a fact-finding mission to examine the success of the establishment of Sirex parasitoids that had previously been released there. Both *Rhyssa persuasoria* (released in 1928–1931) and *Ibalia leucospoides* (released in 1950) were found to have successfully established (Zondag 1969; Nuttall 1989; Bain 2005).

Over a 10 year period (1961–1972) a total of AU$2,258,780 was spent researching control methods, including funding the Sirex Biological Control units in England and Hobart, conducting sawmill surveys and attempting to eradicate Sirex in Victoria and Tasmania. Although Commonwealth funding contributions to the NSF ceased in 1976, the Hobart breeding facility continued to be funded by Commonwealth funds until 1980. The NSF was administered through the Keith Turnbull Research Institute (KTRI) in Victoria from 1976 until 1990. With the closure of that institution, the National Sirex Fund Committee became the National Sirex Coordination Committee (NSCC). The NSF was then funded solely by contributions from State and private softwood growers, via an AU$0.16/ha levy (Insectary maintenance levy $0.06, research levy $0.10) (Anonymous 1990). Following closure of the KTRI Insectary in 1991, the levy was reduced to $0.11/ha (which remains the current rate).

The NSCC continues to oversee and manage the Sirex control program at a state and national level and to conduct and fund research on Sirex and its control. The committee consists of members from each State forest service, including internationally recognised experts in the field of Sirex and Sirex management, and industry and private grower representatives. The NSCC annually reviews Sirex management, and updates the *National Sirex Control Strategy Worksheets* (NSCC 2002, 2007) where appropriate.

The NSCC also collaborates with international colleagues on Sirex control. For example, parasitoids (*I. leucospoides*) have been sent to South Africa over the past few years to supplement South Africa's biological control program, and *Megarhyssa nortoni* and *Rhyssa persuasoria* were sent to Brazil for several years. Sirex larvae and adults, preserved in ethanol, were sent from Tasmania to Nathan Schiff (USDA Forest Service) for DNA sequencing. This identification assisted in developing baseline data for a molecular tool to identify larvae to species level in quarantine detections, and also to differentiate between introduced and native woodwasps in trees killed in the field. Members of the NSCC are also currently members of the USDA Sirex Technical Working Group (formerly Science Advisory Panel) (Bedding et al. 2006).

A main aim of the NSCC are to maintain supplies of the nematode bio-control agent by cryogenically preserving strains of the nematode (*D. siricidicola*) for research comparisons and the Kamona strain for commercial release. In the late 1980s, it was recognised that the original nematode strain had lost its infectivity and was producing very low parasitism of inoculated trees: less than 25% compared to the optimum of greater than 90% (Bedding 1992; Chap. 9). This strain was later dubbed the "defective" strain. A new strain was collected, known as the Kamona strain, which had the originally high infectivity, and has since been used for all subsequent inoculations. The NSCC also funds research to improve Sirex control techniques, annually evaluates the control program to ensure effectiveness of bio-control agents released, monitors the spread of Sirex, and provides information to growers in the form of printed information sheets, popular journal articles, and scientific papers.

14.2.2.2 Sirex Awareness

The NSCC provides and regularly updates information on Sirex detection and management within state forest agencies and private companies. Publications on Sirex are routinely distributed to relevant staff in forestry companies and state agencies, including updated versions of the *National Sirex Control Strategy Operations Worksheets* (NSCC 2002, 2007) and the booklet *Sirex noctilio: The pine-killing woodwasp – Detection, biology and control* (NSCC 2001). A compact disc containing these documents, workshop PowerPoint presentations, and other relevant material is in preparation for release to NSF levy payers in late 2011. Workshops on Sirex detection and management are also conducted by members of the NSCC.

14.2.2.3 Quarantine

Haugen et al. (1990) recommended quarantine measures to restrict the movement of infested logs into uninfested regions. Initially, Sirex was declared a "prohibited pest", and quarantine legislation was enacted to restrict the movement of whole logs between states (Neumann 1979). These quarantine restrictions were removed once Sirex reached – via natural spread – South Australia and New South Wales. Before Sirex was detected in Queensland (again, through natural spread), that state had quarantine measures in place to reduce the chance of Sirex becoming established (Carnegie et al. 2006). Due to the large separation of plantations in Western Australia from those in the eastern states (~2,000 km), the most likely means of Sirex colonising these plantations is from human-assisted transport of infested wood (Carnegie et al. 2006), but there are currently no quarantine restrictions on the import of pine logs into Western Australia. There are also currently no intra-state quarantine restrictions in Queensland or NSW.

14.2.2.4 Sirex Detection

Early detection of Sirex, through surveillance and trap tree plots, provides an opportunity for early implementation of control measures to reduce the chance of Sirex reaching damaging levels (Haugen et al. 1990). Since 1990, detection of the spread of Sirex in Australia has mainly relied on either general forest health surveillance or trap tree plots (Carnegie et al. 2005; Bashford 2008). Aerial and ground surveys of the majority of pine plantations in Australia have been conducted annually since 1996/1997 (Carnegie 2008a; Carnegie et al. 2008; Wotherspoon 2008; Phillips 2008).

Haugen et al. (1990) recommended that "forestry, logging, and sawmill personnel" be trained in Sirex detection, but the New Zealand and Australian experience has shown that *ad hoc* surveys by non-specialists are ineffective in detecting health problems compared with specifically trained personnel (Carter 1989; Bulman et al. 1999; Wardlaw 2008). Operational forestry personnel are often inordinately busy conducting their day-to-day tasks to also look for Sirex. In the early 1990s, surveys to detect Sirex spread were conducted mainly by state government entomologists. Since the mid-1990s these surveys have been conducted by dedicated forest health survey officers (Carnegie et al. 2005). It is recommended that trained and dedicated forest health personnel survey for Sirex, to increase the chance of its early detection.

Haugen et al. (1990) also recommended that trap trees be located in plantations if the nearest source of Sirex is less than 100 km away. Trap tree plots have been placed at the advancing front of Sirex spread in north-eastern NSW and more recently in southern Queensland. This method detected the movement of Sirex into the northern tablelands of NSW in 1997, and the then most northern spread in NSW (Glen Innes) in 2001 (Carnegie et al. 2005). Trap tree plots are established to detect the presence of Sirex in an area, provide an efficient means of introducing the biological control nematode and monitor population levels of Sirex and its biological control agents (Haugen et al. 1990; NSCC 2007). A group of ten trees are poisoned so they die slowly (Fig. 14.2a–c) and are attractive to Sirex females over the Sirex flight season. Nematodes are then inoculated into trees (Figs. 14.3a and b, 14.4 and 14.5a) that have been attacked by Sirex.

The detection of *S. noctilio* in Queensland in 2009 was made via the trapping methodology developed by Bashford (2008) (see Chap. 12), which uses static traps (Intercept® panel traps) containing pinene lures (70% ά-pinene, 30% β-pinene). This method is able to detect Sirex in pine plantations at lower levels than can be consistently detected via aerial surveys (Bashford 2008; Chap. 12). This system now replaces trap tree plots for the detection of Sirex spread into the subtropical coastal regions of NSW, to confirm establishment and further spread in Queensland, and to monitor in Western Australia, where Sirex has not yet been found. This method is also used extensively in South Africa, the United States and Canada (Evans et al. 2006; Pennsylvania Department of Agriculture 2007; Croft 2009). Sirex is usually not detected by aerial or ground surveillance or by trap trees until 2–3 years after it has become established in the area (e.g., Carnegie et al. 2005). However, the use of strategically placed insect traps is likely to detect Sirex 1 year after it reaches a new area.

Static traps can also be used for the early detection of Sirex in young plantations (Bashford 2008, Chap. 12). In Tasmania, traps are routinely placed in plantations aged 7–10 years-old, generally on second rotation sites. Traps are placed on the perimeter or in clearings within the plantation during the Sirex flight season, positioned at least 100 m apart (with a maximum of four traps per 100 ha compartment), and serviced every 3–4 weeks. Once Sirex has been detected in static traps, the surrounding plantation is surveyed for Sirex-struck trees and trap tree plots are established where appropriate for the introduction of the nematode.

14.2.2.5 Sirex Monitoring

Haugen et al. (1990) stressed the importance of knowing the levels of Sirex-associated tree mortality so that appropriate control measures can be planned should

Fig. 14.2 Trap tree establishment methodology: (**a**) holes are drilled at a 45° angle, 10 cm apart around the circumference of the tree, approximately 0.5–1 m above ground level; (**b**) herbicide is injected into drilled holes; (**c**) trees should die slowly over the Sirex flight season, as shown here – fading trees 1 months after poisoning

Fig. 14.3 Nematode inoculation of trap trees or naturally struck trees: (**a**) once the tree has been felled and branches cleared, an operator walks along the tree hammering holes with the Sirex rebound hammer; (**b**) once nematodes have been mixed in the acrylic gel they are applied to each hole from a sauce bottle

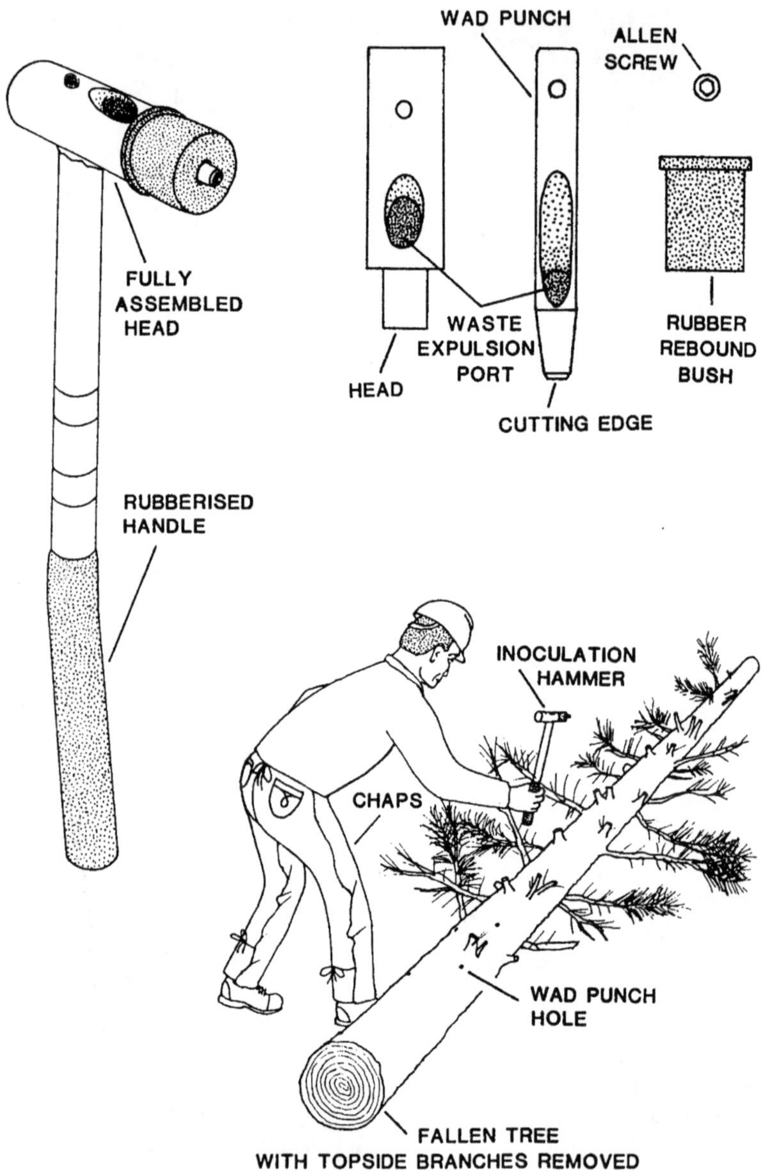

Fig. 14.4 Nematode inoculation rebound hammer and technique (Illustration courtesy of David Spolc)

these levels escalate. However, they highlighted that options for monitoring Sirex populations, including repeatable transects, trap tree plots, aerial surveys in fixed-wing aircraft, and aerial photography, were either expensive, inaccurate or required further investigation and validation. They reported that aerial surveys from fixed-wing

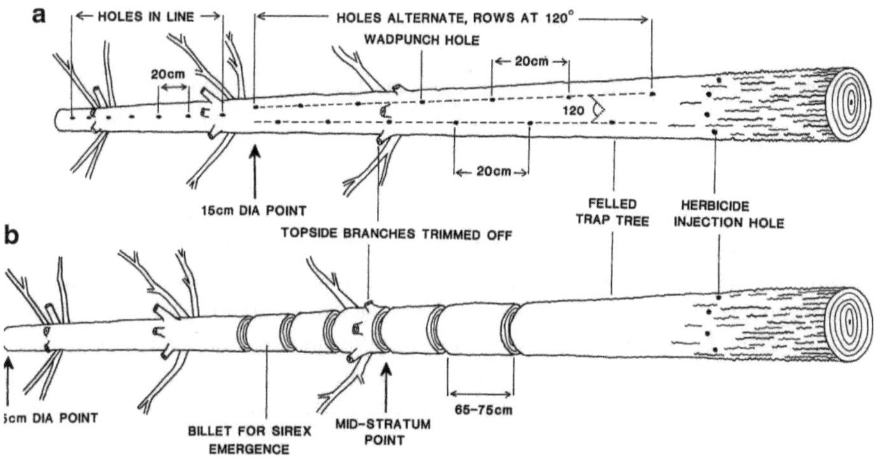

Fig. 14.5 (**a**) Location and distribution of holes along tree stem for nematode inoculation; (**b**) recommended area for cutting billets to then place in cages to monitor emergence and parasitism levels (Illustration courtesy of David Spolc)

aircraft lack accuracy in estimating tree mortality, especially in quantifying tree mortality below 5%. Subsequent research indicates that aerial surveys in a helicopter, flying at 100–300 m.a.g.l. (metres above ground level) and at 60–80 knots, are sensitive enough to detect levels of individual dead and dying co-dominant/dominant trees below 1% (NSCC 2002, 2007; Carnegie et al. 2008; Wotherspoon 2008). However, due to the high cost of helicopter flights, fixed-wing aircraft are also used for forest health surveys in some states (Carnegie 2008a; Phillips 2008). Follow-up ground surveys are conducted to confirm the presence of Sirex or of other agents (Eldridge and Taylor 1989). Used as part of an overall forest health surveillance program (i.e. surveys for all damaging agents, not only Sirex), aerial surveys become a cost-effective procedure to detect and map Sirex infestation (Carnegie et al. 2008). These methods are employed in most states of Australia to detect and map the extent and severity of Sirex-associated tree mortality (NSCC 2002, 2007).

Advances in remote sensing technologies have great potential to improve the accuracy of the location and incidence of dead trees within a pine plantation (Stone et al. 2008). Research in South Africa indicates that high spatial resolution imagery is able to relatively accurately detect Sirex infestation in a pine stand (Ismail et al. 2007; Dye et al. 2008). Such imagery can also provide information on net stocked area, roads, drainage, etc. This technology (digital multispectral imagery) is currently being tested for detection of forest health issues, including Sirex, in Queensland (Forestry Plantations Queensland, unpublished), although further research is required to make this technique more efficient than manned aerial surveillance. However, the current cost of obtaining and processing remote sensing imagery/data is more expensive than aerial surveillance methods. Both methods still require ground truthing.

14.2.2.6 Silvicultural Control

Because healthy, vigorously growing stands are less susceptible to Sirex, the key to silvicultural control is to ensure "on-time" thinning (Eldridge and Taylor 1989; Haugen et al. 1990). Although this is a prescription for Sirex management in Australia, it is not routinely practiced as a "Sirex-management" tool. Thinning operations are scheduled based mainly on "forestry" operational prescriptions and limitations and are generally market driven. In some cases, however, forest health is taken into account when developing thinning schedules. Delayed thinning still occurs though, due mainly to market forces and lack of plantation access (e.g., during wet weather), but not necessarily because of poor management practices. Haugen et al. (1990) also recommended that non-commercial thinning be considered if markets are not available for on-time thinning, but non-commercial thinning is rare in Australian pine plantations.

There have been few outbreaks of Sirex in Australia since the early events in Tasmania (Madden 1975), Victoria (Neumann and Minko 1981) and the Green Triangle (Haugen 1990; Haugen and Underwood 1990). Carnegie et al. (2005) reported sporadic and short-lived outbreaks in NSW since 1980; none have been reported in Victoria since the late 1980s (Collett and Elms 2009) and no significant outbreaks have been reported for South Australia in the past two decades (NSCC unpublished). This is a result of effective implementation of the national Sirex strategy, focusing on biological control, surveillance and silviculture.

There have, however, been several minor outbreaks in NSW in recent years (2006–2009) in southern NSW; one outbreak expanded from ~100 ha in 2006 to ~2,000 ha in 2009 (Carnegie unpublished). Although this does not appear to be a large area in comparison to the area of pine plantation in the region (~130,000 ha), the levels of damage within this area (up to 10%) mean this is one of the most significant outbreaks in Australia in the past two decades. These outbreaks were in unthinned stands, but in general these stands were not overdue for thinning (most were between 10 and 16 years old, with thinning generally carried out at 15–16 years old on such second-rotation sites). These areas have subsequently been targeted with extra trap tree plots and inoculation of naturally struck trees, as well as thinning. Affected stands that have since been thinned, now have low to negligible levels of Sirex.

Several of these stands were, however, overstocked, due to uncontrolled wildlings emerging during the second rotation establishment phase. In such cases, it is often difficult to detect killed suppressed or co-dominant trees during aerial surveys as these trees tend to sit below the canopy. Therefore, there is potential for an outbreak to go undetected for some years. It is recommended that overstocked stands be targeted for Sirex management via trap trees, static traps, or thinning, particularly if wildling control in later rotation plantings is not possible.

Since the outbreaks reported by Madden (1975) there have been only isolated outbreaks in Tasmania, mainly in privately managed stands. These include an outbreak in 1985 in northern central Tasmania, where a 400 ha area had 5% tree mortality; an unthinned compartment on Flinders Island, where 3% of trees died in 1990 and 10% of trees in 2001; in 2002, approximately 40% of trees were killed in a pruned

Fig. 14.6 Sirex killed trees in unthinned plantation in northern Tasmania detected during aerial surveys in 2007

and thinned 200 ha stand in southern Tasmania; and in 2007, a localised outbreak in a private, unmanaged, overstocked plantation in north-western Tasmania where 45% of trees were killed (Fig. 14.6) was identified during aerial surveys. This finding has necessitated control operations in adjacent plantations, including the use of static traps and trap tree plots.

Sustained drought can predispose whole plantations to Sirex infestation (Eldridge and Taylor 1989). Prolonged drought in many pine-growing regions in south-eastern Australia over the past decade has resulted in many stressed trees, and subsequent increased susceptibility to Sirex (a likely factor in the outbreaks discussed above). Serious outbreaks of Sirex in New Zealand in the late 1940s were associated with severe drought conditions and overstocked plantations (Rawlings 1948, 1955; Bain 2005; Chap. 13). With climate change, increased incidence of drought (Hennessy et al. 2006) may warrant a change in thinning regimes to reduce tree mortality and increased susceptibility to Sirex associated with water stress.

14.2.2.7 Culture and Supply of Biological Control Agents

Haugen et al. (1990) recommended a nationally coordinated program to produce and supply the biological control agents. They also recommend that duplicate cultures of nematodes be maintained. Initial mass production of the nematode was conducted at Keith Turnbull Research Institute (KTRI) in Victoria. Production was taken over by CSIRO in Canberra in 1991 when KTRI closed, and by a private company, Ecogrow Australia Pty Ltd (now Ecogrow Environment Pty Ltd), in 2000. Contributors to the National Sirex Fund received a discount on the price of nematodes (currently AU$295 per million). Ecogrow Environment Pty Ltd has also supplied nematodes overseas (e.g., to South Africa). Currently, cultures of Kamona strain are kept at Ecogrow, with secure duplicate cultures held by CSIRO, Division of Entomology (Canberra), and also by the Department of Forest and Ecosystem Science, The University of Melbourne.

A large number (21) of Sirex parasitoids were brought into Australia from many countries, including from Europe, India, Japan and North America, and reared to establish cultures in quarantine (Taylor 1976, Chap. 8). *Ibalia leucospoides, M. nortoni* and *Rhyssa* spp. were the most successful in becoming established (Taylor 1967, 1976; Carnegie et al. 2005; Collett and Elms 2009). Large numbers were released in Tasmania, Victoria, South Australia and NSW during the 1970s and 1980s (Taylor 1976; Haugen 1990; Carnegie et al. 2005; Collett and Elms 2009), but there has been no formal rearing or releases since the late-1980s, following the conclusion of the wasp bio-control breeding program (Carnegie et al. 2005; Collett and Elms 2009). However, parasitic wasps emerging from billets collected during monitoring for Sirex are released back into the forest.

14.2.2.8 Nematode Introduction

The nematode *D. siricidicola* is the primary biological control agent for Sirex in Australia (Haugen et al. 1990; Carnegie et al. 2005; Collett and Elms 2009), and in most of the Southern Hemisphere (Hurley et al. 2007; Bedding 2009). The trap tree plot (Neumann et al. 1982) and nematode inoculation (Bedding and Akhurst 1974) technique discussed by Haugen et al. (1990) is still the main method of introducing the nematode into the Sirex population. Most inoculations are into trap trees, due to the lack of naturally struck trees when Sirex levels are low. However, there have been several instances in recent years where Sirex outbreaks have resulted in inoculation of naturally struck trees, and this is becoming more routine in South Australia.

Haugen et al. (1990) and the *National Sirex Control Strategy Operations Worksheets* (NSCC 2002, 2007) recommended a prescriptive approach for selection of trap tree plot density within a plantation: four plots/100 ha of susceptible stand (10–20 years-old, unthinned) once Sirex is detected, or two plots/100 ha when annual tree deaths are less than 1%. This prescription was appropriate when the original strategy was developed for higher Sirex populations. However, when Sirex numbers are generally low, this would lead to more trap tree plots being established than is necessary. Current selection of trap tree plot locality and density in Australia has changed from the prescriptive approach (a certain number of plots per 100 ha) to an adaptive approach with the use of (1) forest health survey and trap tree plot data on Sirex activity, (2) trap tree plot emergence data, and (3) plantation area/size/age and silvicultural history to determine the number and location of trap tree plots.

Expanding on this adaptive approach (Fig. 14.7), if Sirex has not been recorded in an area for two or more years, either in forest health surveys or from trap tree plots, it is not necessary to put a plot in that area. Similarly, if there is satisfactory nematode parasitism of Sirex from trap tree plots (as determined by dissection of Sirex emerging from billets), and moderate (1–3%) to high (>3%) numbers of Sirex-killed trees have not been identified during surveys, then it is not necessary to put plots in that area, as infected females will spread the nematodes. However, if moderate-to-high levels of Sirex are observed during surveys, or if large numbers of emerging wasps from trap trees have low parasitism, then trap tree plots are put in those areas. Furthermore, based on distances flown by nematode-infected females – up to 7 km (Bedding 1993) – plots can be spaced more efficiently: plots can be located up to 7 km distant from a plot that had good parasitism the previous year (Fig. 14.7). This current process is adaptive to local knowledge and needs, but relies on annual monitoring of Sirex via forest health surveys and emergence data from trap tree plots. Static trapping to detect the presence of Sirex in young stands is a useful tool that assists in the decision making process as to whether to proceed with surveys for suppressed struck trees and subsequent trap tree establishment (Bashford 2008, Chap. 12). Thus, the use of a combination of techniques reduces the overall number of plots that need to be established, reducing costs, without increasing the risk of Sirex outbreaks.

14.2.2.9 Parasitoid Releases

There has been no formal parasitoid rearing program since the closure of KTRI in the late 1980s. Currently, the only releases are wasps captured from billets collected either during forest health surveys or from the trap tree monitoring program. These are then released into areas within the State where they were collected (to supplement local populations), or have been sent overseas or to Queensland where Sirex has recently been detected. Most within-state re-locations have been with *M. nortoni*, as *I. leucospoides* appears to be able to disperse and colonize readily. Recent overseas re-locations (to South Africa) have been with *I. leucospoides*, because these are the most common species emerging from collected billets. In summer 2009/2010, both *M. nortoni* and *I. leucospoides* were released in Queensland (from NSW) for the first time.

14.2.2.10 Evaluation of Biological Control Agents

Haugen et al. (1990) recommended "evaluation of biological agents" but also "*continued* evaluation of biological control agents". Evaluation of the biological control agents is important to determine the establishment, distribution and population

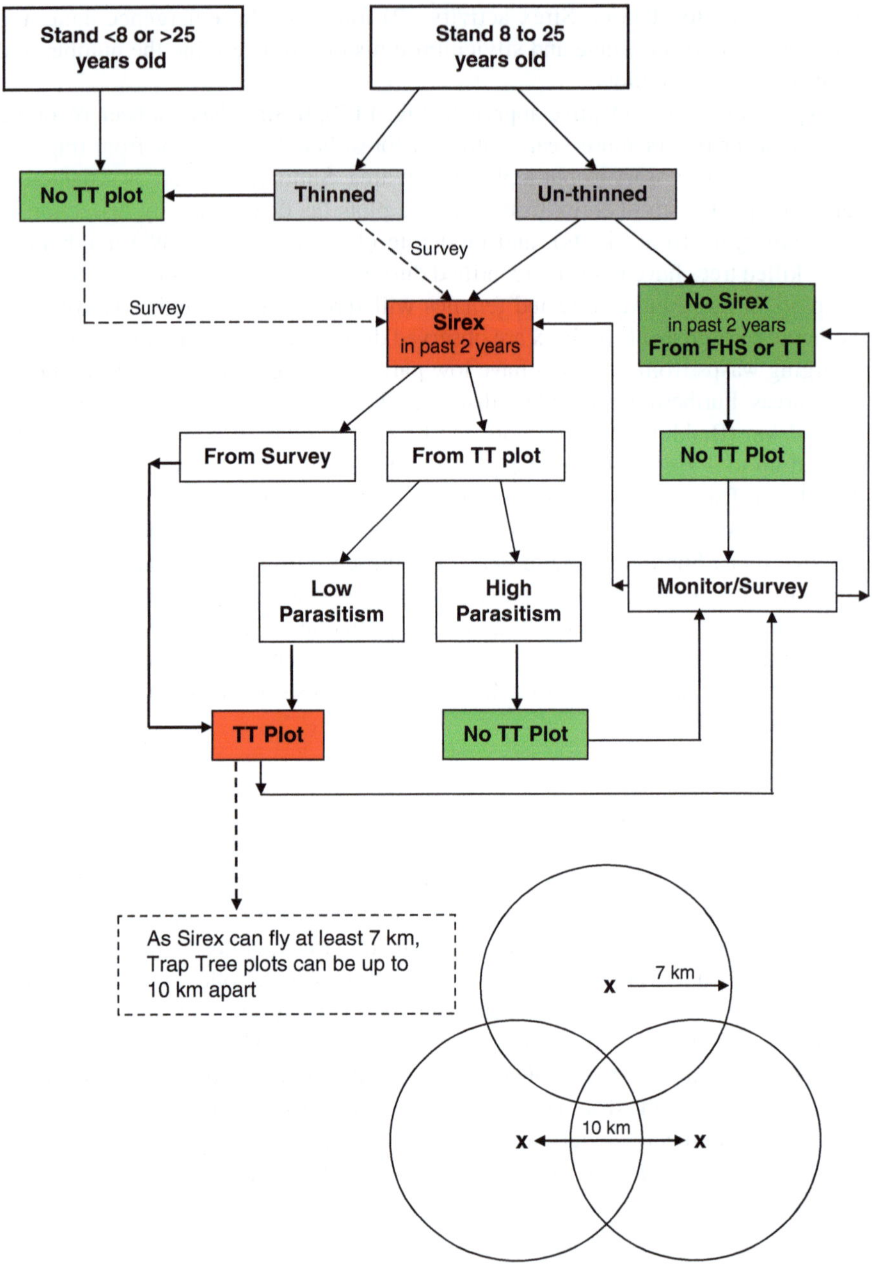

Fig. 14.7 Flow chart for selection of trap tree plot locality (and density) within a plantation (see text for detail)

Fig. 14.8 Sirex biological control monitoring cages: billets cut from trap trees or naturally struck trees are placed in 44 gal drums and monitored regularly for emergence of Sirex and parasitoids. Sirex males and females are later dissected in the laboratory to determine nematode parasitism levels

levels of each agent (Haugen et al. 1990). Evaluations are done annually (mostly by operational staff in the regions) via dissection of emerging Sirex (nematode parasitism), and by collating emergence of parasitoids, from billets collected from trap trees and naturally struck trees (Figs. 14.5b and 14.8). In theory, nematode-inoculated trap trees should yield over 90% of emerging Sirex infested with nematodes (Bedding and Akhurst 1974). If parasitism levels are below this, then investigation into why this is so is warranted. The cause may be incorrect inoculation technique, competition from the defective strain, or some other factor, such as trees drying out too quickly, or attack by bark beetles (e.g., *Ips grandicollis*) (also see Chap. 9).

Analysis of nematode strains using Random Amplified Polymorphic DNA (RAPD) profiles (Calder and Bedding, unpublished) allowed the comparison of infectivity (entry of nematodes into the larval host) levels of individual strains. This analysis showed there were a number of nematode strains present in the wild (especially in Victoria), including the Kamona strain, "defective" strain and several "other" strains. These "other" strains are likely the result of earlier releases of strains from Corsica, Greece and New Zealand almost 40 years ago (Bedding 1993, 2009). They have variable infectivity, and along with the "defective" strain, may be limiting the effectiveness of the more recent Kamona strain inoculations. Results of testing in 2001 indicate that in NSW, the Kamona strain predominated, with infectivity levels between 75% and 97%; in Victoria and the Australian Capital Territory (ACT), only the "defective" and several "other" strains were identified (infectivity 7–30% and 15–38% respectively) – no Kamona strain nematodes were identified at all. Annual monitoring of infectivity rates since 2001 showed infectivity levels of between 74% and 96% determined for Victoria and between 67% and 82% for NSW (Smith et al., unpublished). Samples have also been preserved for later RAPD profile analysis. In Tasmania infectivity levels of 87–94% were obtained from Kamona nematodes introduced to one area (Retreat) in 1995. However, infectivity testing of adult wasps collected from this area each year since 2004 has not detected the presence of any nematodes. Sirex caught in traps in 2009 from 24 sites across the state, including Retreat, confirmed the absence of nematodes from any site in Tasmania. This result indicates that nematodes are unlikely to survive within a very low Sirex population for more than a decade without further releases (NSCC unpublished).

Long-term data sets on parasitism rates have been examined, revealing regional and seasonal variation (Carnegie et al. 2005; Collett and Elms 2009); however, DNA analysis is required to barcode preserved nematode strains from various sites where multi-strain releases have been made to determine if some strains are able to disperse within trees more rapidly than others, and to be able to determine the natural geographical dispersal of different strains.

Analysis of long-term emergence data has shown that for Victoria and NSW, *I. leucospoides* is the most abundant and effective parasitoid (Carnegie et al. 2005; Collett and Elms 2009). It is possible that *M. nortoni* abundance is quite high in some areas but it has not been monitored optimally for several decades. The timing of billet collection (generally October) does not give an accurate picture of *M. nortoni* presence because most adults emerge in the spring following the summer Sirex emergence (Taylor 1967), and billets would need to be held for a second year to get accurate data on *M. nortoni* parasitism. The current system of monitoring, however, works well for *I. leucospoides,* providing an accurate picture of abundance of *I. leucospoides* compared with Sirex population numbers at each site.

14.2.2.11 Research Priorities

Haugen et al. (1990) identified several areas for research to increase efficiency of control procedures. The most important of these was investigating the problem of low parasitism rates of Sirex in inoculated logs. This problem was identified in the late 1980s in the Green Triangle (Haugen and Underdown 1993). As suspected by Haugen and Underwood (1993), the cause was found to be due to continual subculturing of *D. siricidicola* in the free-living, fungal-feeding form, which led to selection of a strain "reluctant" to form the parasitic stage (Bedding and Iede 2005; Bedding 1992, 2009, also see Chap. 9). This strain was dubbed the "defective" strain. Procedures have now been put in place to ensure this loss of infectivity does not occur again with the new Kamona strain (Bedding 2009).

14.3 Current Issues and Research

The NSCC has identified several issues that appear to be having a detrimental effect on the Sirex biological control program and has developed a research program to investigate these issues. The primary focus of this research is to reduce the impact on the productivity of Australia's pine plantations by maintaining or increasing the efficiency and efficacy of the biological control program but also ensuring the program is operationally relevant.

14.3.1 Reduced Nematode Parasitism of S. noctilio

Over recent years there has been an inconsistency, and in some cases a gradual decline, in nematode parasitism rates of Sirex emerging from inoculated trap trees in several regions. For example, in NSW, parasitism rates from trap trees (male and female data combined) in Hume Region have declined from 81% in 2004 to 45% in 2008; in Macquarie Region, from 80% in 2006 to 60% in 2008; and in Monaro Region from 82% in 2006 to 39% in 2009 (A. Carnegie, unpublished). This has been coupled with an increase in Sirex activity in Hume Region and some areas of Macquarie Region. The cause of this (and methods of addressing the problem) is being investigated by the NSCC, including projects outlined below.

14.3.2 Ips grandicollis *Attacking Sirex Trap Trees*

Since 2005, there has been an increase in the numbers of *Ips* bark beetles in several pine growing regions in Australia. Like Sirex, *Ips* bark beetles are attracted to stressed trees (Erbilgin et al. 2002), and in NSW and South Australia have been attacking trap trees poisoned for the Sirex biological control program, which is likely to create a problem for Sirex control. Climate modelling by Bashford (unpublished) using CLIMATE (Pheloung 1996) predicts the distribution of *I. grandicollis* to include all of the *P. radiata* growing areas in mainland Australia, but establishment would be unlikely in Tasmania and the south-western corner of Western Australia. *Ips grandicollis* has been found in several areas of southern subtropical Queensland, where several different species of *Pinus* are grown. It is not yet known if all of the species of *Pinus* grown are susceptible to *I. grandicollis* or if Sirex can successfully develop in all of them. *Ips* populations peaked in NSW in 2006–2007, resulting in large areas (over 17,000 ha) of tree mortality associated with a corresponding drought and *Ips* attack (Carnegie 2008b). It is believed that milder winter temperatures enabled more generations over winter than in more normal years with colder winters. In parallel with the increase in *Ips* numbers, surveys have revealed that the majority of trap trees in Hume Region (NSW) have been attacked by *Ips* over the past few years. Several research projects have been initiated to investigate this problem.

The first project was intended to quantify the efficacy of anti-aggregation pheromones (deterrents) in reducing *Ips* numbers and damage to Sirex trap trees (Loch and Carnegie, unpublished; Carnegie and Loch 2010). It was thought that *Ips* attacks on trap trees may be reducing the attractiveness of trees to Sirex and/or the efficacy of the biological control agents. Paired trap tree plots were established in mid-November 2007 in plantations in two State Forests in Hume Region, NSW, where high *Ips* activity had been observed. North American colleagues recommended a push-pull strategy for *Ips* control (K. Dodds and D. Miller, USDA Forest Service). The "push" part of the strategy involved putative anti-aggregation pheromones (verbenone and *Ips*dienol lures) placed on trees in treatment plots, two of each per tree per month. Early field observations indicated that *Ips*dienol was not an effective anti-aggregant for *I. grandicollis* (and in fact may be an aggregant/attractant), so its use was discontinued. Several researchers have also shown that *Ips*dienol is an aggregant (Dodds and Miller, unpublished) while others showed it to be an anti-aggregant (Wilson 1995). The "pull" part involved intercept panel traps, baited with *Ips*enol and ά-pinene, placed near the plots. Monthly assessments were made of *Ips* and Sirex attack and the rate of tree morbidity (i.e. how quickly poisoned trees were dying), as well as emergence data (in 2008/2009).

The main finding (Loch and Carnegie, unpublished; Carnegie and Loch 2010) was that the push-pull strategy did not work; *Ips* attacked treated and untreated (control) trees similarly, with all but a few trees being attacked by *Ips* by May 2008. *Ips* appeared to attack trees once the foliage changed to less than 50% green and greater than 50% red, which occurred 2 months after trees were poisoned. Results also showed that 50% of trees were also attacked by Sirex. Observations (and evidence) of female Sirex ovipositing indicated they preferred *Ips*-free trees and trees with some green foliage. Nematode parasitism of Sirex emerging from *Ips*-attacked trees was variable and inconsistent, and low (7–44%). Similarly, parasitism by *I. leucospoides* was variable (15–90%, mean 50%). In summary, it appears that *Ips* attack does not stop Sirex, or *I. leucospoides*, developing in trees, but may be affecting the length of time (window of opportunity) that Sirex can oviposit in trees, and also may affect nematode parasitism. Research on this aspect of the study is continuing.

The second project ("Protecting Australia's pine plantations from exotic pests"), funded by the Australian Research Council (ARC) (http://www.arc.gov.au) and the NSCC, is a 3 year research project on various aspects of the disruptive potential of *I. grandicollis* to the Sirex biological control program. This project, which commenced in mid-2010, has three aims, outlined below.

1. To undertake a large scale survey of pine growing regions in Australia, to quantify the magnitude of the threat facing Sirex trap tree establishment by *I. grandicollis* and to identify the geographical regions where *Ips* suppression is most urgently required. The project will identify the abiotic and biotic factors that influence *Ips*-attack and result in disruption of the Sirex biological control program and develop a predictive model and decision support system to assist growers.
2. To elucidate interactions between insects, nematodes and fungi within trap trees by investigating the interactions between Sirex and *I. grandicollis* and their respective fungi (*Amylosterum areolatum* and *Ophiostoma ips*) and nematodes, which for *I. grandicollis* includes not only *Contortylenchus grandicolli,* but also potential phoretic species of nematodes and mites that have previously been identified in Australia (Stone and Simpson 1987, 1990). The main aim is to determine whether these other species (insects, nematodes and fungi associated with *I. grandicollis*) are affecting the ability of *D. siricidicola* to parasitise Sirex. A secondary objective is to investigate whether drought (specifically the increased drying rate of trees) has an influence on these species, and hence on the effectiveness of the Sirex biological control program.

3. To develop and test new management strategies to maximise biological control of Sirex and minimise damage by *I. grandicollis*. The objective of this project is to identify operational procedures that can be used to reduce the impact or effect of *I. grandicollis* on the Sirex biological control program. Some of these procedures are discussed in more detail below, but will include: optimising the timing of trap tree establishment; optimising the timing of nematode inoculation; the use of chemical repellents to protect trap trees from *I. grandicollis*; and the use of *Ips*-resistance provenances of *P. radiata.*

14.3.3 Optimise Timing of Nematode Inoculation of Trap Trees

Bedding and Akhurst (1974) determined that inoculation with *D. siricidicola* was most successful in trees with moisture content greater than 50%. Neumann et al. (1982) recommended that trees be inoculated after the end of the Sirex flight season, which is generally late summer to mid-autumn in south-eastern Australia. Haugen et al. (1990) recommended May–July for inoculation; the recommendation was later altered to April–August (NSCC 2002), because before April many dying naturally struck trees may not be obvious and by the end of August the moisture content may be too low to ensure success. In NSW, Spolch (Forestry Commission of NSW, unpublished) recommended April–May for inoculation, as colder temperatures after May will restrict fungal growth and nematode activity and breeding. Timing is different in Tasmania, with inoculations in mid-winter due to cooler summer temperatures slowing the demise of trap trees and trees retaining high moisture content levels for many months. In the past decade, inoculations in NSW, South Australia and Victoria have been carried out in June–July. Reasons for this change in time of year are unclear, but may be due to competing priorities, such as fire management, in the April–May period.

The recent reduction in parasitism of Sirex in inoculated trees could be due to several factors associated with timing of the inoculations, including (1) colder temperatures in winter restricting fungal growth and nematode activity and breeding, and/or (2) prolonged drought resulting in moisture content of trees reducing to below 50% by winter. To address this, the NSCC now recommends that inoculations occur in April–May. The effectiveness of this change in timing is being monitored. Future research, using paired plots, will compare April–May and June–July inoculations.

14.3.4 Trap Tree Establishment Techniques

The trap tree establishment technique was developed in *P. radiata* plantations in temperate south-eastern Australia (Neumann et al. 1982). There are indications from subtropical Australia (NSW and Queensland), where other *Pinus* species dominate (*P. elliottii, P. taeda, P. elliottii x caribaea* hybrids), that the technique may need to be modified to suit these species and environment, as trees are either dying too quickly or are being attacked by *I. grandicollis*. Furthermore, there have been indications that the rate of chemical used in trap tree establishment is not effective in hot dry areas. A pilot study is being conducted by members of NSCC looking at the effectiveness of two rates of herbicide treatment, using Dicamba 500®, in killing trees slowly. The aim is to determine the most effective rate of herbicide in hot dry sites, and in subtropical Australia, compared with the standard herbicide application as laid down in the *National Sirex Control Strategy Worksheets* (2 ml of undiluted chemical [Dicamba 200®] applied by injection for every 10 cm of tree diameter, NSCC 2007). The current permit for use of Dicamba to poison trap trees describes rates for both Dicamba 200® and Dicamba 500®. However, Dicamba 200® has been phased out in recent years, so current testing is with Dicamba 500® only. An expanded program, within the previously mentioned ARC project, will test a range of different herbicide products, rates of application, and timing of application in a number of different climatic zones.

14.3.5 Storage Life of Nematodes

A major issue raised by field staff conducting inoculations of trap trees was the "shelf life" of nematodes. Nematodes generally need to be inoculated within a week of arriving from the producer (Ecogrow). New packaging and storage methods for nematode transport have now been developed by Bedding and Lacey (unpublished) with partial funding by NSCC. Nematodes are extracted from selected flasks, stored in undiluted medium extract at one million per 20 ml in sealed, thin-walled polypropylene freezer bags. These bags are then kept horizontal on corrugated cardboard within small Styrofoam coolers and kept under refrigeration at 3–5°C. This method results in 80% survival after 2 months and greatly increases the time frame for field delivery and inoculation.

14.3.6 Spread of Sirex into Subtropical Australia

Using the climate matching program CLIMEX (Sutherst and Maywald 1985), Carnegie et al. (2006) predicted that Sirex could establish in the majority of pine growing regions in Australia, including the subtropical regions of NSW and Queensland. Based on the observed rate of spread (30–40 km a year), Eldridge and Taylor (1989) predicted Sirex would reach the Queensland border by 2008. Similarly, based on the rate of spread reported by Eldridge and Taylor (1989) and Carnegie et al. (2005), and distribution of susceptible hosts, Carnegie et al. (2006) predicted Sirex would reach Queensland and north-eastern NSW by 2010. Sirex was detected in Queensland in 2009 but has not been detected in north-eastern NSW. However, a potential limiting factor for Sirex establishing in subtropical Australia is that the temperature within trees in the subtropics may not be suitable for the adequate growth and survival of *A. areolatum,* the reproduction and survival of the nematode *D. siricidicola,* or survival of Sirex larvae. This question is being addressed by the NSCC research on the effect of temperature on growth of *A. areolatum*. Early results indicate that high internal tree temperatures may not be sustained long enough (24 h) to cause fungal mortality at 36°C; however, sustained temperatures of 30°C or higher do prevent fungal growth (Ramsden, unpublished). Thus Sirex may survive, but emerging adults could be small and therefore less fecund. Research in 2010 will examine the survival and reproduction of the nematode and development of Sirex larvae at temperatures above 30°C.

14.3.7 Future Research

The NSCC has identified several other issues that need to be understood to improve the efficiency and effectiveness of the current control program. Future research funded by the NSCC may include:

- Investigation into whether there are other strains of *A. areolatum* that are suitable for culturing the Kamona nematode but which grow faster within trap trees at a greater range of temperatures and wood moisture levels than the current strain.
- Investigation into the mechanism causing low infectivity within trap trees caused by 'wild' defective nematode strains.
- Replacement of "defective" nematode strains with Kamona strain and to determine if it is possible to flood-out defective strains with high volume introductions of Kamona.
- Development of DNA techniques to identify 'strains' of all Sirex-associated organisms to determine what strains are present in plantations in Australia and for use as a monitoring tool.
- Investigation into the population levels of Sirex necessary to ensure viable nematode populations and whether it is necessary to inoculate regularly (say every 5 years) to maintain a background level of nematodes in the case of an outbreak. High levels of nematode inoculations result in generally low Sirex populations as compartments (trees) mature. Harvesting also removes any residual nematode population within that compartment. In isolated areas, several rotations may result in Sirex becoming nematode free. There is some evidence in Tasmania that this is occurring, where nematodes are no longer present in wild Sirex populations.

14.4 Conclusions

Sirex remains a significant pest of exotic *P. radiata* plantations in Australia, requiring continual intensive monitoring and management. Annual expenditure on operational Sirex control in Australia is difficult to accurately quantify, but includes both the trap tree program and forest health surveillance. The cost of the current trap tree program, which includes trap tree plot establishment, nematode inoculation, and monitoring of the efficacy of the biological control agents (and includes the use of static traps for early detection), is approximately AU$400,000 per annum. This cost varies according to the number of trap trees established (averages 350–450 per year) and the number of static traps used (125–150 in recent years). Forest health surveillance is carried out in most of Australia's almost one million ha of *Pinus* plantations, at a cost of approximately AU$1.10 to AU$1.50 ha^{-1} (Carnegie et al. 2008), but determining what proportion of this is directly related to Sirex control is difficult as aerial surveillance includes all health issues, not just Sirex. If AU$100,000, as a conservative estimate for surveillance costs related directly to Sirex control, is added to the above trap tree program costs, the total annual cost of operational Sirex control in Australia is approximately AU$500,000.

Bedding and Iede (2005) quoted M.G. Underdown as having calculated that if mortality levels as observed in the Green Triangle in the late 1980s occurred across Australia, then Sirex had the potential to cause between US$16 and US$60 million (approx. AU$21–AU$78 million) in loss of timber per year. Several authors have attempted to calculate the cost-benefit of research on Sirex control in Australia. Marsden et al. (1980) calculated that the AU$5.21 million spent on Sirex research (by CSIRO) to 1975 had a benefit of AU$12.8 million per rotation to the year 2000. It was estimated that the cost of Sirex control and research in Victoria up to 1974 was AU$4.1 million and the benefits were between AU$4.0 million and AU$6.5 million (Marsden et al. 1980). These benefits would be mirrored in future rotations, thus greatly increasing the overall benefit. Another way to calculate the benefit of Sirex research and control is that given that pine plantations in Australia represent 57% of Australia's AU$3.3 billion per annum forest plantation activity (Parsons et al. 2006), a conservative estimate of avoiding 10% loss to the industry would have a value of AU$188 million pa. Whichever way the cost of Sirex control is calculated, the benefits far outweigh the cost of operational control and/or research.

One of the reasons for the Green Triangle outbreak was that there was no coordinated approach to monitoring or managing Sirex (Haugen 1990). The NSCC was established to ensure continual monitoring of Sirex and its control. The NSSC are able to quickly identify problems as they arise, and develop, coordinate and fund research programs to address these problems. This rapid response is possible because members of the NSCC are directly involved in conducting surveys and managing the biological control program, and therefore have a real-time understanding of the situation.

Despite five decades of research in Australia specifically aimed at the management of Sirex, the dynamic nature of pest, host, symbiotic fungus, parasites, competing agents and environment mean there are still a number of issues that require understanding to improve the control program and make it more effective and efficient. There is no room for complacency once a control strategy has been developed and ongoing vigilance is necessary. The need for continuing research also highlights the importance of groups such as the NSCC. The Australian experience is a good lesson for overseas countries where Sirex is relatively new and shows that continual monitoring and ongoing research is a necessity, not an option.

Acknowledgements The authors are grateful to Charlma Phillips, Robin Bedding and Robert Eldridge for constructive input into earlier drafts of this chapter, and to other members of the NSCC for their input. David Spolc is thanked for the use of his illustrations.

References

Anonymous (1990) Biological control of Sirex in Victoria and the role of Keith Turnbull Research Institute. Austral Forest Grower 13(2):29

Bain J (2005) *Sirex noctilio* (Hymenoptera: Siricidae) – the New Zealand experience. In: Gottschalk KW (ed) Proceedings, 16th U.S. Department of Agriculture interagency research forum on gypsy moth and other invasive species 2005. 18–21 Jan; Annapolis, MD. General Technical Report. NE-337. Newtown Square, PA, US: Department of Agriculture, Forest Service, Northeastern Research Station

Bashford R (2008) The development of static trapping systems to monitor for wood-boring insects in forestry plantations. Aust For 71:236–241

Bedding RA (1967) Parasitic and free-living cycles in entomophagous nematodes of the genus *Deladenus*. Nature 214:174–175

Bedding RA (1972) Biology of *Deladenus siricidicola* (Neotylenchidae) an entomophagous-mycetophagous nematode parasitic in siricid woodwasps. Nematol 18:482–493

Bedding RA (1992) Strategy to overcome the crisis in control of Sirex by nematodes. Aust For Grow 14(4):15–16

Bedding RA (1993) Biological control of *Sirex noctilio* using the nematode *Deladenus siricidicola*. In: Bedding RA, Akhurst RJ, Kaya HK (eds) Nematodes and the biological control of insect pests. CSIRO Publications, East Melbourne

Bedding RA (2009) Controlling the pine-killing woodwasp, *Sirex noctilio*, with nematodes. In: Hajek AE, Glare TR, O'Callaghan M (eds) Use of microbes for control and eradication of invasive Arthropods. Springer Science+Business Media BV, Rotterdam

Bedding RA, Akhurst RJ (1974) Use of the nematode *Deladenus siricidicola* in the biological control of *Sirex noctilio* in Australia. J Aust Entomol Soc 13:129–135

Bedding RA, Iede ET (2005) Application of *Beddingia siricidicola* for sirex woodwasp control. In: Grewal PS, Ehlers R-U, Shapiro-Ilan DI (eds) Nematodes as biological control agents. CAB International, London, pp 385–399

Bedding RA, Carnegie AJ, de Groot P, Dodds K, Germain R, Hain F, Haugen D, Iede E, Lance D, Mastro V, Reardon R, Rogers D, Schiff N, Tumlinson J, Williams D (2006) Sirex Science Advisory Panel. Report 9–10. Jan 2006. Annapolis, MD. http://www.aphis.usda.gov/plant_health/plant_pest_info/sirex/downloads/sirex-sap-report.pdf. Accessed 12 Dec 2009

Bulman LS, Kimberley MO, Gadgil PD (1999) Estimation of the efficiency of pest detection surveys. N Z J For Sci 29:102–115

Carnegie AJ (2008a) A decade of forest health surveillance in Australia: an overview. Aust For 71:161–163

Carnegie AJ (2008b) Exotic bark beetle turns nasty in pine plantations. Bush Telegraph, 13(Autumn)

Carnegie AJ, Eldridge RH, Waterson DG (2005) The history and management of sirex wood wasp, *Sirex noctilio* (Hymenoptera: Siricidae), in New South Wales, Australia. N Z J For Sci 35:3–24

Carnegie AJ, Loch AD (2010) Is *Ips grandicolis* disrupting the biological control of *Sirex noctilio* in Australia? In: McManus KA, Gottschalk KW (eds). Proceedings 21st U.S. Department of agriculture interagency research forum on invasive species 2010, January 12–15, Annapolis, MD. Gen Tech Rep NRS-P-75, pp 8–11.

Carnegie AJ, Matsuki M, Hurley BP, Ahumada R, Haugen DA, Klasmer P, Sun J, Iede ET (2006) Predicting the potential distribution of *Sirex noctilio* (Hymenoptera: Siricidae), a significant exotic pest of *Pinus* plantations. Ann For Sci 63:119–128

Carnegie AJ, Cant RG, Eldridge RH (2008) Forest health surveillance in New South Wales, Australia. Aust For 71:164–176

Carter PCS (1989) Risk assessment and pest detection surveys for exotic pests and diseases which threaten commercial forestry in New Zealand. N Z J For Sci 19:353–374

Collett NG, Elms S (2009) The control of Sirex wood wasp using biological control agents in Victoria, Australia. Agric For Entomol 11:283–295

Commonwealth Scientific Industrial Research Organisation. Division of Entomology Annual Reports (1965/1966 to 1979/1980), Canberra, ACT

Croft P (2009) South Africa Sirex control programme. Institute for commercial forestry research. http://www.icfr.ukzn.ac.za/collaboration/sirex/monitoring-sirex/. Accessed 14 Apr 2010

Dye M, Mutanga O, Ismail R (2008) Detecting the severity of woodwasp, *Sirex noctilio,* infestation in a pine plantation in KwaZulu-Natal, South Africa, using texture measures calculated from high spatial resolution imagery. Afr Entomol 16:263–275

Eldridge RH, Simpson JA (1987) Development of contingency plans for use against exotic pests and diseases of trees and timber. 3. Histories of control measures against some introduced pests and diseases of forests and forest products in Australia. Aust For 50:24–36

Eldridge RH, Taylor EE (1989) Sirex woodwasp – a pest of pine in N.S.W. Forest Protection Series No. 1. Forestry Commission of New South Wales, Sydney

Erbilgin N, Nordheim EV, Aukema BH, Raffa KF (2002) Population dynamics of *Ips pini* and *Ips grandicollis* in red pine plantations in Wisconsin: within- and between-year associations with predators, competitors, and habitat quality. Environ Entomol 31:1043–1051

Evans HJ, Hopkins AA, Scarr TA (2006) Status of important forest pests in Ontario in 2006. Natural Resource Canada and Ontaria Ministry of Natural Resources, Sault Ste Marie, ON Report

Gilbert JM, Miller W (1952) An outbreak of *Sirex noctilio* F. in Tasmania. Aust For 16:63–69

Haugen DA (1990) Control procedures for *Sirex noctilio* in the Green Triangle: review from detection to severe outbreak (1977–1987). Aust For 53:24–32

Haugen DA, Underdown MG (1993) Reduced parasitism of *Sirex noctilio* in radiata pines inoculated with the nematode *Beddingia siricidicola* during 1974–1989. Aust For 56:45–48

Haugen DA, Underwood MG (1990) *Sirex noctilio* control program in response to the 1987 Green Triangle outbreak. Aust For 53:33–40

Haugen DA, Bedding RA, Underdown MG, Neumann FG (1990) National strategy for control of *Sirex noctilio* in Australia. Austral Forest Grower 13(2): Special lift-out section No. 13

Hennessy KJ, Macadam I, Whetton PH (2006) Climate change scenarios for initial assessment of risk in accordance with risk management guidance. CSIRO Marine and Atmospheric Research, Aspendale, VIC, Australia. Available at http://www.greenhouse.gov.au/impacts/publications/pubs/risk-scenarios.pdf. Accessed 12 Dec 2009

Hurley BP, Slippers B, Wingfield MW (2007) A comparison of control results for the alien invasive woodwasp, *Sirex noctilio*, in the southern hemisphere. Agric For Entomol 9:159–171

Irvine CJ (1962) Forest and timber insects in Victoria 1962. Victoria's Resour 4:40–43

Ismail R, Mutanga O, Bob U (2007) Forest health and vitality: the detection and monitoring of *Pinus patula* trees infected by *Sirex noctilio* using digital multispectral imagery. South Hemisph For J 69:39–47

Madden JL (1971) Some treatments which render Monterey pine (*Pinus radiata*) attractive to the woodwasp *Sirex noctilio* F. Bull Entomol Res 60:467–472

Madden JL (1975) An analysis of an outbreak of the woodwasp *Sirex noctilio* F. (Hymenoptera: Siricidae), in *Pinus radiata*. Bull Entomol Res 65:491–500

Marsden JS, Martin GE, Parham DJ, Risdill Smith TJ, Johnson BG (1980) Returns on Australian agricultural research. The Joint Industries Assistance Commission – CSIRO benefit-cost study of the CSIRO Division of Entomology. CSIRO, Melbourne

National Sirex Coordination Committee (2001) *Sirex noctilio* – the pine killing woodwasp: detection, biology and control. CSIRO, Canberra

National Sirex Coordination Committee (2002) National Sirex control operations worksheets. National Sirex Coordination Committee, Melbourne

National Sirex Coordination Committee (2007) National Sirex control operations worksheets. National Sirex Coordination Committee, Melbourne

Neumann FG (1979) Insect pest management in Australian radiata pine plantations. Aust For 42:30–38

Neumann FG, Minko G (1981) The sirex wood wasp in Australian radiata pine plantations. Aust For 44:46–63

Neumann FG, Harris JA, Kassaby FY, Minko G (1982) An improved technique for early detection and control of Sirex wood wasp in Radiata pine plantations. Aust For 45:117–124

Neumann FG, Morey JL, McKimm RJ (1987) The sirex wasp in Victoria. Bulletin No. 29. Department of Conservation, Forests and Lands, Victoria

Nuttall MJ (1989) *Sirex noctilio* F., sirex wood wasp (Hymenoptera: Siricidae). In: Cameron PJ, Hill RL, Bain J, Thomas WP (eds) Review of biological control of invertebrate pests and weeds in New Zealand 1874–1986. Technical Communication No. 10. CAB International Institute of Biological Control, Wallingford, pp 299–306

Parsons M, Gavran M, Davidson J (2006) Australia's plantations 2006. Bureau of Rural Sciences, Canberra

Pennsylvania Department of Agriculture (2007) Eurasian wood wasp (*Sirex noctilio*). www.dcnr.state.pa.us/FORESTRY/fpm_invasives_woodwasp.aspx

Pheloung PC (1996) CLIMATE: a system to predict the distribution of an organism based on climate preferences. Agriculture Western Australia, Perth

Phillips CL (2008) Forest health surveillance in South Australia. Aust For 71:196–201

Rawlings GB (1948) Recent observations on the *Sirex noctilio* population in *Pinus radiata* forests in New Zealand. N Z J For 5:411–421

Rawlings GB (1955) Epidemics in *Pinus radiata* forest in New Zealand. N Z J For 7:53–58

Spradbery JP (1970a) The immature stages of European Ichneumonid parasites of Siricine woodwasps. Proc R Entomol Soc Lond (A) 45:14–28

Spradbery JP (1970b) Host finding by *Rhyssa persuasoria* (L.), an Ichneumonid parasite of siricid woodwasps. Anim Behav 18:103–114

Stone C, Simpson JA (1987) Effect of six chemicals on the insects, mites, nematodes and fungi associated with *Ips grandicollis* (Eichhoff) (Coleoptera: Scolytidae) in northeastern New South Wales. J Aust Entomol Soc 30:21–28

Stone C, Simpson JA (1990) Species associations in *Ips grandicollis* Eichhoff galleries in *Pinus taeda* L. N Z J For Sci 20:75–96

Stone C, Turner R, Verbesselt J (2008) Integrating plantation health surveillance and wood resource inventory systems using remote sensing methodologies. Aust For 71:245–253

Sutherst RW, Maywald GF (1985) A computerised system for matching climates in ecology. Agric Ecosyst Environ 13:281–299

Taylor KL (1967) The introduction, culture, liberation, and recovery of parasites of *Sirex noctilio* in Tasmania, 1962–67. Technical paper No. 8. CSIRO Division of Entomology, Canberra

Taylor KL (1976) The introduction and establishment of insect parasitoids to control *Sirex noctilio* in Australia. Entomophaga 21:429–440

Wardlaw T (2008) A review of the outcomes of a decade of forest health surveillance of state forests in Tasmania. Aust For 71:254–260

Wilson IM (1995) Antiaggregants for the Mountain Pine Beetle, Masters thesis, Simon Fraser University, Canada

Wotherspoon KP (2008) Forest health surveillance in Tasmania. Aust For 71:182–187

Zondag R (1969) A nematode infection of *Sirex noctilio* (F.) in New Zealand. N Z J Sci 12:732–747

National Sirex Coordination Committee (2007) National Sirex control operations worksheets. National Sirex Coordination Committee, Melbourne
Neumann FG (1979) Insect pest management in Australian radiata pine plantations. Aust For [illegible]
Neumann FG, Minko G (1981) The sirex wood wasp in Australian radiata pine plantations. Aust [illegible]
[illegible]

Chapter 15
The Ecology and Biological Control of the Woodwasp *Sirex noctilio* in Patagonia, Argentina

Paula Klasmer and Eduardo Botto

Abstract The woodwasp *Sirex noctilio* has become the most important forest pest affecting pine plantations in Argentina. Sirex was first detected in the northeastern provinces in 1985, and it has spread from there throughout most of the commercial pine plantations of the country causing great losses and tree mortality. In 1993, *S. noctilio* and its parasitoid *Ibalia leucospoides* became established in the Western Andean Patagonia, near the city of San Carlos de Bariloche (Patagonia, Argentina). A management program that includes trap trees for early detection, thinning and biological control with the parasitic nematode *Deladenus* (=*Beddingia*) *siricidicola* and the entomophagous parasitoid *I. leucospoides* has since been developed. An overview of the research carried out to date regarding Sirex management in Patagonia is presented and discussed.

15.1 Introduction

The woodwasp, *Sirex noctilio* F. (Hymenoptera: Siricidae) is a well-known forest insect pest with a world-wide distribution. This insect is endemic to Eurasia and North Africa (Spradbery and Kirk 1978, Chap. 5) and has become one of the most important forest pests of pine plantations in the Southern Hemisphere.

P. Klasmer (✉)
Campo Forestal General San Martín, EEA INTA Bariloche, C.C. 26, 8430 El Bolsón, Río Negro, Argentina
e-mail: vklasmer@bariloche.inta.gov.ar

E. Botto
Insectario Investigaciones Lucha Biológica, IMYZA. CICVyA. INTA, 1712 Castelar, Buenos Aires, Argentina

B. Slippers et al. (eds.), *The Sirex Woodwasp and its Fungal Symbiont: Research and Management of a Worldwide Invasive Pest*, DOI 10.1007/978-94-007-1960-6_15,
© Springer Science+Business Media B.V. 2012

Sirex noctilio was detected in New Zealand in 1900 (Hudson 1919; Miller and Clark 1935), in Australia in 1950 (Madden 1968), in Uruguay in 1980 (Rebuffo 1988), in Brazil in 1988 (Iede et al. 1988) and in South Africa in 1994 (Tribe 1995). It was first detected in Argentina in 1985 (Espinoza et al. 1986) (details described in Chaps. 13, 14, 16–18).

Sirex noctilio is a primitive and solitary wasp that parasitizes weakened and/or stressed trees. In unmanaged or poorly managed plantations and in the absence of biological control agents Sirex can cause up to 80% tree mortality (McKimm and Walls 1980; Neumann et al. 1987), and is thus a serious threat to the forest industry. Many pine species (*Pinus* spp.) are susceptible to attack by this woodwasp. In Argentina, the woodwasp attacks *Pinus bandsman, P. contorta var. latifolia, P. elliottii, P. jeffreyi, P. ponderosa, P. radiata*, and *P. taeda* (Klasmer et al. 2000). Different tactics and strategies have been implemented worldwide to reduce Sirex damage in commercial pine plantations. These have included silvicultural practices such as pruning and selective thinning prior to the woodwasp flight season to reduce the number of stressed trees and to maintain healthy plantations, and the use of biological control agents using the parasitic nematode *Deladenus siricidicola* or the parasitic wasps *Ibalia* spp., *Rhyssa* spp. and *Megarhyssa* spp. This chapter describes the establishment, spread and distribution of *S. noctilio* in Argentina and the research into its ecology and management.

15.2 Establishment, Spread and Distribution of *Sirex noctilio* in Argentina

Sirex was first recorded in Argentina in 1985 in Gualeguaychú province of Entre Rios (Espinoza et al 1986), an area where pine plantations are extensive. Since then, the woodwasp has spread to different localities within the country. By 1988, it appeared that the woodwasp had colonized pine plantations located in the southwest of Buenos Aires province. In 1993 and 1995, Sirex was reported in the northern provinces of Jujuy (Quinteros and Vilte 1995) and Salta (Quinteros and Vilte 1996). In the central province of Cordoba, Sirex was first reported in 1994 (Lopez et al. 2002) whereas in the northeastern province of Misiones it has been present since 1995 (Echeverría 1996). By the end of 1989 Sirex was recorded at a sawmill near the city of San Carlos de Bariloche (Aguilar et al. 1990; Klasmer and Fritz 1994). Three years later, the first pine plantations attacked by the woodwasp were observed nearby. Between that time and 2001, Sirex spread into the region from north (San Martin de Los Andes, Neuquén, province) to south (Esquel, Chubut province) along the Andean Patagonia, a distance of nearly 300 km and an area with noncontiguous plantations (Fig. 15.1). In the northeastern provinces of Entre Ríos, Corrientes and Misiones, the northwestern province of Jujuy, and the province of Cordoba located in the center of the country, *S. noctilio* attacks *P. elliottii* and *P. taeda,* the most common commercially planted pine species in Argentina and in Brazil. In Western Patagonia, all pine species used in afforestation, *Pinus ponderosa*, *P. radiata*,

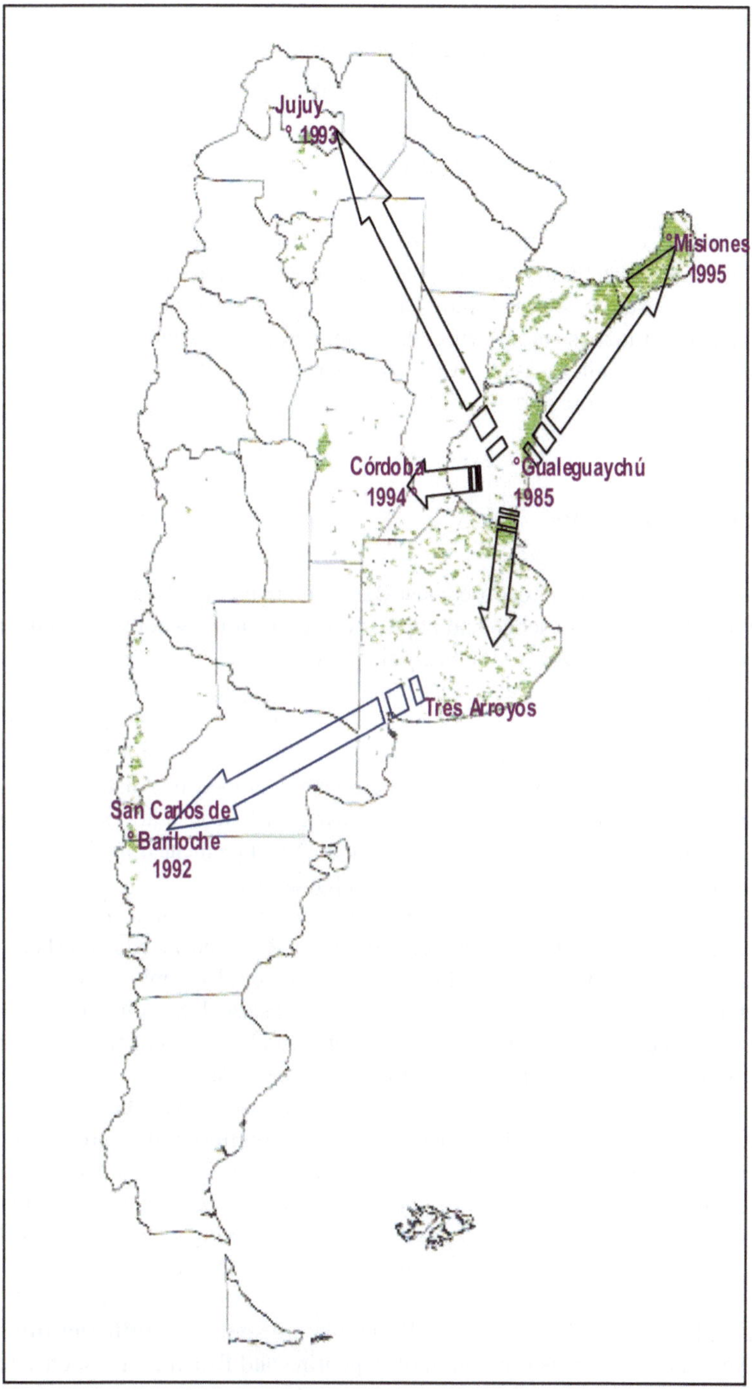

Fig. 15.1 Distribution of pine plantations and dispersal of *Sirex noctilio* in Argentina (1,150,000 ha). Secretaría de Agricultura Ganadería Pesca y Alimentación (SAGPyA) December 2001

P. contorta var. latifolia, *P. jeffreyi* and *P. bandsman*, are susceptible to infestation by the woodwasp (Klasmer et al. 2000).

15.3 Management Strategies for *Sirex noctilio* in Patagonia

In Argentina, several attempts have been made to develop management strategies for *S. noctilio*, particularly in Patagonia where efforts have been focused on field and laboratory studies on the biology and ecology of Sirex and on the development of management tactics (monitoring, thinning, and pruning and biological control). The aim of these studies was to develop simple management tools that could be easily transferred to local foresters.

15.3.1 Monitoring

Soon after Sirex was first recorded near Bariloche in Patagonia, a survey was undertaken to evaluate the presence and population status of the woodwasp in the region (Klasmer and Fritz 1994). In the beginning, most of the surveys conducted in areas with epidemic populations were based on "visual inspection" of the tree stand and recording the symptomatic trees attacked by Sirex on a census basis. Monitoring practices were developed, including the use of trap trees (Klasmer and Fritz 1994; Klasmer et al. 1997a) following the methods developed in Australia and Brazil (Neumann and Morey 1984; Spradbery and Kirk 1981; Iede et al. 1993). Monitoring techniques based on sequential sampling procedures implemented in Brazil (Penteado 1995) were used in some commercial pine plantations.

Trap trees used to monitor the presence of Sirex were felled by the end of the flight season and cut into one meter sections and placed in wire-meshed (1 m^3) cages. These cages were checked on a daily basis until no emergence of Sirex or potential natural enemies of the woodwasp was observed. Seasonality, abundance, sex ratios, generation time and the presence of natural enemies of Sirex (particularly parasitic wasps) were estimated from the rearing cage data.

The capacity of *S. noctilio* to develop a diapause has allowed the wasp to become readily adapted to the cold temperate climatic conditions that characterize the Andean Patagonian Region (see also discussion in Chap. 4). This insect displays a 1, 2 or 3 year generation life cycle. The first adults emerge in January and the flight season can last until May, with a peak in February or March (Fig. 15.2). Although some adults can emerge during the cold days in June, their activity is very low and they probably have a low contribution to Sirex population growth. Both, adult emergence and flight activity patterns in Patagonia are somewhat different from those observed in the Northeastern region of Argentina and Brazil as a response to local climatic conditions. Therefore, long life cycles (2–3 years) are more common than short ones (3–4 month to 1 year generation time) (Carvalho et al. 1993). This pattern

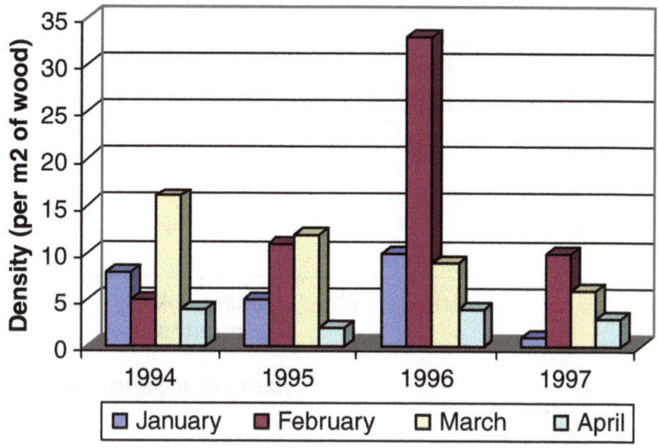

Fig. 15.2 Phenology of *Sirex noctilio* (1994–1997)

of adult emergence and activity has also been observed in other countries, such as has been reported for August and September in the Northern Hemisphere (Spradbery and Kirk 1978).

Monitoring studies allowed understanding the dispersal of Sirex in Western Patagonia. By the end of 1992, *S. noctilio* was detected in the Andino Patagonian Region in a mixed conifer plantation at Dina Huapi a locality near San Carlos de Bariloche (71°W – 41°S) in the province of Río Negro. Apparently the wasp arrived in Patagonia with a cargo of pine logs from an infested forest located at Tres Arroyos, province of Buenos Aires (Aguilar et al. 1990). In 1995, Sirex was already established at Estancia San Ramón, 40 km northeast from Bariloche. In April 1996, the woodwasp was recorded in Villegas (41° 34′ 54″ S – 71° 30′ W), 80 km south of Bariloche in a mixed pine plantation of mainly *P. radiata* and *P. contorta* var. *latifolia* (Klasmer 1996). In 2000, this pest was detected in a trap tree at the Campo Forestal General San Martin, INTA (Instituto Nacional de Tecnología Agropecuaria, Estación Experimental Agropecuaria Bariloche) located in Las Golondrinas, Chubut province, (42° S – 71° 3′ W), 2 km south from El Bolsón (Villacide and Klasmer 2001). At the same time, *S. noctilio* was observed further south in the province of Chubut near Epuyen Lake, Esquel and Trevelin. In March 2001, the wasp was found in Valle de Meliquina (40° 26′16″ S – 71° 14′05″ W) and in wooden houses in the city of San Martín de Los Andes (40° 9′23″ S – 71° 21′21″ W) in the province of Neuquén (northeast Patagonia) (Klasmer 2001). The infested wood probably came from the Sirex-infested plantations of Estancia San Ramón in Bariloche. These data confirm the great capacity that *S. noctilio* has for rapid dispersal.

Studies conducted in Australia have shown that females of *S. noctilio* are able to fly up to 30 km/year (Neumann et al. 1987). Similarly in South Africa, a dispersal rate of 48 km/year was recorded (Tribe and Cillié 2004). More recently Villacide and Corley (2008) conducted comparative studies on the flight capacity of healthy vs. nematode parasitized female Sirex wasps in Argentina. These studies demonstrate that healthy individuals are able to fly over 30 km whereas the parasitized individuals fly only 16.1 km. In addition to this great natural capacity of Sirex to

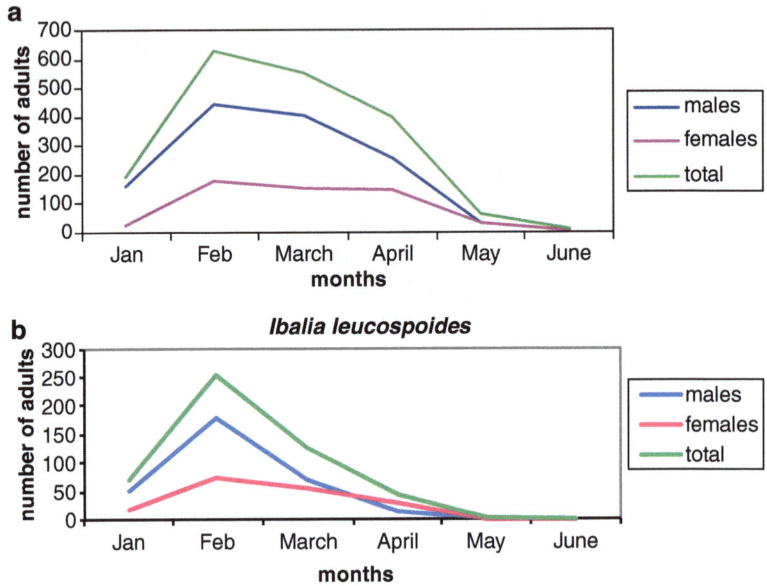

Fig. 15.3 (**a**) Seasonal abundance of *Sirex noctilio* (2005); (**b**) Seasonal abundance of *Ibalia leucospoides* (2005)

disperse naturally via flight anthropic activities i.e., transportation of wood or wooden products from infested to uninfested areas is undoubtedly a major dispersal factor.

Monitoring studies have provided useful data, not only for Sirex, but also for its natural parasitoid *I. leucospoides*. Population trends of both Sirex and *I. leucospoides* in the field seem to be well synchronized irrespective of the length of Sirex generations (Klasmer et al. 1997b) (Fig. 15.3a, b).

15.3.2 Biological Control with Parasitoids

In 1993 the entomophagous wasp *I. leucospoides* was detected in pine plantations in the Andean Patagonian Region, apparently introduced together with its host (Klasmer and Fritz 1994). *Ibalia leucospoides* was the only biological control agent found in the region. Biological attributes of this parasitoid such as high survival and fecundity and a good dispersal capacity are some of the main features accounting for its success as an efficient natural enemy of the *S. noctilio*. Research on pest-parasitoid relationships in Patagonia showed that *I. leucospides* is well adapted to the woodwasp as shown by the excellent synchronization between the host and the parasitoid life cycles. Parasitism levels observed in the region ranged in average between 22%

during 1993–1994 (Klasmer and Fritz 1994) and 45% in 1997 (Klasmer et al. 1997a). Similar results were observed in Brazil by Reis (1999).

15.3.3 Biological Control with Entomophagous Nematodes

The introduction of the parasitic nematode *D. siricidicola* was the key factor in the biological control strategy used against Sirex in Argentina. *Deladenus siricidicola* was introduced into Argentina from Brazil (Encruzhilada do Sul strain) in 1996 and 1999 (Eskiviski et al. 2003). A mass rearing facility for the entomophagous nematode was built at INTA Montecarlo, in the province of Misiones that produced the nematodes for mass inoculation throughout the country. In May 1999, *D. siricidicola* was introduced into the Patagonian Region (Klasmer 1999). Previous to this inoculation, laboratory and field studies were carried out to evaluate the nematode's adaptive response to a completely new environment. Releases began in 1999 in local pine plantations of San Carlos de Bariloche following the Brazilian technique (Iede et al. 1993). The nematode was released during different seasons to determine the optimal colonization time for its establishment; its adaptability was estimated by assessing its survival in the field after each inoculation. Special care was taken during the fall-winter season to avoid the detrimental effect of low temperatures, which could preclude nematode establishment. Therefore, no inoculations were conducted when temperatures were below 5°C.

The presence of *D. siricidicola*, its survival and relative abundance in a *S. noctilio* infested plantation were evaluated before Sirex emergence season by the woodchip sampling method (Haugen and Underdown 1991). After each inoculation, the presence of the nematodes was estimated by cutting three (5 cm wide) wood disk samples obtained from the base, middle and upper part of each tree. The bark was removed and woodchips (6 × 5 × 1.5 cm) were taken from each disk. Woodchips were labelled and placed into a Petri dish and covered with water to extract the nematodes. The number of nematodes was counted and levels of relative abundance were assigned following the method described by Haugen and Underdown (1991). The presence of live nematodes in woodchips after the initial inoculations ranged from moderate to good. It was observed that the nematodes moved from the bark to the core of the tree, through the tracheids, resin channels and parenchyma and they were able to successfully parasitize *S. noctilio* larvae (Klasmer et al. 2004). *Deladenus siricidicola* was able to survive the low temperatures that characterize Northwestern Patagonia during the winter season as suggested by the average parasitism levels of 50–60%.

In 2001, similar studies on the nematode were conducted on 40 year-old pine plantations (ca. 105 ha) in Campo Forestal San Martin, INTA, El Bolsón, province of Río Negro (71° 30′W 42° S), (Klasmer et al. 2000). Since then, seasonal inoculations have been carried out in the area throughout the year to ensure the establishment of the nematode. To evaluate the success of this biological control practice, log billets were collected periodically throughout the seasons and placed within mesh cages until the emergence of Sirex adults. The emerged wasps were dissected to determine the presence of the nematode. It was very important to check for the presence of nematodes inside the eggs of the woodwasp females and thus to confirm the parasite-induced sterilization. This is important because it is sometimes observed that the parasitic nematodes are in the haemocelle, but not within the reproductive organs of adult woodwasps.

The performance of the nematode gave positive results considering the levels of average parasitism recorded after the first inoculations carried out in 2001 (30%) and subsequent records of field parasitism: 58% (2002), 75% (2003), 95% (2005) and nearly 100% in 2007 (Klasmer 2007, 2008; Klasmer et al. 2004). Parasitism rates did not show significant differences among the inoculations made in autumn, winter or spring season, which indicates that the inoculation procedure was carried out properly and that the nematode was able to survive well under the environmental conditions in the El Bolsón area.

Parasitism rates of *S. noctilio* by the nematode in infested logs reached its optimum when the water content of the wood ranged between 50% and 70% (Klasmer et al. 2004). These results agreed with those reported by Bedding and Akhurst (1974) who emphasized the importance of the water content during the inoculation of the nematodes to assure a successful establishment of the natural enemy. With humidity contents over 70%, the *S. noctilio* symbiotic fungus *Amylostereum areolatum* will grow very slowly and at moisture contents lower than 20%, only mature larvae will survive.

The majority of the *S. noctilio* population was found in the middle and upper parts of trees (Fig. 15.4a) whereas the parasitoid *I. leucospoides* was found mostly in bottom portions along the bole (Fig. 15.4b). The nematodes were observed throughout the trees, but were mostly concentrated in the (middle and upper) parts (Fig. 15.4c). It was also observed that the second and third Sirex generation from the same tree emerges from the bottom portions (Klasmer and Fritz 1994).

15.4 Summary and Conclusions

Sirex noctilio has been established in the Southern Hemisphere for more than a century. Despite the efforts made in Argentina to eradicate Sirex soon after its first recorded appearance in 1985, the woodwasp has become the most important forest pest in pine plantations nation-wide. This situation triggered the development of pest management strategies based mainly on the use of environmentally safe methods such as biological control and silvicultural practices. In Northern Patagonia, several biological studies have been conducted to develop an integrated pest management strategy that could contribute to reduce Sirex population levels and control its expansion in the region.

Studies on *S. noctilio* in Argentina have focused on the development of monitoring systems using traps trees and sequential sampling methods, which have provided basic information about the presence, relative abundance, dispersion and natural

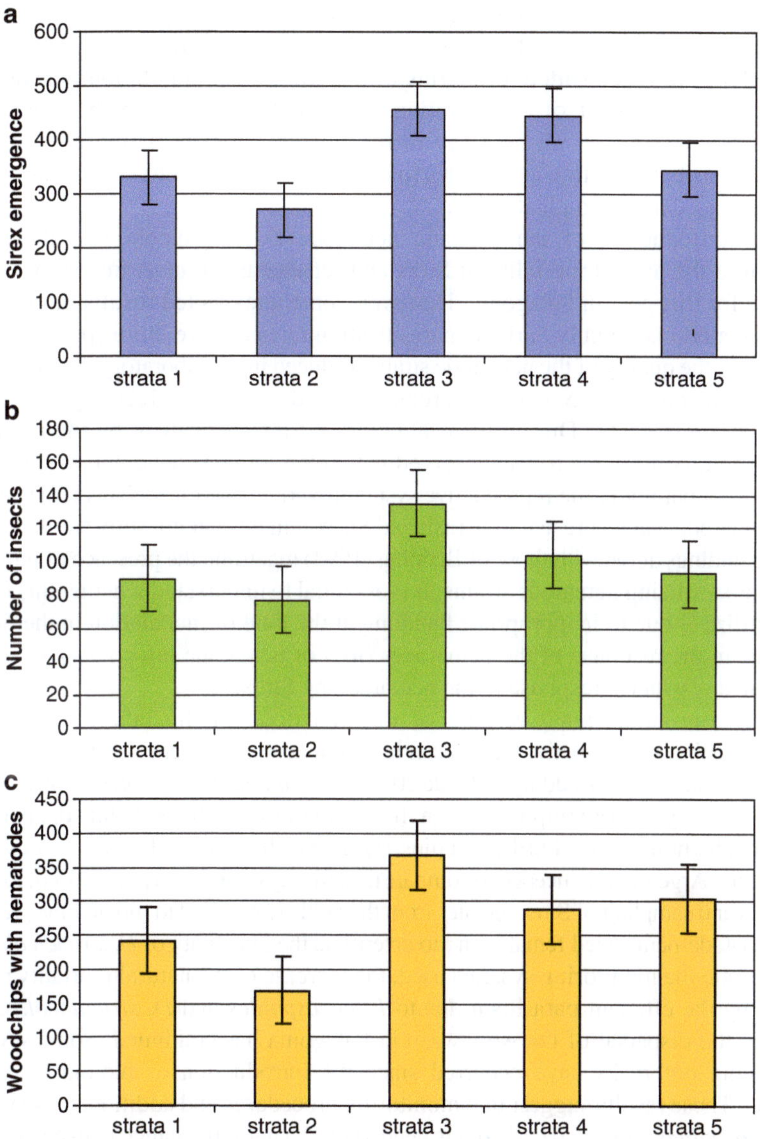

Fig. 15.4 (**a**) Distribution of *Sirex noctilio* larvae within the trees (2005). Trees were divided into five approximately equal strata, with strata 1 representing the bottom section; (**b**) Distribution of *S. noctilio* larvae parasitized by *Ibalia leucospoides* within the trees; (**c**) Distribution of *Deladenus siricidicola* within the trees (2005)

enemies of the woodwasp. As a consequence of these studies, it has been shown that *S. noctilio* has a complex life history in Patagonia which includes a 2–3 year life cycle as a diapause-mediated response to winter low temperatures. The fact that the woodwasp infests a broad group of pine species in the region was also shown.

The occurrence of natural biological control of *S. noctilio* by the exotic parasitoid *I. leucospoides* in most of the areas of the Patagonia constitutes an important biological fact to be considered in terms of the woodwasp population management and the development of conservation biological control with this parasitoid. *I. leucospoides* might not be a key factor regarding the regulation of *S. noctilio* populations but its action contributes to delay Sirex dispersal trends and also minimizes Sirex damage to plantations.

The introduction of the parasitic nematode *D. sirici*dicola into the region improved the natural mortality of Sirex and represents the most important control option for this pest in Patagonia. However, some unexpected results in the use of the nematode i.e. highly variable parasitisation rates require more specific studies regarding the quality of the nematode strain produced in the laboratory. *D. siricidicola* efficacy in controlling *S. noctilio*, strongly depends upon the infectivity of the nematode strain released. During the evaluation of Sirex parasitism by the nematode under field conditions, it was observed that some females have nematodes in the haemocoel but not in the reproductive system (ovaries, ovarioles). This indicates that the nematode can be present in parasitized Sirex females but not fully sterilize them. These findings agree with those of Bedding (1993) regarding the presence of infective-but-not-sterilizing nematode strains. It is essential to understand whether this failure to sterilize is due to inappropriate handling at the time of inoculation in the field, a change in the response of the nematodes (loss of nematode infectivity), or a host-mediated resistance response to the nematode by Sirex.

The continuous mass rearing of the nematode in the laboratory without re-introducing new infective strains can lead to a genetic bottleneck ending in loss of nematode infectivity (Bedding and Iede 2005). For this reason, mass rearing of an annual field sampling is more important than the origin of the strain (S. Penteado, personal communication 1997). Further studies regarding the nematode strains present in Patagonia Argentina are needed to compare the efficacy of different populations. Regular surveys and sampling of Sirex females from the field are required to monitor the presence of nematode-parasitized females and to determine the infectivity of *D. siricidícola.*

Despite the monitoring systems used, the increase of the natural mortality rate of Sirex by the efficient parasitism due to *I. leucospoides* and *D. siricidocola* in the region, the dispersal of the woodwasp in Patagonia has continued. Some severely damaging outbreaks have occurred since the introduction of biological control agents. These results suggest that monitoring procedures and additional research on *I. leucospoides* and *D. siricidicola* is needed until a natural balance between the pest and its natural enemies is reached.

Introductions of new entomophagous parasitoids (i.e. *M. nortoni*) to enrich the guild of natural enemies of *S. noctilio* in the region are required. This could provide a more robust Integrated Pest Management strategy for *S. noctilio.* The introduction of *M. nortoni* into the North Patagonia, provinces of Neuquén and Rio Negro in 2005 and 2009 (Sandra Ide, Servicio Agricola y Ganadero-SAG-Chile, personal communication 2010) has been successful in terms of parasitoid establishment and dispersal. This research was conducted between Argentina and Chile (SAG-SENASA), according to the bilateral Plan de Contingencia 2003–2006.

Silvicultural practices such as those implemented by Villacide and Corley (2006), including sanitary thinning of infested trees, proved to be a useful for local wasp control. This strategy is particularly important under epidemic situations were Sirex infested trees can be removed, especially those that could not be inoculated with the parasitic nematodes.

Acknowledgements Research presented in this review was partially funded by grants Proyecto Nacional Forestal PNFOR 2212 (Instituto Nacional de Tecnologia Agropecuaria), Proyecto de Investigacion Aplicada (PIA) 13/97, PIA 09/98 and PIA 03/04. Secretaria de Agricultura, Pesca y Alimentacion (SAGPyA) and PNFOR MIP de Forestales (INTA 2006–2009). We thank SENASA (Servicio Nacional de Sanidad Animal) and Servicio Forestal Andino from Bariloche, province of Rio Negro for their support.

References

Aguilar VA, Lanfranco D, Puentes O (1990) *Prospección para la detección de Sirex noctilio (Hymenoptera: Siricidae) en Bariloche – República Argentina.* (Informe de Convenio N° 180). Valdivia, Chile: Universidad Austral de Chile, Facultad de Ciencias Forestales, Serie Técnica, Convenio CONAF/UACH

Bedding RA (1993) Biological control of *Sirex noctili*o using the nematode *Deladenus siricidícola*. In: Bedding RA, Akhurst RJ, Kaya HK (eds) Nematodes and the biological control of insect pests. CSIRO, East Melbourne

Bedding RA, Akhurst RJ (1974) Use of the nematode *Deladenus siricidícola* in the biological control of *Sirex noctilio* in Australia. J Aust Entomol Soc 13:129–135

Bedding RA, Iede ET (2005) Application of *Beddingia siricidicola* for *Sirex* woodwasp control. Grewal, Ehlers and D.I. Shapiro-llan. Nematodes as Biocontrol Agents, London

Carvalho AG, Pedrosa-Macedo JH dos Santos HR (1993) Bioecologia de *Sirex noctilio* Fabricius, 1793 (Hymenoptera: Siricidae) em povoamentos de *Pinus taeda* L.Conferencia Regional da Vespa-da-Madeira, *Sirex noctilio,* na América do Sul (1992: Florianópolis). Anais. Colombo. EMBRAPA-CNPF/FAO-ONU/USDA/FUNCEMA

Echeverría N (1996). Problemática de Sirex noctilio F. XI Jornadas Forestales, Concordia, Entre Rios, Argentina. Sección I:1–3

Eskiviski E, de Coll O, Faraldo GS, Bennesch MA (2003) Efficiency of new isolations of *Deladenus siricidicola* B. in the biological control of *Sirex noctilio*. Jornadas tecnicas forestales y ambientales. FCF UNM – EEA Montecarlo, INTA, Argentina

Espinoza Zúñiga HA, Lavanderos V, Lobos C (1986) Reconocimiento de la plaga *Sirex noctilio* en plantaciones de pinos de Uruguay y Argentina. S.A.G, Santiago

Haugen DA, Underdown MG (1991) Woodchip sampling for the nematode *Deladenus siricidícola* and the relationship with the percentage of *Sirex noctilio* infected. Austral For 54:3–8

Hudson GV (1919) *Sirex juvencus* in New Zealand. Entomologist's Monthly Magazine 55:181

Iede ET, Penteado SR, Bisol JL (1988) Primeiro registro de ataque de *Sirex noctilio* em *Pinus taeda* no Brasil. EMBRAPA-CNPF, Circular tecnica 20:1–12

Iede ET, Silva SN, Penteado SRC, Machado DC (1993) Seleçao e instalaçao de árvores armadilha para detecçao da vespa-da-madeira. Embrapa, Curitiba

Klasmer P (1996) Aparición de *Sirex noctilio* en la zona de Villegas (Prov. de Río Negro). INTA: Comunicación Técnica No. 3, Recursos Naturales Forestales, Plagas

Klasmer P (1999) *Sirex noctilio* (Fab.). La avispa de los pinos en la región Andino-patagónica. Guía práctica para su reconocimiento, detección y prevención. Secretaría de Agricultura, Ganadería, Pesca y Alimentación – INTA – Proyecto Forestal de Desarrollo. Buenos Aires

Klasmer P (2001) Presencia de *Sirex noctilio* Fabricius en la provincia de Neuquén. EEA INTA Bariloche
Klasmer P (2007) The woodwasp, *Sirex noctilio* (Hymenoptera: Siricidae): ecology and biological control in Argentina. International Sirex symposium and workshop, Pretoria & Pietermaritzburg, Pretoria, 9–16 May
Klasmer P (2008) Actualizaciones Biológicas: Principales plagas forestales en la Región Andino-patagónica de Argentina. Primer Curso Internacional de Protección Fitosanitaria Forestal, Medellín, Colombia
Klasmer P, Fritz G (1994) Acciones de detección y control de *Sirex noctilio* en la región cordillerana andino patagónica argentina (Hymenoptera, Siricidae, Siricinae). Informe de la Temporada 1993–1994. INTA – Campo Forestal Gral. San Martín (El Bolsón RN)
Klasmer P, Fritz G, Corley JC, Botto EN (1997a) Current status of research on *Sirex noctilo* F. in the Andean-Patagonian region in Argentina. Proceedings of a conference: training in the control of Sirex noctilio by the use of natural enemies. Forest Health Technology Enterprise Team-USDA, Curitiba, Brazil
Klasmer P, Corley JC, Botto EN (1997b) Presencia de la avispa barrenadora de los pinos *Sirex noctilio* F. (Hymenoptera: Siricidae) en la región Andino-patagónica de Argentina. Estado actual de las investigaciones para su control biológico. Congreso Internacional de Plagas Forestales. Pucón, Chile. Sesión Nº 2: Detección, muestreo y evaluación de plagas de insectos forestales
Klasmer P, Botto EN, Corley JC, Villacide JM, Fernandez Arhex V (2000) Avances en el control biológico de *Sirex noctilio* en la región patagónica de argentina. Primer Simpósio do Cone Sul sobre Manejo de Pragas e Doenças do Pinus. Conferencia. Serie Técnica IPEF 13:21–30, Curitiba/PR, Brasil, 15 a 17 de marzo
Klasmer P, Botto EN, Corley JC, Villacide JM (2004) Evaluación del nematodo *Deladenus siricidícola* Bedding (Nematoda: Neotylenchidae) como potencial agente para el control de *Sirex noctilio* F. (Hymenoptera: Siricidae) en la Patagonia Argentina. Investigación Forestal al Servicio de la Producción II. 1997–2000. SAGPyA – Proyecto Forestal de desarrollo
Lopez A, Demaestris M, Zupan E, Barotto O (2002) Antecedentes del *Sirex noctilio* (Hymenoptera: Siricidae) en el Valle de Calamuchita, Córdoba, Argentina. Nota Técnica Bosque 23:111–114
Madden JL (1968) Physiological aspects of host tree favorability for the Woodwasp *Sirex noctilio* F. Proc Ecol Soc Aust 3:147–149
McKimm RJ, Walls JW (1980) A survey of damage caused by the Sirex woodwasp in the Radiata pine plantation at Delatite, North-eastern Victoria, between 1972 and 1979 (Forestry Technical Papers, Nº 28). Victoria: Forests Commission
Miller D, Clark AF (1935) *Sirex noctilio* (Hym) and its parasite in New Zealand. Bull Entomol Res 26:149–154
Neumann FG, Morey JL (1984) Influence of natural enemies on the *Sirex* wood wasp in herbicide-treated trap trees of radiata pine in north-eastern Victoria. Austral For 47:218–224
Neumann FG, Morey JL, McKimm RL, Neumann FG, Morey JL, McKimm RL (1987) The *Sirex* wasp in Victoria. Department of Conservation, Forests and Lands, Victoria, Bulletin N° 29
Penteado S, Do RC (1995) Métodos de amostragem para avaliacao populacional de *Sirex noctilio* F. (Hymenoptera: Siricidae) e de seus inimigos naturais, em *Pinus taeda* L. *E* aspectos do controle biologico. Tesis de Po-Graduacao, Univ. Federal do Paraná, Curitiba, Brasil
Quintana de Quinteros SL, Vilte H (1995) Detección de *Sirex noctilio* en *Pinus* spp., en la provincia de Salta. Libro Resumen. Pág. 50. IX Jornadas Fitosanitanias Argentinas, Mendoza, Argentina
Quintana de Quinteros SL, Vilte H (1996) Primer registro del ataque de *Sirex noctilio* (Hymenoptera – Sinicidae) en coníferas de la Provincia de Jujuy. Revista Científica. Facultad de Ciencias Agrarias, Universidad Nacional de Jujuy 1:27–33
Rebuffo S (1988) La avispa de la madera *Sirex noctilio* F. en el Uruguay. Montevideo, Ministerio de Ganadería Agricultura y Pesca, Montevideo, Uruguay
Reis FW (1999) Fatores biologicos e comportamentais de *Ibalia leucospoides* Hochenw. (Hymenoptera: Ibaliidae) e de seu hospedeiro *Sirex noctilio* Fabricius, 1793, (Hymenoptera:

Siricidae), visando a otimização do control biológico natural. Tesis de Pos-Graduacao em Ciências Biológicas, Universidad de Paraná, Brasil
Spradbery JP, Kirk AA (1978) Aspects of the ecology of siricid woodwasps (Hymenoptera: Siricidae) in Europe, North Africa and Turkey with special reference to the biological control of *Sirex noctilio* F. in Australia. Bull Entomol Res 68:341–359
Spradbery JP, Kirk AA (1981) Experimental studies on the responses of European siricid woodwasps to host trees. Ann Appl Biol 98:179–185
Tribe GD (1995) The woodwasp *Sirex noctilio* Fabricius (Hymenoptera: Siricidae), a pest of *Pinus* species, now established in South Africa. Afr Entomol 3:215–217
Tribe GD, Cillié JJ (2004) The spread of *Sirex noctilio* Fabricius (Hymenoptera: Siricidae) in South African pine plantations and the introduction and establishment of its biological control agents. Afr Entomol 12:9–17
Villacide JM, Klasmer P (2001) Primer Registro de la avispa de los pinos, *Sirex noctilio*, en la Comarca Andina del Paralelo 42°. Comunicación técnica N° 18. CFGSM. El Bolsón, Río Negro
Villacide JM, Corley JC (2006) Control de niveles poblacionales endémicos de la avispa de los pinos *Sirex noctilio* (Hymenoptera: Siricidae) mediante el raleo sanitario de hospederos atacados. Revista de Investigaciones Agropecuarias 35:121–134
Villacide JM, Corley JC (2008) Parasitism and dispersal potential of *Sirex noctilio* implications for biological control. Agric For Entomol 10:341–345

Chapter 16
The Woodwasp *Sirex noctilio* in Brazil: Monitoring and Control

Edson T. Iede, Susete R.C. Penteado, and Wilson R. Filho

Abstract Until 1988, a small number of outbreaks of native pests had been recorded in the 1.84 million ha of pine plantations in Brazil. Most of these stands were planted with little species diversity, and at high density and received inadequate management. In 1988, an outbreak of *Sirex noctilio* was first recorded in southern Brazil and the insect currently infests 450,000 ha in the southern and southeast regions. The presence of Sirex has the potential to cause losses of up to $US30 million annually. The discovery of this insect resulted in an immediate change in silvicultural practices used in forest plantations, with the introduction of Integrated Pest Management (IPM) as a part of the Forest Management Programs. In 1989, the National Programme of Woodwasp Control (PNCVM) was established. This programme has a broad scope and includes the early detection and monitoring of the spread of Sirex through the use of trap trees (trees artificially stressed with herbicides) and ground inspections. Appropriate silvicultural practices, especially stand thinning, preventing competition and providing healthy growing conditions are also an important aspect of control programs. Curative measures used are based on biological control using the nematode parasite *Deladenus* (=*Beddingia*) *siricidicola*, introduced from Australia, which sterilizes the females and reaches up to 100% parasitism. The egg parasitoid *Ibalia leucospoides* has also been introduced and can reach about 25% parasitism. Two other parasites, *Rhyssa persuasoria* and *Megarhyssa nortoni* were introduced into Brazil between 1996 and 1998, and in 2003, from Australia. Establishment of these two parasites has not been confirmed. This combination of biological control and appropriate silvicultural practices are essential to keep the Sirex population under control in Brazil.

E.T. Iede (✉) • S.R.C. Penteado • W.R. Filho
Embrapa Florestas, Estrada da Ribeira km 111, Cx. Postal 319, CEP 83.411-000
Colombo, PR, Brazil
e-mail: iedeet@cnpf.embrapa.br

B. Slippers et al. (eds.), *The Sirex Woodwasp and its Fungal Symbiont: Research and Management of a Worldwide Invasive Pest*, DOI 10.1007/978-94-007-1960-6_16,
© Springer Science+Business Media B.V. 2012

16.1 Introduction

Sirex noctilio F. (Hymenoptera: Siricidae) is a minor pest in its regions of origin, Europe, Asia and North Africa (Chap. 5). In contrast, it has become the main pest of pine plantations where it has been introduced, in countries such as New Zealand, Australia, Uruguay, Argentina, Brazil, Chile and South Africa (Chaps. 13–18).

In South America, *S. noctilio* detected for the first time, in Uruguay in 1980 infesting several species of *Pinus*, with *P. taeda* being the most susceptible (Rebuffo 1988). Its presence in Argentina was first recorded in the province of Entre Rios, in the northwest of the country, between 1985 and 1986, where it was found infesting *P. taeda* and *P. elliottii* (Espinoza et al. 1986, Chap. 15). The first record of *S. noctilio* in Brazil was in February 1988, in a *P. taeda* plantation, in the municipal districts of Gramado, Canela and São Francisco de Paula, in Rio Grande do Sul State (Iede et al. 1988). In the following year, the insect was detected in trap trees, in a *P. taeda* plantation located in the municipal district of Lages, in Santa Catarina State. In Paraná State, two interceptions occurred in 1993 and 1994, but there was no evidence of pest establishment. However, in July 1996, its establishment was verified in *P. taeda* plantations in several municipal districts of Paraná State (Iede et al. 1998). The first record in the Southeast Region was in 2005, in São Paulo and Minas Gerais States. The woodwasp can now be found in 450,000 ha of pines throughout the states of Rio Grande do Sul, Santa Catarina, Paraná, Sao Paulo and Minas Gerais.

In Brazil there are 1.84 million ha of pine plantations. Most of the areas where *S. noctilio* is established have historically been planted with little species diversity and at a high density, and in some cases subject to inappropriate silvicultural practices. These conditions are ideal for pest and disease outbreaks. In the Southern and Southeastern regions, there are approximately 1.2 million ha of mainly *P. taeda* (loblolly pine) and *P. elliottii* (slash pine) established in plantations.

The presence of *S. noctilio* in Brazil awakened the Brazilian forest sector to the need for control strategies, prevention and monitoring actions because this exotic woodwasp has caused significant economic losses, mainly in the *P. taeda* and *P. elliottii* plantations. The National Fund for Woodwasp Control (FUNCEMA) was thus created. FUNCEMA involves public and private organizations working together to support the National Programme of Woodwasp Control (PNCVM).

Monitoring with trap trees is the most important activity of this program. In areas where the pest has not yet been detected, it is possible to carry out early detection and immediate release of biological control nematodes when necessary. It has been recommended that trap tree monitoring can cease when the number of attacked trees is 1%, and at that point to start using the sequential sampling technique described by Penteado et al. (1998, 2002a) (see discussion below). Biological control is the most efficient means to keep *S. noctilio* under control, especially when the nematode, which sterilizes up to 70% of the females, is used. It is also important to use forest management to maintain high levels tree vigor and health in the stands.

16.2 Biology and Ecology of *Sirex noctilio*

In Brazil, *S. noctilio* infests *Pinus taeda, P. elliotti, P. patula* (mainly in Southern region of Brazil) and *P. oocarpa, P. caribaea caribaea, P. caribaea bahamensis, P. caribaea hondurensis, P. kesiya* and *P. strobus chiapensis* (in the Southeast Region). In Brazil, the normal life cycle of the insect usually takes 1 year; however, it can have a short cycle, also called a summer cycle, which varies from 3 to 4 months. This cycle can occur in dominated or even in co-dominant trees (Iede et al. 1988). In the South of Brazil, Carvalho et al. (1993) and Iede et al. (1993) verified that *S. noctilio* adults emerge between October and April, with peaks in emergence during the months of November and December.

According to Carvalho (1993), the longevity of adults in the field is of 4 and 5 days for females and males, respectively. Iede et al. (1988) obtained an average of 8 days survival for adults, in climatic chambers (25 ± 1°C; RH 70 ± 10%, and photoperiod of 12 h). The males emerge before the females and the male/female ratio varies from 1.5:1 to 32:1.

After the initial flight period, females penetrate the tree trunk with their ovipositors and lay their eggs in the sapwood; they can lay up to four galleries in each hole, with an average of 2.2 eggs per hole. Each female lays an average of 210 eggs in approximately 5 days (Carvalho et al. 1993). During this process, females introduce spores (arthrospores) of the symbiotic fungus (*Amylostereum areolatum*) and a mucous secretion; both the fungus and the toxin is thought to contribute to the death of trees. The pathogenic fungus, which is also a source of nutrients for the larvae, dries up the wood and causes rot. The wood quality is also affected by the larvae constructing tunnels and by the entrance of secondary agents that help damage the wood, limiting its use or ruining it for the wood market. Once trees are dead, the wood degrades rapidly and it must be used within 6 months of infestation. Penteado et al. (2000) verified that most of the larvae are distributed in the top half of the tree.

Plantations that are more susceptible to *S. noctilio* are generally up to 8 years-old and under stress. Stands that have not been thinned are more susceptible to infestation than those that have been thinned. The symptoms of infestation begin right after the insect population peaks, in November and December, but become more visible after March. The external symptoms of attack are a progressive yellowing of the crown, which later becomes brownish-red, wilting of the foliage, loss of needles, resin drops on the bark (due to the oviposition holes) and emergence holes. Internal symptoms are brown spots along the inner bark caused by the fungus *A. areolatum* and tunnels drilled by larvae, which affect wood quality.

16.3 The National Program of Woodwasp Control in Brazil

Sirex noctilio is present in 450,000 ha, in the Southern and Southeast regions of Brazil. The Brazilian forest sector went through a difficult phase due to the damage by this pest. It is estimated that the potential losses caused by this insect may reach

US$ 30 million annually. Urgent and efficient strategies were required to monitor, control and delay the spread of *S. noctilio*. The seriousness of the situation resulted in an immediate change in the handling of forest plantations, with introduction of Integrated Pest Management (IPM) as part of the general forest management programs. Biological control associated with prevention measures is the most effective and economical method to combat damage due to this insect, because as an exotic pest, it was introduced without its complex of natural enemies (Iede et al. 1998).

In June 1989, FUNCEMA was created to manage the *S. noctilio* problem in Brazil. FUNCEMA is a not-for-profit civil organization formed by private and public institutions whose main objective is to generate funds for the development of the National Programme of Woodwasp Control (Iede et al. 1989). This program includes research and development activities such as: (a) monitoring for the early detection of the pest using trap trees intentionally stressed with herbicides to attract the insects; (b) adopting prevention strategies to improve the phytosanitary conditions of forest stands using silvicultural practices, especially thinning, to prevent and minimize attacks; (c) adopting quarantine strategies to control and slow-down dispersal; (d) mass rearing and releasing the nematode *Deladenus siricidicola* (Nematoda: Neotylenchidae); (e) mass rearing and releasing the parasites *Rhyssa persuasoria* and *Megarhyssa nortoni* (Hymenoptera: Ichneumonidae) collected in Australia; (f) field-collecting and releasing the parasite *Ibalia leucospoides* (Hymenoptera: Ibaliidae), to increase the range of natural enemies; (g) developing an extensive training program for technical personnel and forestry workers to improve pest management; and (h) providing information to the public (Iede et al. 1989).

The National Programme of Woodwasp Control is an excellent example of the integration of a research and development policy nationwide, as more than 120 private companies in Southern Brazil are involved together with public organizations. The companies also offer technical assistance to small private forest owners so the control strategies will reach all plantations attacked by the woodwasp (Iede et al. 1989).

16.3.1 Prevention Strategies

Trees resistant to *S. noctilio* are those that remain free of injuries and continue to grow vigorously on good sites and in well-managed blocks. The level of mortality of the trees is significantly related to their diameter at breast height (DBH). Trees with small DBH are more likely to be attacked than larger individuals in the same stand. However, during the event of a broad abiotic disturbance dominant trees can also be attacked.

Effective forest management practices limit rotation lengths and encourage forest managers to consider the composition, structure, age and vigor of the plantation. According to Davis (1966), more effective pest control may be obtained through silvicultural practices in the long run, creating reasonable insect-resistance in the forest. Complete control will never be attained this way, but the loss caused by insects can be reduced.

Thinning is one of the most important silvicultural practices in pine plantation forestry. It is conducted to accelerate or modify the course of competition. The position of the crown is an important characteristic in deciding which trees to cut and which to retain. Vigorous trees that outgrow their neighbors dominate the canopy and usually have a greater chance of surviving future competition than less dominant ones, which take lower positions in the plantation. Most of the thinning practices reduce loss due to damaging agents not only because thinning is a prevention strategy but also because it increases the vigor and resistance of the remaining trees. Thinning can enhance the susceptibility of trees to insect attack only under certain circumstances, such as when it is carried out during the flight period of the pest. Pruning should also be avoided during the flight season because pruning-related injuries release some chemical compounds, like monoterpenes, that attract the female woodwasps to oviposit in the damaged trees. Even in well-managed stands, e.g., a thinned, 8 year-old, low density (650 trees/ha) stand with trees 28 cm DHB, was observed tree mortality of 8% because pruning was done during the *S. noctilio* flight season.

16.3.2 Early Detection and Monitoring of S. noctilio

16.3.2.1 Early Detection Using Trap Trees

The use of trap trees stressed by herbicide injection is the most appropriate and efficient technique for the early detection of the pest, and for monitoring Sirex dispersal. Detecting *S. noctilio* during its early stages of development and colonization helps to define locations for release of biological control agents and allows thinning practices to be carried out before the pest reaches high levels of damage. Maintaining a trap tree system can greatly increase the efficiency of the biological control of this woodwasp.

At the beginning of the IPM Sirex project in Brazil, the herbicide Dicamba was used to stress trees, but after some years, the product was no longer sold in Brazil. Some experiments were carried out with other products; Padron, which showed efficacy similar to Dicamba, has been used since that time.

The detection method, as well as the intensity with which it should be applied, must be based on a risk analysis of the introduction and dispersal of the pest in each region. Penteado et al. (2002b) recommended that the trees intended as trap trees should have a DBH of between 10 and 20 cm and be installed in groups of five trees. In areas where Sirex is already present, as well as in areas up to 10 km away from the infestation foci, the recommendation is to install groups of five trap trees every 500 m, which means, four groups of five trees in each 100 ha. At distances between 11 and 50 km from the infestation, the trap tree groups should be spaced 1,000 m apart.

The installation of trap trees in Brazil must be done from August 15 to September 30, 2 months before the population of the adult woodwasps peak, which generally occurs between November and December. During February to May, the groups of trap trees should be revisited to check for insect attack (Penteado et al. 2002b). Trap trees should be installed in susceptible plantations, close to saw mills, along the main wood transportation routes and on the borders of the area of natural dispersal of the pest. After detection, trap trees must be installed annually in areas with percentages of attack less than 1%. When more than 1% of the trees are attacked, the trap tree method is changed to the sequential sampling method.

16.3.2.2 Remote Sensing

The monitoring of *S. noctilio* attack is an essential component of an integrated pest management program. The national program originally intended to have all pine plantations in the country mapped from satellite images with the attacked trees identified on the map. Although it was possible to identify pine plantations, it was not possible to conclusively identify trees that had symptoms of Sirex attack.

Aerial sketch mapping was tested for monitoring and early detection of the pest, but because the pest initially attacks subdominant trees under the forest canopy, the method was not appropriate. According to Haugen et al. (1990), infested trees are not always visible in an aerial inspection, and this kind of evaluation is not recommended when the level of attack is lower than 5% (also see Chap. 14).

16.3.2.3 Sequential Sampling

Haugen et al. (1990) observed that methods for monitoring *S. noctilio* in Australia at the time (see Chap. 14 for current methods in Australia) were characterized by low precision and high operational costs, emphasizing the need for the development of more practical and economical methods. It was, therefore, necessary to establish a damage sampling technique that is practical and takes into consideration the vast stands of pines, the levels of attack, and the cost of the activity.

Sequential sampling, developed by Penteado et al. (1998, 2002a), in which the size of the sample is defined according to the results obtained during the application of the procedure, has proven to be a good alternative for evaluating the attack intensity of *S. noctilio* in pine plantations. This type of sampling reduces the time and cost of application and gives immediate results. The use of a sequential sampling Table (Table 16.1) reduced the size of the sample between 45% and 86%, depending on the level of attack and in comparison to the fixed-size procedure generally recommended. Sequential sampling saves time and money and increases precision when reduced sample sizes are used.

The sequential sampling technique is used as follows:

1. Start by sampling 68 trees;
2. Note the number of trees attacked in the sample in the second column of the table and compare it to the number of trees attacked shown in the third column of the table, in this case 34;

Table 16.1 Table used to determine sample size for evaluating levels of attack of *Sirex noctilio* in *Pinus* stands, based on sequential sampling

	Number of trees attacked	
Number of trees sampled	Trees attacked in sample	Minimum to interrupt sampling
68	–	34
74	–	36
80	–	37
87	–	38
94	–	39
102	–	41
111	–	42
121	–	44
132	–	45
145	–	46
159	–	48
175	–	49
194	–	50
215	–	52
241	–	53
272	–	54
272	–	49
272	–	44
272	–	38
272	–	27
272	–	22
272	–	16
272	–	11
272	–	5

3. If the number of trees attacked in the sample is equal to or greater than 34, the sampling procedure is complete;
4. If the number of trees attacked in the sample is lower than 34 the procedure must be continued by sampling six more trees, totalling 74 sampled trees;
5. If the number of trees attacked is 36 or more, the sampling should be interrupted;
6. If the number of trees attacked is lower than 36, the sampling must continue until the number of attacked trees shown in the third column of the table is attained or until 272 trees have been sampled;
7. Calculate the percentage of trees attacked as a function of number of trees sampled.

When establishing the sampling path in the plantation for this type of inspection, consider the distance covered and the amount of time spent, and maximize the coverage of the area. To start the sampling procedure, select a row (path) in which 40 trees will be checked. When this number is achieved in one row, skip five rows and begin again on 40 trees of the fifth row, continuing until the sampling is completed.

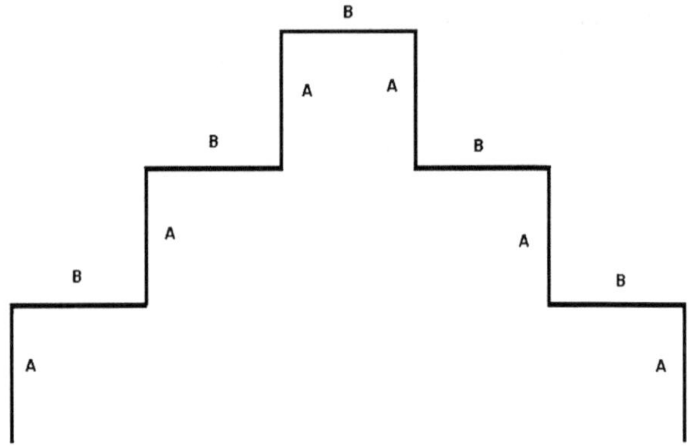

Fig. 16.1 Pathway to conduct sequential sampling. (**a**) Evaluation of *S. noctilio* attack on 40 trees, (**b**) Five or ten rows are skipped

In blocks larger than 10 ha, however, it is advisable to skip ten rows instead of five to guarantee greater coverage of the area under evaluation. This method of sampling in a plantation is similar to a step-like pathway (Fig. 16.1). We have observed that Sirex attacks are usually randomly distributed (not in groups or clusters) regardless of their intensity. This has also been noted by the National Sirex Co-ordination Committee (1991), and contributes to the generation of more precise sampling results.

It is recommended that sequential sampling be applied in areas up to 50 ha in size. The criteria that define the number of necessary samples in a forest block, is characterized by the soil type, relief, management practice, age and density, among other factors, can be used to define the appropriate sampling intensity for each area according to its size. The stands should, therefore, be sampled block by block, as these are considered homogeneous. If, however, there are biotic or abiotic conditions within a block that may stress the trees, at least one extra sampling unit is recommended.

The best time for sampling must be defined by its purpose. In Brazil, *S. noctilio* adults from the normal life cycle are found between the second half of October and the first half of January. Thus sampling done between March and April will indicate the number of trees that should be inoculated with nematodes. The nematode inoculations are done between March and August. Sampling conducted later than that, between June and July, will indicate the total mortality in the period but may interfere with the planning of the activities of inoculation with nematodes, as has also been observed by the National Sirex Co-ordination Committee (1991).

16.3.3 Biological Control

Biological control, along with prevention strategies, is the most efficient and economical method for controlling *S. noctilio* in Brazil. A system of control similar to that adopted in other countries (see for example Chaps. 8, 9, 13–18) where Sirex has been introduced was developed in Brazil. These biological controls include the nematode *D. siricidicola* and the parasites, *I. leucospoides, R. persuasoria* and *M. nortoni.*

16.3.3.1 Nematodes

The most effective biological control agent of the woodwasp is the nematode *D. siricidicola*, which sterilizes the females. This nematode has two life cycles: a free-living cycle, when it feeds on the same symbiotic fungus as the woodwasp and a parasitic-life cycle, which occurs inside larvae, pupae and adults of *S. noctilio*. Because its free-living cycle feeds on the fungus *A. areolatum,* the nematode is easily bred under laboratory conditions and then released in the field by application into trees attacked by *S. noctilio.* The 20 ml doses of nematodes contain approximately one million nematodes (Bedding 1972). The inoculation of nematodes into trees is done with a specially developed hammer that is used to make holes in the trunks of infested tree at a 30 cm spacing (see images in Chap. 14). The doses are mixed in a 10% solution of gelatine and introduced by a syringe into the holes in the wood made with the hammer (Penteado et al. 2002b). After inoculation, the nematodes penetrate the wood in search of the fungus they feed on and reproduce in free-living cycle. However, when they locate Sirex larvae, they develop into infective adults and penetrate the larvae, leaving scars in their integuments. When the host pupates, the juvenile nematodes are released into the body of the host and move to the reproductive organs. They then penetrate the eggs, sterilizing the female. The infected adult female Sirex emerge from the trees and lay their eggs in susceptible trees but these eggs are not fertile, and can contain between 100 and 200 nematodes (Bedding 1972).

In Brazil, the first introduction of nematodes was in 1989 and 1990, from the Australian *Commonwealth Scientific and Industrial Research Organization* (CSIRO). The inoculations were done between March and August, when most of the larvae are inside the tree, and the nematode can be more effective. Three years after the first inoculations, it was observed that parasitism levels were very low. The average levels of parasitism obtained in Australia with the nematode was 70%. The level of parasitism verified for the nematode in attacked areas in Rio Grande do Sul and Santa Catarina states has been quite variable. It was found to be as high as 70% in a 12,000 ha *P. taeda* plantation, in Encruzilhada do Sul, Rio Grande do Sul, but in some places we have observed parasitism levels ranging from 10% to 90%.

It was discovered that the introduced strain of nematode became defective because it had been cultured in the free-living form for over 20 years in Australia without the intervention of the parasitic life cycle, which led to the selection of a strain that rarely formed the infective stage. Unfortunately the situation did not become apparent until this defective strain had been released for many years (Bedding and Iede 2005; also see Chap. 9). Two strategies were used to solve the problem in Brazil. These included importing (in 1994) the Kamona strain, which was re-isolated in Tasmania, from CSIRO, Australia, and releasing it in 1995; and also re-isolating the nematode from the field, in Brazil every year, and introducing the field-captured nematodes into the mass rearing programme. Every year, the nematode parasitism levels are evaluated and parasitism levels are currently above 70%. In most places where the nematode was released, it is difficult to find attacked trees because the Sirex population is under control.

16.3.3.2 Parasites

The parasite *Ibalia leucospoides* was recorded in Brazil for the first time in December 1990, in a pine plantation attacked by the woodwasp around the city of São Francisco de Paula, Rio Grande do Sul (Carvalho et al. 1991). It is now established in all areas where the woodwasp occurs with parasitism as high as 39%, but averaging around 25%.

Megarhyssa nortoni and *Rhyssa persuasoria* were introduced to Brazil from Tasmania in 1996, 1997 and 1998, in a cooperative project between Embrapa Florestas, International Institute of Biological Control (CABI-Bioscience) and the USDA (United States Department of Agriculture) Forest Service. The first introduction of *M. nortoni* was done in 1996 with 133 individuals (88 females and 45 males). In October 1997, an experimental release of *M. nortoni* was conducted in Correia Pinto, in Santa Catarina State. In this place 18 previously mated females were released next to trap trees. In October 1998, to verify the establishment of the parasite, traps were installed on 10 trees near the points of release and pine logs were collected from ten trees located near the point. However, the presence of the insect was not recorded. In 1998, five field releases were carried out between October and December, with a total of 97 females and 134 males of *M. nortoni* and ten females and two males of *R. persuasoria.* Areas selected for release had been attacked by *S. noctilio*, were close to native forests, and had not been thinned to allow the establishment of parasites. Also in 1998 parasites were introduced from Tasmania, however, due to problems during the collection of insects in the field, the number of insects was very low, with only 15 females of *M. nortoni* and one female of *R. persuasoria* being released.

In 2000, a total of 86 males and 67 females of *M. nortoni* and 35 males and 70 females of *R. persuasoria* were used to determine longevity at different temperatures. Average longevity for *M. nortoni* was 17.9 and 68.87 days, at 23°C and 12°C, respectively, and for *R. persuasoria,* average longevity was 20.11 and 50.28 days at 23°C and 12°C, respectively. In 2005, *M. nortoni* adults were released in one area located in Rio Branco do Ivaí, Paraná, but its establishment has not yet been confirmed.

16.3.4 Quarantine Strategies

Sirex noctilio can spread naturally between 30 and 50 km/year. However, the transportation of infested wood to uninfested areas increases the possibility of dispersal.

Probably this is the way the pest was introduced into Brazil from Uruguay. For that reason, monitoring affected areas and a quarantine system that prohibit the transportation of infested wood outside defined quarantine areas are needed.

In Brazil, moving wood from a Sirex-infested area to a Sirex-free area is permitted with an Origin Phytosanitary Certificate provided by the National Plant Protection Organization. In areas near the border of areas known to contain Sirex, trap tree monitoring is required each year. Fumigation of wood or heat treatment (minimum of 56°C throughout the profile of the wood for a minimum of 30 min) to control insects inside the wood is also required.

16.4 General Recommendations

Sirex noctilio is essentially a secondary and an opportunistic pest. The use of prevention measures in pine plantations is the most appropriate way to avoid economic losses. These preventive measures can be achieved by monitoring the plantations and by the use of appropriate silvicultural practices, such as thinning. After the initial damage by *S. noctilio* in Brazil, it was estimated that most plantations showed low levels of mortality and only a small portion severely infested. However, this status will change if monitoring, prevention and control strategies are not carried out.

Further research is needed to understand the genetics of the nematode *D. siricidicola* and its symbiotic fungus *A. areolatum* in different regions Brazil. This is especially because differences in levels of parasitism between different locations and also differences in the development of different isolates of the nematode and fungus in the laboratory have been observed. Cryopreservation of isolates of the fungus and nematode is needed to prevent loss of nematode infectivity and to be able to preserve isolates selected as promising for biological control.

References

Bedding RA (1972) Biology of *Deladenus siricidicola* (Neotylenchidae) an entomophagous-mycethophagous nematode parasitic woodwasps. Nematology 18:482–493

Bedding RA, Iede ET (2005) Application of *Beddingia siricidicola* for sirex woodwasp control. In: Grewel PS, Ehlers R, Shapiro-Ilan DI (eds) Nematodes as biocontrol agents. CAB International, Wallingford

Carvalho AG (1993) Parasitismo de *Ibalia* sp. (Hymenoptera: Ibaliidae) em *Sirex noctilio* Fabricius, 1793 (Hymenoptera: Siricidae) em São Francisco de Paula, RS. Boletim de Pesquisa Florestal, Colombo, no. 26/27

Carvalho AG, Berti Filho E, Vadas J, Maia SJLS (1991) Ocorrência de *Ibalia* sp.(Hymenoptera, Ibaliidae) em *Sirex noctilio* F. (Hymenoptera, Siricidae) no Rio Grande do Sul. Revista de Agricultura da USP, vol 66. USP (Escola Superior de Agricultura Luiz de Queiroz), Piracicaba – São Paulo

Carvalho AG, Pedrosa-Macedo JH, Santos HR (1993) Bioecologia de *Sirex noctilio* F. 1793 (Hymenoptera: Siricidae) em povoamentos de *Pinus taeda* L. In: Conferência Regional da

Vespa-da-Madeira, *Sirex noctilio*, na América do Sul, Florianópolis, 1992. *Anais*. Embrapa-CNPF/FAO-ONU/USDA – Forest Service, Colombo
Davis KM (1966) Forest management: regulation and valuation, 2nd edn. McGraw-Hill, New York
Espinoza ZH, Lavenderos A, Lobos C (1986) Reconocimiento de la plaga *Sirex noctilio* en plantaciones de pinos de Uruguay y Argentina. Corporación Nacional Forestal, Santiago
Haugen DA, Bedding RA, Underdown MG, Neumann FG (1990) National strategy for control of *Sirex noctilio* in Australia. Aust For Grow 13:1–8, special liftout section
Iede ET, Penteado SRC, Bisol JC (1988) Primeiro registro de ataque de *Sirex noctilio* em *Pinus taeda* no Brasil. EMBRAPA-CNPF, Circular Técnica, Colombo
Iede ET, Bedding RA, Penteado SRC, Machado DC (1989) Programa Nacional de Controle da vespa-da-madeira-PNCVM. EMBRAPA-CNPF, Curitiba
Iede ET, Penteado SRC, Gaiad DCM, Silva SMS (1993) Panorama a nível mundial da ocorrência de *Sirex noctilio* F., (Hymenoptera:Siricidae). In: Conferência Regional da Vespa-da-Madeira, *Sirex noctilio*, na América do Sul, 1992, Florianópolis. Anais. EMBRAPA-CNPF, Colombo
Iede ET, Penteado SRC, Schaitza EG (1998) *Sirex noctilio* problem in Brazil - detection, evaluation and control. In: Iede ET, Schaitza EG, Penteado SRC, Reardon R, Murphy T (eds) Training in the control of *Sirex noctilio* by use of natural enemies. USDA Forest Service, Morgantown
National Sirex Co-ordination Committee (1991) National Sirex control strategy - operations worksheets. Canberra, Australia
Penteado SRC, Oliveira EB, Iede ET (1998) Sampling methods for evaluating *Sirex noctilio* attack levels in *Pinus taeda* stands and for monitoring the efficiency of its natural enemies. In: Iede ET, Schaitza EG, Penteado SRC, Reardon R, Murphy T (eds) Training in the control of *Sirex noctilio* by use of natural enemies. USDA Forest Service, Morgantown
Penteado SRC, Oliveira EB, Iede ET (2000) Distribuição da vespa-da-madeira e de seus inimigos naturais ao longo do tronco de *Pinus*. Boletim de Pesquisa Florestal, Colombo
Penteado SRC, Oliveira EB, Iede ET (2002a) Aplicação da amostragem sequencial para monitoramento dos níveis de ataque de *Sirex noctilio* em povoamentos de *Pinus taeda*. Embrapa Florestas, Boletim de Pesquisa e Desenvolvimento, Colombo
Penteado SRC, Iede ET, Reis Filho W (2002b) Manual de controle da vespa-da-madeira (*Sirex noctilio*) em plantios de *Pinus* spp. Embrapa Florestas, Colombo
Rebuffo S (1988) La "avispa de la madera" *Sirex noctilio* F, en el Uruguay. Montevideo: Ministerio de Ganaderia, Agricultura y Pesca, Direccion Forestal. Serie Proteccion Forestal 1:1–14

Chapter 17
Surveillance and Control of the Sirex Woodwasp: The Chilean Experience

Marcos Beèche, Dolly Lanfranco, Mario Zapata, and Cecilia Ruiz

Abstract *Sirex noctilio* was first detected in Chile in the Valparaíso Region in 2001. This led to an intensive attempt to eradicate the pest. The same year and after the detection of the pest in Los Lagos Region and its expansion to other regions north to Los Lagos, actions to deal with the pest were focussed on its containment and suppression. With this goal, Chile implemented quarantine and biocontrol measures with the parasitic nematode *Deladenus* (*=Beddingia*) *siricidicola* and the insect parasitoids, *Megarhyssa nortoni* and *Rhyssa persuasoria. Ibalia leucospoides* had been independently introduced earlier. Research and technical programs financed mainly by the Chilean Government in collaboration with private forestry companies and the regulations and measures implemented by the Servicio Agrícola y Ganadero made it possible to maintain the Sirex populations at tolerable limits. An important result of the control programs have been the successful establishment of various biological control agents and growing levels of parasitism of *D. siricidicola.*

17.1 Introduction

Continental Chile is approximately 4,300 km long and 427 km wide at its widest point. Chile is 75.5 million ha in size; approximately 11% of this area is forested. This forested area is comprised of about 4.75 million ha of unprotected native forest,

M. Beèche (✉) • M. Zapata
Subdepartamento Vigilancia y Control Oficial Forestal, Divisiόn de Protecciόn Agrícola y Forestal, Servicio Agrícola y Ganadero, Paseo Bulnes 107, Off. 24, Santiago, Chile
e-mail: marcos.beeche@sag.gob.cl

D. Lanfranco • C. Ruiz
Instituto de Silvicultura, Facultad de Ciencias Forestales y Recursos Naturales, Universidad Austral de Chile, Campus Isla Teja, Valdivia, Chile

B. Slippers et al. (eds.), *The Sirex Woodwasp and its Fungal Symbiont: Research and Management of a Worldwide Invasive Pest*, DOI 10.1007/978-94-007-1960-6_17,
© Springer Science+Business Media B.V. 2012

1.4 million ha of protected native forest and approximately 7.5 million ha of commercially productive native forest. The south-central part of Chile (35° S–42° S) is considered the most important area for commercial forestry because it provides the most suitable conditions for the growth of *Pinus radiata* and other *Pinus* species. Extensive establishment of commercial pine plantations began in the 1970s with the implementation of government incentives for plantation development. These incentives promoted the expansion of the forest products industry and subsidized the majority of costs for reforestation and plantation establishment. The forest sector in Chile currently comprises 2.2 million ha of plantations lying between 32° and 42° latitude south (the south-central region), 65% of which are planted to *P. radiata*. Plantations of *P. ponderosa* and *P. contorta* have also recently been established in Chile mainly south of the 45th parallel (INFOR 2007).

Forestry in Chile has experienced marked growth in the last few decades and has evolved both in technological and economic terms. This growth is reflected in the strong presence of the Chilean forest sector in the world's most demanding markets, which require high quality products and increasingly strict environmental standards of production. Large companies met this challenge early by adopting environmental management systems. They currently use the Chilean System for Sustainable Forest Management Certification, which grants a seal of approval required mainly by European and North American markets (Comisión Nacional de Buenas Prácticas Agrícolas 2006).

In light of the detection of *Sirex noctilio* in various South American countries (Argentina, Uruguay and Brazil) during the 1980s, a National Program for the Early Detection of *S. noctilio* in Chile was implemented in 1990 under the national coordination of Servicio Agrícola y Ganadero (SAG). The program included the full participation of the primary Chilean forestry companies through an agreement of understanding, which included surveillance and control under the standardization of work protocols which are still in operation.

The National Programme for the Early Detection of *S. noctilio* involved specific forest health surveys, the establishment of a network of trap trees and a monitoring program using funnel traps to detect the pest and initiate early control measures. As a result, the first outbreak of *S. noctilio* was detected in Chile in January 2001 in a trap baited with alpha-pinene and ethanol located near Guardia Vieja, Valparaíso Region. This outbreak was associated with a nearby hydroelectric facility that had imported large quantities of equipment from Europe in wood packaging materials. In March of the same year, a *S. noctilio* female was captured in Ensenada, Los Lagos Region, about 1,000 km south of the first outbreak. In both cases the presence of immature individuals of the pest in *P. radiata* trees was confirmed. The most recent discovery is believed to have been a result of migration of the pest from southern Argentina.

As a consequence of the first detections of *Sirex* in 2001, an Official Control Program for *S. noctilio* in Chile was developed and implemented in the same year. The chief objective was to avoid damage to commercial *P. radiata* plantations in the country. The program includes a strategy for containment and suppression of outbreaks of the pest, including surveillance, outbreak control and quarantine measures in Chile. It also includes cooperative measures with neighbouring countries to reduce the infestation of new areas in accordance with the phytosanitary guidelines set by the International Plant Protection Convention (IPPC).

Given the importance of the approximately 1.5 million ha of commercial *P. radiata* plantations and the damage that *S. noctilio* can potentially cause to the Chilean forest industry, the pest was declared under official control as a part of the national strategy for its detection and control. This strategy includes: eradicating or suppressing new outbreaks to reduce infestations to protect commercial pine plantations; increasing the surveillance in pest risk areas for the quick detection of the pest; and developing a biological control program. Chilean forestry companies have taken part in the program from the beginning by applying technical surveillance and control protocols in their forested lands. They have also participated in pest control activities in neighbouring countries to reduce the risk of introduction into new areas.

This review outlines the Chilean experience relating to the surveillance and control of *S. noctilio*. We begin by briefly providing some background on the biology of *S. noctilio* and the history of forest health practices in Chile.

17.2 Forest Health in Chile

Countries such as Australia, New Zealand and Chile have established extensive, continuous plantations of exotic species of the genus *Pinus*, particularly *P. radiata*. These plantations are homogenous not only in terms of species but also in terms of age, density and management. A similar situation exists in Brazil, Argentina, Uruguay and South Africa with species such as *P. taeda*, *P. elliotti* and *P. patula*. However, in those areas, the size and diversity requirements of plantations are standard. All of these countries try to maximize production by planting the species best suited to the forest soil types in their geographic region. The motivation for this approach is decidedly economic, and it seems that there has been no real concern for risks such as pests and fire that are inherent in the establishment of extensive monoculture plantations. However, increased international trade during the past decades, particularly involving wood products used as packaging material for the transportation of goods, have increased the risk of new pests being introduced. These new pests can cause serious damage because they normally arrive without their natural enemies.

Sirex noctilio is an economically important pest in areas dominated by plantations of *P. radiata, P. taeda, P. elliotti* and *P. caribaea.* Plantations are more susceptible when established on poor sites, at high densities, in extensive stands of the same age and without adequate silvicultural planning and management. *Sirex noctilio* seems to prefer codominant or suppressed trees with a diameter at breast height of greater than 12 cm. Abiotic factors such as prolonged drought, wind and snow can also greatly influence the establishment and dispersal of the pest.

Frequent monitoring during the rotation of a forest crop will ensure that infestations are discovered. Abiotic factors can also influence the establishment of a pest, and must be considered. The predictability of these factors increases with the availability of long-term data, facilitating decision-making and simulation modelling of different disturbance parameters. The importance of forest health information should not be underestimated because its absence can lead to huge losses. The Green Triangle outbreak in Australia is a good example, where use of the nematode *Deladenus siricidicola* failed after some years of mass rearing, losing its effectiveness as a control method. Thus, the lack of sufficient surveillance allowed a new and serious outbreak to occur (Haugen 1990; Haugen and Underdown 1990).

Many of the factors that favour the arrival of a pest can be avoided through effective forestry practices and the maintenance of forest health. In terms of silviculture, preventive actions can include quarantine, sanitary thinning and wood waste management, such as is practiced in Chile and Argentina. These practices are effective for the control of *S. noctilio* at the local level (Villacide and Corley 2006, 2007).

17.3 *Sirex noctilio*: A Remarkable Taxon

Anyone who has studied *S. noctilio* must recognise that they are dealing with a remarkable taxon and this is especially true for the female wasps. From the biological point of view, various characteristics are fundamental to the life history strategy of *S. noctlio*. These insects have mycangia that have evolved with the fungus *Amylostereum areolatum* and whose hyphae provide food for the larvae during their first stages of development. They also have reservoirs for the phytotoxic mucus that prepares the tree for oviposition. The presence of pro-ovigenic females with a haplodiploid reproductive system is remarkable as is the ovipositor adapted to drill into the bark and deposit its eggs and the insect has is able to disperse very effectively. Their reproductive strategy is perfectly synchronized between the pest, the fungi and the susceptible trees.

For effective *S. noctilio* infestation, the quality or quantity of susceptible trees is important, depending on population levels. Under low population levels of the pest, high density stands and suppressed, codominant trees, broken by wind are most susceptible, while at high population levels all trees are susceptible. Sex ratio has been considered a good indicator of population age of *S. noctilio*. When *Sirex* first arrives in a new area, there are more males than females, while in areas where the pest is well established the sex ratio tends to be closer to 1:1. Pest management decisions were made based on this knowledge of the biology and ecology of this pest in areas of southern Chile where it has been established since 2001. For example, the population level and sex ratio may help to explain the time of the infestation, its probable origin and allow forest managers to make the best decision in a specific area (e.g., sanitation versus biological control, declaration of a new quarantine area or selected areas for intensive surveillance).

The symbiotic fungus, *A. areolatum* has recently received special attention. Molecular techniques have provided news tools for identification, understanding the mutualistic associations of *Amylostereum* species with different species of siricids, and the phylogenetic relationships among its species. Advances have also been made in the understanding of the ecology, biology, behaviour and functions of these species (Slippers et al. 2000, 2002, 2003). A collaborative project between the University of Pretoria (South Africa) and Uppsala University (Sweden) investigated the evolutionary biology of the mutualistic symbiosis of siricids and fungi, as well as the monitoring and control of forest pests (Slippers et al. 2006).

The phytotoxic mucus of *S. noctilio* has received less attention apart from the mention of its mode of action in weakening attacked trees and causing their eventual mortality. In general, little research has been done on the mucus (Wong and Crowden 1976). Its effect is a key factor in successful establishment of *Sirex* and it is deposited before the process of oviposition. It is clear that the fungus and the mucus together are responsible for the phytotoxicity and subsequent mortality of living trees (Spradbery and Kirk 1978) (see Chap. 3 for further discussion).

17.4 Surveillance of *Sirex noctilio* in Chile

17.4.1 Surveillance Program

The objective of the surveillance program for *S. noctilio* is to detect outbreaks in endangered areas and to determine its distribution throughout the country during each season. This program allows for control of the outbreaks, the collection of information to define quarantine areas in the country, and the assessment of the pest's status in the area, according to International Standards for Phytosanitary Measures (ISPM) No. 6: Guidelines for Surveillance (IPPC/FAO 1997).

The program first assesses the risk of pest introduction into endangered areas so that the distribution of available resources can be prioritized. The intensity of surveillance activities to be carried out annually is then determined, with emphasis placed on areas of greatest risk. The specific phytosanitary surveillance program for the detection of *S. noctilio* in endangered areas includes three different detection strategies applied in a complementary manner: (1) detection via ground surveys; (2) trap trees, and (3) monitoring via funnel traps.

17.4.1.1 Ground Surveys

An annual ground survey is carried out in Chile in *Pinus* spp. plantations in *S. noctilio* endangered areas. Teams of surveyors search for suspect trees that present signs and/or symptoms of attack by the pest and then trees are cut and sawn into logs. Log samples are sent for laboratory analysis to determine the presence or absence of the pest through a search for immature *S. noctilio* or adults close to emergence that may be present in the wood. Approximately 10,000 survey stations throughout the country are set up each year.

Table 17.1 Number of locations where *Sirex noctilio* was detected in Chile from 2001–2009

Region	2001	2002	2003	2004	2005	2006	2007	2008	2009
Valparaíso[a]	13	0	0	0	0	0	0	0	0
Bio-Bio	0	0	0	0	0	0	0	0	12
La Araucanía	0	11	12	96	18	70	72	14	8
Los Ríos and Los Lagos	2	19	11	70	234	152	8	12	17
Total	15	30	23	166	252	222	80	26	37

[a]Outbreak eradicated

17.4.1.2 Trap Trees

A network of trap trees treated with herbicides is set up in *Pinus* spp. plantations in *S. noctilio* endangered areas, according to the basic methodology developed by Neumann et al. (1982). The purpose of this activity is to artificially predispose trees to attack by the pest. Detection plots each consist of five trees. The trees are cut, sawn into logs and analyzed during the spring according to specified procedures. Samples of wood are then extracted and sent to the laboratory for analysis to determine the presence or absence of the pest. A network of approximately 1,000 trap tree detection plots is set up annually according to a specific work plan involving SAG and forestry companies.

17.4.1.3 Funnel Traps

A country-wide network of 8–12 unit funnel traps baited with the attractants alpha-pinene and ethanol is set up and checked every 10–15 days. This phytosanitary surveillance activity is oriented towards the capture of *S. noctilio* adult females during the flight of the insect in *S. noctilio* endangered areas of the country where plantations, woodlots or isolated trees of *Pinus* spp., *Abies* spp., *Picea* spp., and *Larix* spp. are present. Special emphasis is given to areas close to international border access roads where goods in wood packaging materials arrive from abroad.

17.4.2 Distribution of Sirex noctilio *in Chile*

After the outbreak in the Valparaíso Region was eradicated, the distribution of the pine woodwasp in Chile has been restricted to the Regions of Bio Bio, La Araucanía, Los Ríos and Los Lagos (SAG 2005a). The majority of commercial *P. radiata* plantations located in endangered areas are currently Sirex free. *Sirex noctilio* was detected at 37 points in 2009, including the detection of the pest in a new region; the wasp was detected in *P. radiata* plantations in the Alto Bio Bio area (Table 17.1).

17.5 Control of *Sirex noctilio* in Chile

Control of *S. noctilio* in Chile is carried out using a combination of quarantine measures, sanitation and biological control. The aim of these measures is to reduce the possibility of infestation in susceptible areas throughout the country that are still pest free, which make up close to 95% of the area occupied by *P. radiata* plantations in Chile, and to reduce the prevalence of this pest in quarantine areas where its presence has been reported.

17.5.1 Quarantine Control

The Official control program for the Sirex woodwasp includes the application of internal and external quarantine phytosanitary measures. These are aimed at reducing the possibility of pest spread in endangered areas. They also seek to lower the chances of reintroduction of *S. noctilio* from abroad, which could broaden the genetic diversity of the pest.

17.5.1.1 Internal Quarantine

The main purpose of this activity is to reduce the risk of spread of the pest in Chile when it is detected within the country. The application of this quarantine measure is based on Resolution No. 2.758 (SAG 2009). This resolution defines the area under quarantine as the entire area where *S. noctilio* is detected plus a 20 km radius around this area. The endangered area is defined as the zone surrounding the area under quarantine where host species of the pest are present (Fig. 17.1).

The phytosanitary procedures of the internal quarantine prohibit the movement of timber from *S. noctilio* host species from quarantine and endangered areas, unless this timber has been subjected to heat treatment, chipping or fumigation with methyl bromide. The movement of timber is controlled through supervision of the harvest and highway inspections of lumber transport trucks. Companies that do not comply with these internal phytosanitary regulations are penalized and the timber is confiscated, destroyed, or returned to the place of origin (Fig. 17.2).

17.5.1.2 External Quarantine

This activity is carried out mainly through the application of the ISPM No.15: Guidelines for regulating wood packaging material in international trade (IPPC/FAO 2009) as well as Resolution No. 133 (SAG 2005b). This establishes specific phytosanitary requirements for the entry of wood materials used to package goods imported from abroad.

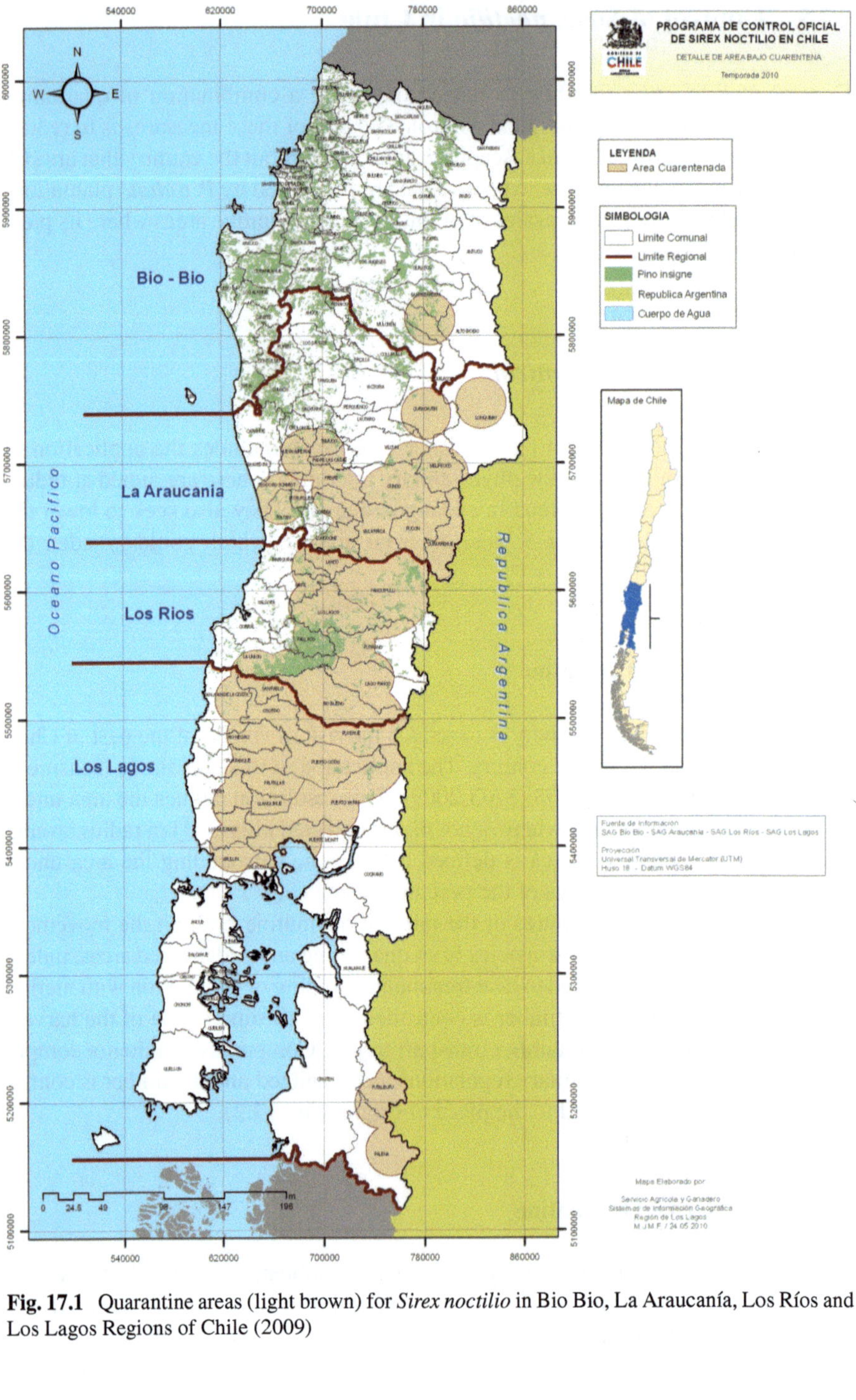

Fig. 17.1 Quarantine areas (light brown) for *Sirex noctilio* in Bio Bio, La Araucanía, Los Ríos and Los Lagos Regions of Chile (2009)

Fig. 17.2 Quarantine control of transportation of pine logs in Chile

In practice, the fulfilment of these regulations is verified through the inspection of 7–10% of shipments of foreign goods packaged in wood material. These materials must be heat-treated or subjected to fumigation with methyl bromide in the country of origin and must come with the mark specified by ISPM No.15. Inspections are carried out at international border controls in Chile and at the good's final destination, mainly industrial sites where machinery is imported. Wood packaging that does not comply with the phytosanitary regulations is destroyed, returned to the country of origin or subjected to quarantine, while any detected pests are sent to a SAG laboratory for taxonomic analysis. As a result of the above procedures, *S. noctilio* has been intercepted in wood packaging material from abroad 40 times between 1995 and 2009, highlighting the risk of spread of this pest via international trade of goods in wood materials.

17.5.2 Sanitation

Between 2001 and 2006, the control of *S. noctilio* was carried out through the phytosanitary treatment of all pest outbreaks. The pest was detected 708 times during this period. Affected trees were cut down and either burned, buried, fumigated or

Table 17.2 Parasitism levels of *Deladenus siricidicola* in adult females of *Sirex noctilio* in the provinces of Río Negro, Neuquén and Chubut (República Argentina), evaluated during the periods 2002–2003 to 2006–2007

	Parasitism in females (%) by year				
Province	2002–2003	2003–2004	2004–2005	2005–2006	2006–2007
Río Negro	13.8	48.7	64.1	81.8	60.1
Neuquén	30.2	38.8	32.5	53.4	90.7
Chubut	–	–	–	70.5	85.3

heat-treated to eliminate immature or adult pests that could be present in the timber. The primary objective was local eradication. However, because eradication is not always possible, sanitation reduces the pest populations thus decreasing the risk of spread to endangered areas, and allowing sufficient time for the parallel development and implementation of a biological control program.

17.5.3 Biological Control

The aim of the biological control program is to suppress the pest with the nematode *D. siricidicola*, and insect parasitoids *I. leucospoides*, *M. nortoni* and *R. persuasoria*. The program strategy involves obtaining these biological control organisms, placing them under quarantine in the SAG laboratories, mass rearing them in Chile and then releasing them in the area under quarantine.

Chile and Argentina have a bilateral agreement, Servicio Nacional de Sanidad y Calidad Agroalimentaria (SENASA)/Argentina and SAG/Chile for the biological control of *S. noctilio.* The objective of this bilateral plan is to use biological control methods to reduce the populations of *S. noctilio* in the Patagonian provinces of Neuquén, Río Negro and Chubut in southern Argentina, close to the Chilean border. The plan involved the production of *D. siricidicola* in two laboratories (one in Chile and one in Argentina) for release in those provinces during the period 2001–2007, and the quarantine and breeding of the parasitoid wasps *M. nortoni* and *R. persuasoria* in the SAG/laboratories for release during the period 2004–2009 using the biosecurity protocols of SAG/Chile and SENASA/Argentina.

In Argentina from 2001 to 2007, 13,877 pine trees were inoculated with *D. siricidicola* with 3,918 dosages of the nematode. The evaluation of parasitism levels showed that this organism is established in the three provinces of Argentina, with parasitism levels in adult females between 60.1% and 90.7% (Table 17.2). From 2004 to 2009, 93 females and 97 males of *M. nortoni* were released in Neuquén and Río Negro provinces of Argentina, where the parasitoid has been established successfully in both Argentinean areas. *Rhyssa persuasoria* has been released at the same places (23 females and 59 males), without recovery from field until recently (2009).

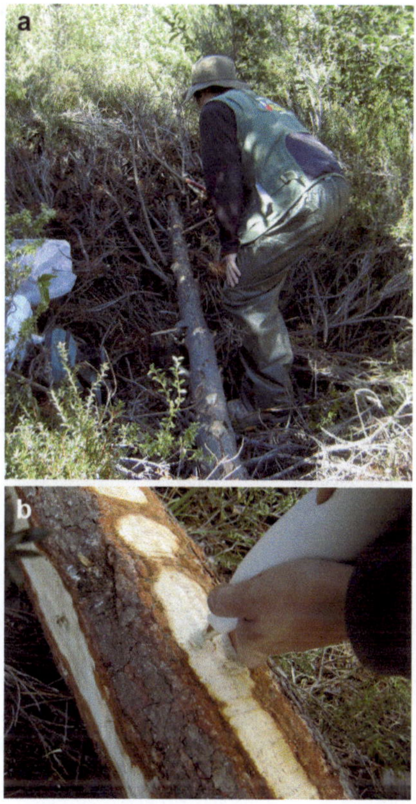

Fig. 17.3 Inoculation of *Deladenus siricidicola* in radiata pine trap trees in Chile. (**a**) Pruning and cleaning of the trap tree; (**b**) Gel inoculation

17.5.3.1 Quarantine, Rearing and Inoculation of *Deladenus siricidicola*

Two strains of *D. siricidicola* are currently available in Chile: Encruziliada do Sul, from Brazil; and Tangoio, from New Zealand. Both of these strains have been subjected to quarantine protocols according to Chilean legislation; the Encruziliada do Sul strain has been authorized for field release. After quarantine, the nematode is produced in two Chilean laboratories: SAG laboratory in Osorno and Controladora de Plagas Forestales (CPF S.A.) laboratory located in Los Angeles. The production at both laboratories assures the sufficient supply of *D. siricidicola* for each season. The nematodes are reared on *A. areolatum* grown on wheat grain using established techniques (Bedding and Iede 2005). They are released in the field only within the quarantine area, and are applied in gel form to inoculate groups of five *P. radiata* trap trees previously treated with herbicide (Fig. 17.3).

Inoculations with *D. siricidicola* were first carried out in Chile in 2006. That year, 2,307 groups of trap trees were inoculated. Almost 50% of the trees were on the property of small forest owners and the remaining trees were on land belonging to forestry companies (Table 17.3).

Table 17.3 Inoculations of *Deladenus siricidicola* in Chile in Bio Bio, La Araucanía, Los Ríos and Los Lagos Regions from 2006 to 2009

Year of inoculation	Number of trap trees inoculated in groups	Number of inoculated trees	Number of dosages of *D. siricidicola* used
2006	847	5,464	1,116
2007	1,460	8,810	1,606
2008	1,329	8,120	1,245
2009	999	5,005	1,015
Total	4,635	27,399	4,972

Table 17.4 Releases of *Megarhyssa nortoni* in Chile from 2005 to 2009

Region	Year of release	Number of places	Number of females	Number of males	Total
Bio Bio	2009	1	29	30	59
La Araucanía	2005	1	23	39	62
	2007	4	124	132	256
	2008	4	136	135	271
	2009	1	0	95	95
Los Ríos	2007	2	60	60	120
	2008	4	120	113	233
Los Lagos	2006	1	32	32	64
	2007	3	74	155	229
	2008	2	54	70	124
	2009	2	53	58	111
Total		**25**	**705**	**919**	**1,624**

17.5.3.2 Collection, Quarantine, Rearing and Release of *Megarhyssa nortoni* and *Rhyssa persuasoria*

These species were collected during the years 2004–2008 in different localities on the North and South Islands of New Zealand in *P. radiata* forests where *S. noctilio* is present. Their post-entry quarantine was carried out in the Quarantine Laboratory for Wood Insects SAG Lo Aguirre in the city of Santiago. The rearing of these insects in quarantine was achieved using *Pinus* spp. logs infested with *S. noctilio* from Argentina. This was because Chile's has a very low *S. noctilio* population and thus, logs sufficiently infested for the rearing of this insect were not available.

The logs were placed in a quarantine room under controlled temperature, relative humidity and photoperiod in an effort to generate the environmental conditions of a forest to allow the development and mating of the insects. After two generations under quarantine, the insects obtained were examined individually, sexed, and released in authorized quarantine areas in *P. radiata* forests where *S. noctilio* existed or was suspected to be present. Between 2005 and 2009, a total of 1,624 adult individuals of *M. nortoni* were released in quarantine areas (Table 17.4). Adult parasitoid releases are carried out in the mornings on sunny days in the Spring. Standardized

Fig. 17.4 Release of *Megarhyssa nortoni* in Chile. (**a**) Transport cages; (**b**) Liberation of an adult female

protocols are followed for every step of the process. The infested *P. radiata* stands where the releases are carried out have previously been evaluated to ensure that releases are carried out where they have a high probability of succeeding (Fig. 17.4).

17.5.3.3 Evaluation of Biological Control Programs

The biological control program for *S. noctilio* is evaluated based on the presence and incidence of both the parasite and the parasitoids released in the country. The establishment, parasitism and dispersion of parasites and parasitoids associated with *S. noctilio* is evaluated through combined systematic and random sampling of previously numbered 50 × 50 km grids in each of the La Araucanía, Los Ríos and Los Lagos Regions where the pest is present. Stands that are infested or suspected to be infested with *S. noctilio* are selected at random from each grid. Standardized samples of pieces of pine are taken from each stand and are placed in an insectarium

Table 17.5 Parasitism levels of *Deladenus siricidicola* in adult females of *Sirex noctilio* in Chile in Bio Bio, La Araucanía, Los Ríos and Los Lagos Regions from 2007 to 2009

	Parasitism (%)		
Region	2007	2008	2009
Bio-Bio			0
Araucanía	10.5	38.6	41
Los Rios	14.8	13.2	30.8
Los Lagos	40.0	22.2	64.5
Total	13.6	29.9	44.0

under field conditions to rear the insects. For a period of a year, all insects that emerge, including both the pest and its parasitoids, are removed from the insectarium, counted and sexed daily. *Sirex noctilio* females are dissected to determine the presence or absence of *D. siricidicola.*

As a result of these studies, it has been shown that *D. siricidicola* has become established in Chile and is present in areas infested by the pest in the La Araucanía, Los Ríos and Los Lagos Regions. The level of parasitism observed in *S. noctilio* females to December 2007 is 13.6%; however, the nematode's distribution is irregular because in some of the areas evaluated no establishment has been observed, while in others present levels of parasitism are close to 80%. During 2009, the national parasitism level was 44%; the Los Lagos Region has the highest parasitism level (Table 17.5).

Ibalia leucospoides introduced into Chile by SAG in 1997 has shown an acceptable level of parasitism (25–30%), close to that observed for the same species in other areas where this organism is present (Ruiz 2006). *Ibalia leucospoides* has dispersed to almost half of the areas infested with *S. noctilio* and, due to its long history of presence in the country, is not expected to significantly increase its level of biological control. *Megarhyssa nortoni* has been observed as established at two localities in Chile, thus rearing and release activities must be continued for this parasitoid.

17.6 Summary and Conclusions

The management of *S. noctilio* in Chile has been an integrated effort that has maintained the populations at low levels for several years. Currently the quarantine areas for *S. noctilio* are between 37.6° and 43.8° south latitude and the control measures are diminishing the points of detection and the population levels. Just recently, the pest was detected in the Bio Bio Region where the main *Pinus* plantations in Chile occur. Here with different sites, climatic conditions and large areas of plantations, it will be important to examine any changes in insect behaviour and the use of biocontrol methods must be closely monitored.

Deladenus siricidicola is the key species used in the control of *S. noctilio*. Chile adopted its use and developed modifications for rearing, preservation, viability, inoculation and final evaluation of this nematode. Some aspects, such as loss of viability after cryopreservation or parasitic capacity after generations of repetitive mass rearing under laboratory conditions using the same strains are currently under investigation. From 2007 the SAG and CPF-S.A. laboratories, located in Los Lagos and Bio Bio Regions, have produced sufficient nematodes to treat the areas under quarantine for *S. noctilio* in southern Chile. Because the procedures for mass rearing nematodes are already well-established probably just a few innovations can be developed; however, the cryopreservation of nematodes to reduce the loss of virulence has been considered a very important issue to maintain the nematode vitality. The levels of biological control by *D. siricidicola* in Chile are considered suitable.

Parasitoids recognized to be effective for *S. noctilio* control have been introduced into some infested localities that have a high enough population of the pest to support the introduction of the parasitoids. *Ibalia leucospoides*, a koinobiont endoparasitoid of eggs, first and second instars, averages control levels between 25% and 30% of parasitism. This parasitoid displays a Type III functional response, which is considered desirable because biological control agents must work well even with variable pest population levels (Fernández-Arhex and Corley 2003; Ruiz 2006). Apart from its effectiveness, its searching behaviour and association with *Amylostereum* have been studied (Villacide and Corley 2002; Fernández-Arhex and Corley 2003; Martínez et al. 2006). Other parasitoid species such as *M. nortoni* and *R. persuasoria*, both idiobiont ectoparasitoids, have been released in Chile from material collected in New Zealand, with the objective of reducing the population levels of the pest and to complement control realized with *Ibalia* and *D. siricidicola*. *Megarhyssa nortoni* was established in southern Chile, which implies correct release techniques in the field. At this time, the Official Program of *S. noctilio* in Chile includes the production of this wasp in insectaries located close to the infested areas to increase the production and distribution of the parasitoids. The techniques for breeding, quarantine and release of parasitoids have been successfully developed in Chile by SAG and CPF-S.A.

There are still many unanswered questions as to how and where *S. noctilio* mate, the number of males, survival rates and more. These questions do not seem to have engaged research interest despite their importance in population monitoring, creating predictive models of dispersal, and generating data to simulate population growth and predict the incidence of the haplodiploid condition in the following season. Unfortunately, quite a few observations are not generally available because they are unpublished or appear in media with limited circulation. Nonetheless important contributions from Argentina, Chile, Uruguay and Brazil regarding the understanding the life cycle and behaviour of *S. noctilio* include Aguilar and Lanfranco (1988); Corley et al. (2002); Villacide et al. (2004) and Corley and Villacide (2005), and are summarized in Chaps. 4, 15–17.

References

Aguilar A, Lanfranco D (1988) Aspectos biológicos y sintomatológicos de *Sirex noctilio* Fabricius (Hymenoptera-Siricidae): Una revisión. Bosque 9:87–91

Bedding R, Iede E (2005) Application of *Beddingia siricidicola* for *Sirex* woodwasp control. In: Grewal P, Ehlers R, Shapiro-Illan D (eds) Nematodes as biocontrol agents. Oxford University Press, New York

Comisión Nacional de Buenas Prácticas Agrícolas (2006) Especificaciones Técnicas de Buenas Prácticas Para Plantaciones Forestales. Ministerio de Agricultura, Santiago

Corley J, Villacide J (2005) Una Visión Dinámica del Manejo de la Plaga Forestal *Sirex noctilio*. IDIA XXI:132–134

Corley J, Bettinelli J, Villacide J (2002) Dinámica poblacional de la plaga forestal *Sirex noctilio*: endemias y epidemias. SAGPyA Forestal 23:2–8

Fernández-Arhex V, Corley J (2003) Biología del parasitoide *Ibalia leucospoides* (Hymenoptera: Ibaliidae), enemigo natural de *Sirex noctilio* (Hymenoptera: Siricidae). Patago For Año IX N° 4, pp 4–5

Haugen D (1990) Control procedures for *Sirex noctilio* in the Green Triangle: review from detection to severe outbreak (1977–1987). Aust For 53:24–32

Haugen D, Underdown M (1990) *Sirex noctilio* control program in response to the 1987 Green Triangle outbreak. Aust For 53:33–40

INFOR (2007) Estadísticas Forestales Chilenas 2006. Boletín Estadístico 117. Ministerio de Agricultura, Santiago

IPPC/FAO (1997) International standards for phytosanitary measures N° 6: guideliness of surveillance

IPPC/FAO (2009) International standards for phytosanitary measures N° 15: regulation of wood packaging material in international trade

Martínez A, Fernández-Arhex V, Corley J (2006) Chemical information from the fungus *Amylostereum areolatum* and host-foraging behaviour in the parasitoid *Ibalia leucospoides*. Physiol Entomol 31:336–340

Neumann F, Harris J, Kassaby F, Minko G (1982) An improved technique for early detection and control of the *Sirex* wood wasp in radiata pine plantations. Aust For 45:117–124

Ruiz C (2006) Razón sexual de *Sirex noctilio* Fabr. y detección de sus potenciales enemigos naturales, mediante el estudio de parcelas cebo, implementadas por el Servicio Agrícola y Ganadero entre los años 2002–2005 en la X Región de Chile. Trabajo de Titulación Lic. en Cs. Biológicas. Universidad Austral de Chile. Facultad de Ciencias, Valdivia, Chile

SAG (2005a) Informe Anual 2004 Ministerio de Agricultura. Servicio Agrícola y Ganadero. Subdepartamento de Vigilancia y Control de Plagas Forestales y Exóticas Invasivas, Santiago

SAG (2005b) Resolución N° 133: Establece regulaciones cuarentenarias para el ingreso de embalajes de madera. Diario Oficial de la República de Chile N° 38.070

SAG (2009) Resolución N° 2.758: Declara el control obligatorio de la avispa de la madera del pino *Sirex noctilio* Fabricius y deroga Resolución N° 2.630 de 2001. Diario Oficial de la República de Chile N° 39.380

Slippers B, Wingfield M, Wingfield B, Coutinho T (2000) Relationships among *Amylostereum* species associated with siricid woodwasps inferred from mitochondrial ribosomal DNA sequences. Mycologia 92:955–963

Slippers B, Wingfield B, Coutinho T, Wingfield M (2002) DNA sequences and RFLP data reflect geographical spread and relationships of *Amylostereum areolatum* and its insect vectors. Mol Ecol 11:1845–1854

Slippers B, Coutinho T, Wingfield B, Wingfield M (2003) A review of the genus *Amylostereum* and its association with woodwasps. S Afr J Sci 99:70–74

Slippers B, Vasiliauskas R, Hurley B, Stenlid J, Wingfield M (2006) A collaborative project to better understand Siricid-Fungal symbiosis. Aktuelt fra skogforsk 1:65–68

Spradbery J, Kirk A (1978) Aspects of the ecology of siricid woodwasps (Hymenoptera: Siricidae) in Europe, North Africa and Turkey with special reference to the biological control of *Sirex noctilio* F. in Australia. B Entomol Res 68:341–359

Villacide J, Corley J (2002) Distribución potencial del parasitoide *Ibalia leucospoides* (Hymenoptera:Ibaliidae) en la Argentina. Quebracho 10:7–13

Villacide J, Corley J (2006) Control de Niveles Poblacionales Endémicos de la avispa de los pinos *Sirex noctilio* (Hymenoptera: Siricidae) mediante el Raleo Sanitario de Hospederos Atacados. RIA 35:121–134

Villacide J, Corley J (2007) Manejo Integrado de la Avispa Barrenadora de los Pinos *Sirex noctilio*. Serie Técnica: Manejo Integrado de Plagas Forestales. INTA-Cambio Rural

Villacide J, Fernández-Arhex V, Klasmer P, Corley J (2004) Estrategias de manejo de la plaga forestal *Sirex noctilio*: definiendo escalas y acciones. SAGPyA Forestal 30:13–18

Wong L, Crowden R (1976) Preliminary studies on the mucus secretion of the wood wasp *Sirex noctilio* F. Physicochemical and biochemical properties. Aust J Biol Sci 29:21–32

Chapter 18
The Control of the Sirex Woodwasp in Diverse Environments: The South African Experience

Brett P. Hurley, Philip Croft, Marcel Verleur, Michael J. Wingfield, and Bernard Slippers

Abstract The woodwasp *Sirex noctilio* was detected in South Africa in 1994, in Cape Town. Within the following 15 years, *S. noctilio* had spread across four of the five main forestry regions. Losses incurred from *S. noctilio* damage increased dramatically as it moved into the Eastern Cape and KwaZulu-Natal provinces, where densely stocked pulpwood stands provided ideal conditions for the population of the wasp to increase. Control strategies developed in Australasia were adopted, including the introduction of the parasitic nematode *Deladenus* (*=Beddingia*) *siricidicola* and the parasitic wasp *Ibalia leucospoides*. Adaptations were required for the techniques used to release *D. siricidicola* due to the poor success obtained in the summer rainfall areas. These adaptations increased the effectiveness of the inoculations and subsequent natural spread, although overall inoculation success remained poor. *Ibalia leucospoides* was successfully released in the Western Cape in 1998, but only recently (since 2006) have there been efforts to release this parasitoid in other infested areas. Monitoring has been crucial for the early detection of infestations and release of biological control agents. Plantation management strategies to combat *S. noctilio* have been used to some extent, but are restricted by economic

B.P. Hurley (✉) • M.J. Wingfield
Department of Zoology and Entomology, Forestry and Agricultural Biotechnology Institute, University of Pretoria, Pretoria, South Africa
e-mail: brett.hurley@fabi.up.ac.za

P. Croft
Forestry South Africa, Pietermaritzburg, South Africa

M. Verleur
Sappi Forests (Pty) Ltd, KwaZulu-Natal, South Africa

B. Slippers
Department of Genetics, Forestry and Agricultural Biotechnology Institute, University of Pretoria, Pretoria, South Africa

B. Slippers et al. (eds.), *The Sirex Woodwasp and its Fungal Symbiont: Research and Management of a Worldwide Invasive Pest*, DOI 10.1007/978-94-007-1960-6_18,
© Springer Science+Business Media B.V. 2012

considerations. Although infestations of *S. noctilio* have recently decreased, the movement of *S. noctilio* into the largest pine-growing province and the current inability to quickly establish biological control due to poor inoculation success emphasizes the fact that *S. noctilio* continues to pose a serious threat to the pine industry in South Africa.

18.1 Introduction

The woodwasp *Sirex noctilio* (Siricidae: Hymenoptera) is a major pest in pine plantations throughout the Southern Hemisphere, where together with its fungal symbiont *Amylostereum areolatum*, it infests and kills *Pinus* spp. *Sirex noctilio* is native to Eurasia and North Africa and was first detected in the Southern Hemisphere around 1900 in New Zealand (Miller and Clarke 1935, Chaps. 5, 13). During the next 100 years, *S. noctilio* spread across the Southern Hemisphere and is now present in Australasia, South America and South Africa (Carnegie et al. 2006; Hurley et al. 2007, Chaps. 14–17). Most recently, an established population of *S. noctilio* was recorded in the United States of America and Canada (Hoebeke et al. 2005; de Groot et al. 2007).

Due to the long history of *S. noctilio* in New Zealand (since 1900) and Australia (since 1952) (Gilbert and Miller 1952), and the extensive damage that it has caused in these countries (see Rawlings 1955; Haugen 1990 as examples), management strategies to reduce the impact of *S. noctilio* were well established before the wasp spread to South America and South Africa. These management strategies were strongly focused on the establishment of biological control, although silvicultural measures such as thinning and timely pruning were also seen as important (Madden 1968; Neumann et al. 1987). Biological control of *S. noctilio*, which had been very successful in New Zealand and Australia (Hurley et al. 2007), was rapidly adopted in South America and South Africa when *S. noctilio* was first detected in these locations.

The control of *S. noctilio* in South America and South Africa has not been uniformly effective, specifically when compared to the success recorded in Australia and New Zealand (Hurley et al. 2007). In a detailed examination of *S. noctilio* infestations and their control in the Southern Hemisphere, Hurley et al. (2007) suggested that the integrated control approach for *S. noctilio* might require significant local adaptation in new areas of introduction for effective control to be obtained. The term local adaptation refers to the changes required to the adopted management program, in response to the difference in conditions that can influence the program's success between the local environment and the environment where the specific approach was developed. The aim of this chapter is to provide an overview of the efforts to control *S. noctilio* in South Africa, where the environment has dictated local adaptation of existing control strategies. Information on pine plantations and management, and the history of *S. noctilio* in South Africa is also provided.

18.2 Pine Forestry in South Africa

South Africa has a low mean rainfall and limited natural forests. Savannah woodland covers 19% of the land area, and closed canopy forest only 0.5% (Owen and van der Zel 2000). The arrival and expansion of European colonies in South Africa, and the consequent demand for timber, rapidly placed a strain on the limited supply of wood resources. Thus, from the early 1700s attempts were made to protect natural forests and exotic tree species were introduced to supplement the demand for wood (Owen and van der Zel 2000). In the late seventeenth century, *P. sylvestris*, *P. pinea* and *P. pinaster* were introduced. Numerous pine species were introduced after this, including *P. radiata* (pre-1865), *P. taeda* (pre-1900), *P. patula* (1907) and *P. elliottii* (early 1900s) (Owen and van der Zel 2000). These are currently the four main pine species planted in South Africa (Poynton 1977). The first commercial plantations were established in 1875, where black wattle (*Acacia mearnsii*) was planted to provide railway sleepers, and later as a superior vegetable tanning material. By 1910, 60,000 ha were planted to *A. mearnsii*. Plantations of non-native tree species increased greatly after World War 2; these included species of *Pinus* and *Eucalyptus*. By 1960, 980,000 ha had been planted to non-native species and by 1975 the planted area had risen to 1.1 million ha (Owen and van der Zel 2000).

Plantation forests currently cover about 1.0% of the land area (about 1.3 million ha) in South Africa, the majority of which is in the Mpumalanga (40.7%) and KwaZulu-Natal (38.5%) provinces. The forest industry produces nearly 18 million tons of roundwood per annum, and forest products exports contribute over R12 billion to foreign trade (third only to metal ores and coal), making forestry a major contributor to the South African economy (Godsmark 2008, based on 2007 figures). *Pinus* species make up about 53% of the planted area, distributed in Mpumalanga (46.6%), KwaZulu-Natal (21.5%) and the Eastern and Western Cape (27.8%). Most of the pine is used for saw timber (68.5%), followed by pulpwood (30.4%) and poles (0.9%) (Godsmark 2008).

South African forests cover a broad range of climatic, soil and altitudinal zones (Figs. 18.1–18.4). Site-species matching is thus a crucial aspect of South African forestry, although the presence and location of processing plants is also a major factor determining which pine species are planted. In the Western Cape, which is a winter rainfall and all-year rainfall area, *P. radiata* is most commonly planted. In KwaZulu-Natal, Mpumalanga and Limpopo, which are summer rainfall areas, *P. patula* is primarily planted, although *P. taeda* and *P. elliottii* are also widely utilised. *Pinus patula* is mainly planted at altitudes higher than 1,000 m.a.s.l. (metres above sea level) due to its snow tolerance and its susceptibility to *Diplodia pinea* at lower altitudes.

Management strategies for *Pinus* spp. differ in the different geographical areas where plantations have been established. In the Western Cape, Eastern Cape and Limpopo, pine trees are predominantly grown for saw timber. In KwaZulu-Natal, pine trees are predominantly grown for pulpwood. In Mpumalanga, there are large areas of pine trees planted for both saw timber and pulpwood. Management for saw timber production includes pruning and thinning, where suppressed trees are

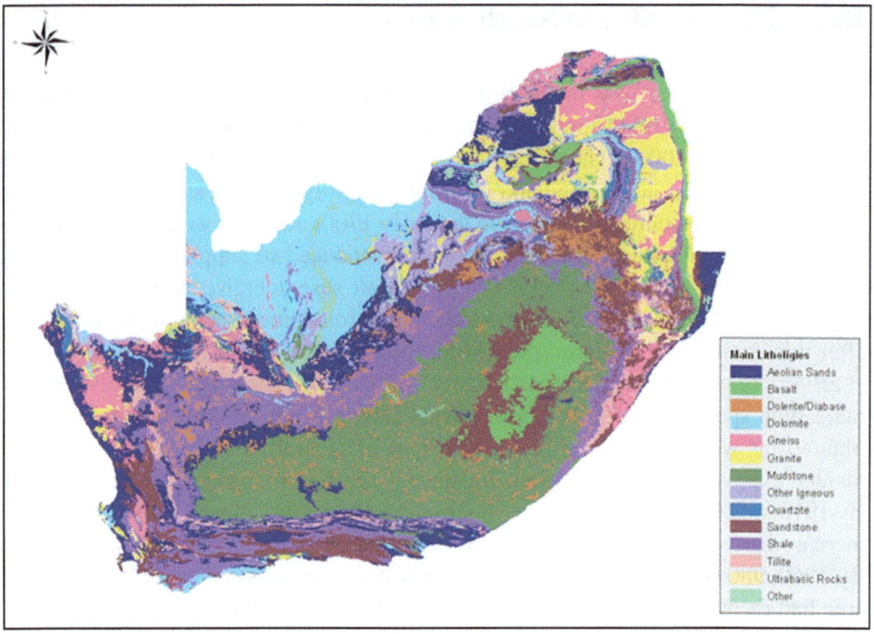

Fig. 18.1 Maps of South Africa (*left* to *right*, starting from the *top*). Lithology (Data sources: Schulze 2007)

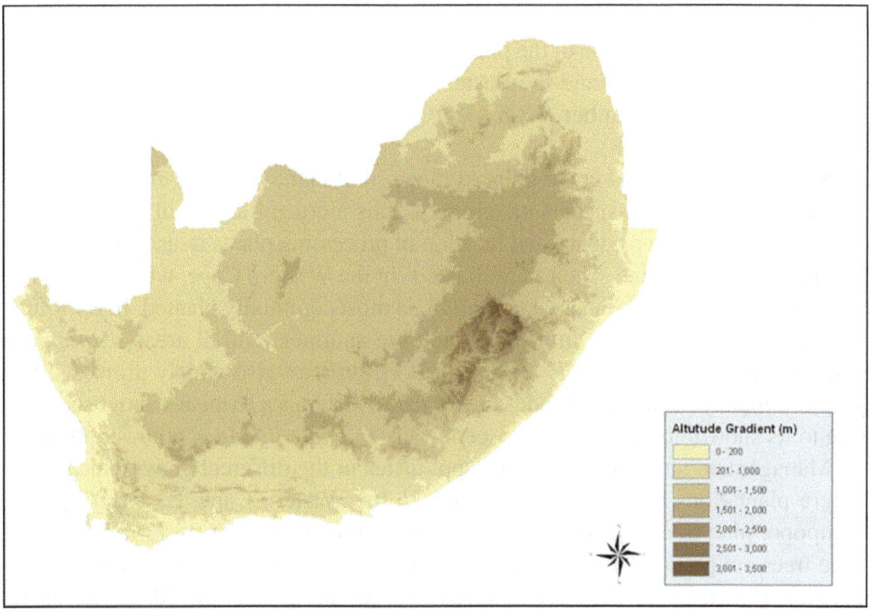

Fig. 18.2 Maps of South Africa (*left* to *right*, starting from the *top*). Altitude (Data sources: Schulze 2007)

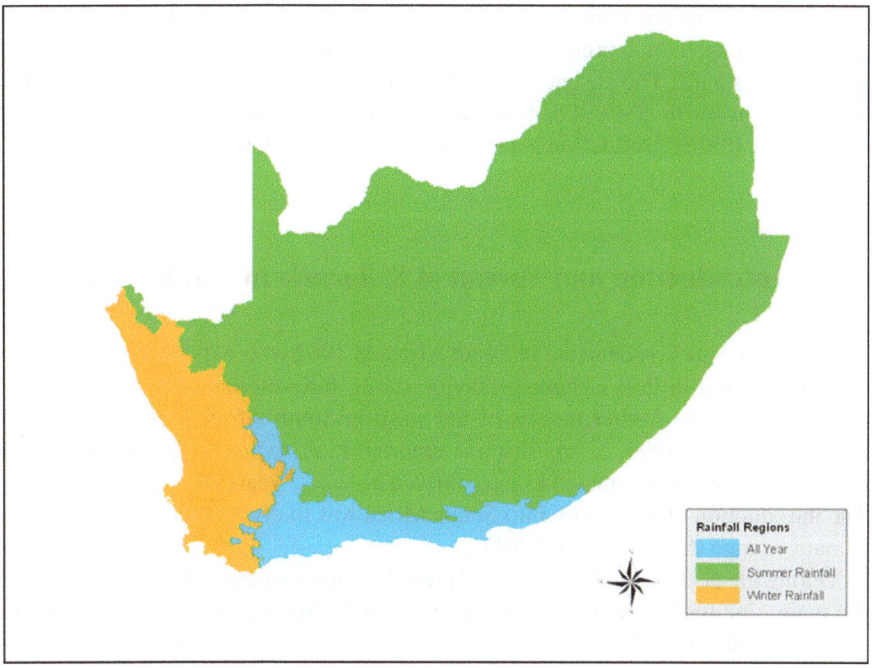

Fig. 18.3 Maps of South Africa (*left* to *right*, starting from the *top*). Rainfall (Data sources: Schulze 2007)

Fig. 18.4 Maps of South Africa (*left* to *right*, starting from the *top*). Distribution of pine forestry (Data sources: Schulze 2007)

removed. In contrast, management for pulpwood production generally does not include thinning, though pruning for access to a height of 2 m is performed, especially in *P. patula* stands. The absence of thinning in pulpwood production stands results in a much higher final stand density (approx. 800–1,250 stems per hectare [spha]) than in saw timber production stands (approx. 250 spha), at the time of harvest.

18.3 Introduction and Spread of *S. noctilio* in South Africa

Sirex noctilio was first detected in South Africa in 1962 where it was found in wood at a timber yard in Port Elizabeth (Taylor 1962). Infested material was destroyed and there were no further reports of the wasp in South Africa for the next three decades. In April 1994, *S. noctilio* was reported in *P. radiata* trees in Cape Town (Tribe 1995). It was estimated to have arrived at least 2 years prior to its detection. The introduction of *S. noctilio* into South Africa was likely from South America or its native range (Slippers et al. 2001).

During the next 15 years, *S. noctilio* spread in a north-easterly direction, following the main pine resources in the country (Fig. 18.5). Surveys in the Cape region in 1994 detected *S. noctilio* as far as 90 km from Cape Town where it was first detected,

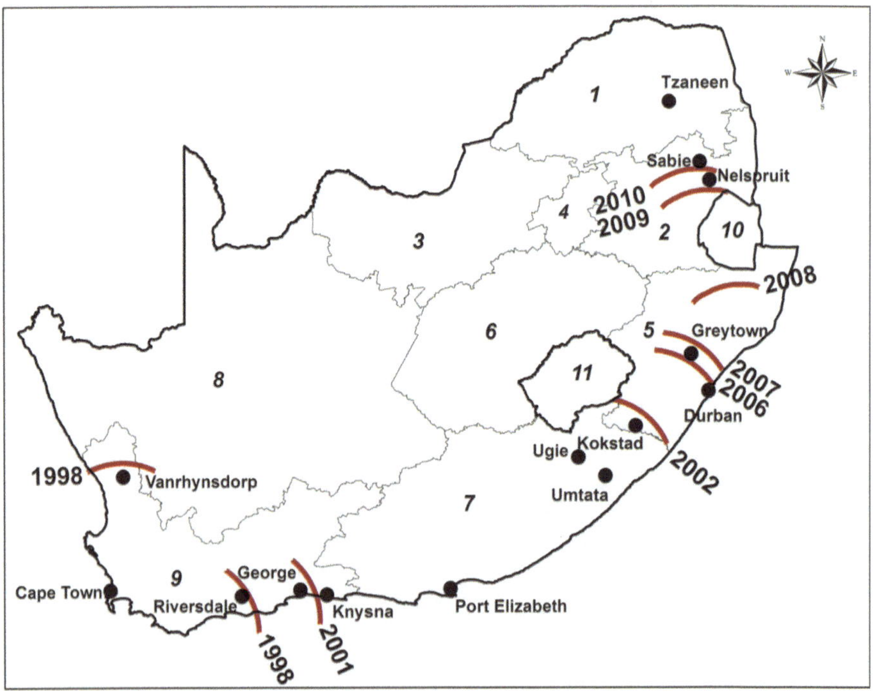

Fig. 18.5 Spread of *S. noctilio* in South Africa from 1994 to 2010

infesting mainly *P. radiata* (Tribe 1995). By 1998, *S. noctilio* had been detected as far as Riversdale in the east and Van Rhynsdorp in the north (Tribe and Cillié 2004). In 2001, the wasp was found near Knysna (Brenton-on-Sea). In 2002, *S. noctilio* was detected in the Eastern Cape and KwaZulu-Natal (at Weza) in *P. patula* stands (Hurley et al. 2007). In these areas, the wasp rapidly reached epidemic proportions and gradually made its way northwards. By 2010, *S. noctilio* had spread throughout KwaZulu-Natal and into Mpumalanga - South Africa's main pine growing province (authors, unpublished data). The spread of *S. noctilio* over the last 15 years has likely been accelerated by accidental human-assisted transport, such as through the movement of infested logs and wood packaging material (Fig. 18.6a).

Although it was considered a major threat to pine forests in the Western Cape after its detection, *S. noctilio* has caused little damage in this province. The exception was an infestation in the George area where *S. noctilio* caused an average of 10% mortality in an overstocked area of 12–13 year-old *P. radiata* (Hurley et al. 2007). In contrast, soon after the detection of *S. noctilio* in the Eastern Cape and KwaZulu-Natal, serious mortality occurred. The damage peaked in 2004–2006, where mortality in some stands was over 35% and the mean mortality was 6% (Hurley et al. 2007) (Fig. 18.6b).

Hurley et al. (2007) suggested that differences in management strategies between the provinces had contributed to differences in infestation levels. In contrast to the Eastern Cape and KwaZulu-Natal, the Western Cape does not grow trees for pulpwood and thus does not have large areas of unthinned, heavily stressed stands. In addition to the differences in management strategies, the slow establishment of biological control agents in the Eastern Cape and KwaZulu-Natal also contributed to the epidemic in this area. Infestation levels of *S. noctilio* have declined in KwaZulu-Natal after 2006, and mean infestation levels in the area were below 1% in 2009 (authors, unpublished data). Nonetheless, *S. noctilio* is still considered the most serious pest of pine currently in South Africa, based on resulting tree mortality and the threat it poses as it moves in to new pine areas.

18.4 Efforts to Control *S. noctilio* in South Africa

18.4.1 Biological Control with Parasitic Nematodes

The nematode *Deladenus* (=*Beddingia*) *siricidicola* Bedding is considered the primary biological control agent for *S. noctilio* (Bedding and Iede 2005). This nematode has a complex life-cycle that includes fungus-feeding and parasitic phases (Bedding 1972). In the fungus-feeding form, *D. siricidicola* feeds on the *S. noctilio* fungal symbiont *A. areolatum*, and has a life-cycle of about 2 weeks at 22°C, which makes it possible to rear the nematode in Petri-dish cultures in a laboratory environment (Fig. 18.6c, Chap. 9). In the parasitic form the nematode parasitizes but does not kill the larvae of *S. noctilio*, and causes sterility in the female wasp. Infested females act

Fig. 18.6 (**a**) *Sirex noctilio* male emerging from untreated wooden packing material bores through empty cement bags sent to Johannesburg, showing ease at which *S. noctilio* can spread within South Africa. (**b**) *Pinus patula* trees killed by *S. noctilio* in the KwaZulu-Natal province. (**c**) The nematode *D. siricidicola* growing on the fungus *A. areolatum* on Petri-dish cultures in a laboratory environment. (**d**) The parasitoid wasp *I. leucospoides* released as a biological control agent for *S. noctilio*. (**e**) Lure-based black panel traps used to detect new infestations of *S. noctilio*

as dispersal agents for the nematode via their parasitized eggs. These attributes of *D. siricidicola*, together with its high specificity and parasitism rates, make it an ideal biological control agent.

Shortly after the detection of *S. noctilio* in Cape Town, efforts were made to introduce the Kamona strain of *D. siricidicola* (Tribe 1995; Tribe and Cillié 2004). In 1995, 250 *S. noctilio*-infested trees were inoculated with 70 million nematodes, and in 1996, 46 *S. noctilio*-infested trees were inoculated with 20 million nematodes

(Tribe and Cillié 2004). Inoculations were done on *P. radiata* trees within a 90 km arc of Cape Town, an area demarcating the boundary of the known occurrence of S. *noctilio* at that time. The standard inoculation method, described in Bedding and Iede (2005) was used. The nematodes were imported directly from Australia. In addition to these inoculations, in 1998 and 2002 *S. noctilio*-infested logs, where *D. siricidicola* parasitism had been confirmed, were transported to *P. radiata* plantations further east, as far as George, with the intention that infected *S. noctilio* females would spread the nematode in that area (Tribe and Cillié 2004). No inoculations were done in the Western Cape from 1997 to 2003.

Results of the nematode inoculations in the Western Cape were reported to be good, although variable. Tribe and Cillié (2004) recorded parasitism of 96.1% within the inoculated area by 1998. But other studies have revealed more variable success in the province, ranging from 0% to 64% (Hurley et al. 2007, data from 2001 to 2002 collections). More recent data reflect average parasitism levels of 53% (authors, unpublished data, based on adult female parasitism) to 64% (Verleur and Kanzler 2007), based on larval scarring.

Prior to 2001, *S. noctilio* was thought to be well controlled in the Western Cape because it had not caused any serious mortality of pine. Consequently, the South African forest industry did not consider the wasps to be an imminent threat to the majority of the country's pine resources in KwaZulu-Natal and Mpumalanga. The perceived threat of *S. noctilio* increased when it was detected in Knysna in 2001 and then in Umtata, north Eastern Cape and KwaZulu-Natal in 2002 (Fig. 18.5). In 2003, *S. noctilio* was detected in various plantations in the Eastern Cape and KwaZulu-Natal and heavy infestations were observed. Unfortunately, neither *D. siricidicola* nor any other biological control agents of the wasp had apparently migrated northwards with *S. noctilio*. Again, *D. siricidicola* (Kamona strain) was seen as the most important biological control agent to release in these new epidemic areas, and the first releases started in 2004. This release required a re-negotiation with a private Australian company because the previous licence allowed release of the nematode only below 32°S latitude.

In 2004, a total of 1,763 *S. noctilio*-infested trees were inoculated in the Eastern Cape and KwaZulu-Natal, with 178 million nematodes (Table 18.1). These inoculations were mainly in pulpwood compartments of *P. patula* where infestations were highest. Standard inoculation techniques (Bedding and Iede 2005) were used. Success from these inoculations was poor, with only 3% female parasitism obtained. Efforts were made to improve the inoculation success the following year. Particular emphasis was placed on inoculation technique and nematode survival during transport and in wood after inoculation. Operating practices involving nematode transport and inoculations were improved and preliminary studies showed that nematodes had survived well during transport at temperatures between 5°C and 10°C and that they were penetrating and surviving in the wood. Despite these efforts, inoculations in 2005 again resulted in low levels of parasitism (Table 18.1).

Field trials were established in 2006 to test possible factors influencing success of *D. siricidicola* inoculations. Data from 2005 trials had indicated that the position (bottom, middle or top) of the inoculation in the trees affected inoculation success, which was thought to be due to differences in moisture content between these sections

Table 18.1 Inoculations of *Deladenus siricidicola* in the Eastern Cape and KwaZulu-Natal provinces of South Africa from 2004 to 2008

Year[a]	Nematodes used (mil.)	Trees inoculated	% female parasitism of *S. noctilio*[b]
2004	178	1,763	3 (n = 548)
2005	480	4,400	7.8 (n = 1,836)
2006[c]	40	396	7.2 (n = 2,897)
2007	4,641	57,522	22 (n = 2,132)
2008	3,115	38,541	35 (n = 1,408)

[a] In 2004–2005 felled trees were inoculated, while in 2007–2008 mainly standing trees were inoculated (fewer nematodes are used per tree). Comparison of parasitism results between years requires caution as different sampling strategies were used, where a greater proportion of bottom logs (where parasitism was highest) was sampled in 2007 and 2008

[b] n is the total number of female wasps dissected from sampled logs

[c] Inoculations in 2006 were part of a trial. All commercial inoculations were stopped in this year pending the result of the trial

(Verleur and Kanzler 2006). These factors, as well as nematode source (those reared in South Africa as opposed to nematodes sourced from Australia), period of inoculation and inoculation method (conventional method of inoculating felled trees compared to inoculating standing trees) were examined (Hurley et al. 2008; Verleur and Kanzler 2008a). The inoculation of standing trees was tested as this targeted the bottom sections of the trees, where inoculation success was highest, was safer and more cost and time effective than having to fell trees for inoculation. Background parasitism, defined as natural spread and establishment of *D. siricidicola*, was also examined.

Results from the 2006 trials confirmed that inoculation success was highest from the bottom section of the tree and in earlier inoculation periods, where moisture content was highest (Hurley et al. 2008; Verleur and Kanzler 2008a). However, the correlation between moisture content and parasitism was low; indicating that moisture content was not the only factor influencing inoculation success. Inoculation of standing trees was more cost effective and safer (see Fig. 18.7). Nematode source was found not to influence inoculation success. Background parasitism was detected at the trial sites, from just over 1% (Hurley et al. 2008) to about 20% (Verleur and Kanzler 2008a). These results were very encouraging as they showed that despite the low inoculation success (7.2%, Table 18.1), the nematodes had, at least to some degree, become established in the plantations.

Mass inoculations of *D. siricidicola* in the Eastern Cape and KwaZulu-Natal were planned for 2007 and 2008. This project was motivated mainly by the evidence of background parasitism, which meant that inoculations, even with low success, could spread the nematode across areas infested with *S. noctilio*, allowing the nematode population to naturally increase over time. Over these 2 years, 96,063 trees were inoculated with about 7.8 billion nematodes (Table 18.1). The majority of inoculations were in pulpwood stands where standing trees were inoculated. In saw timber stands, the conventional method of inoculating felled trees was followed. Average female parasitism was 22–35% from the 2007 and 2008 inoculations,

Fig. 18.7 Techniques used to inoculate *D. siricidicola* into *S. noctilio* infested trees. (**a–c**) Standard procedure of felling trees to inoculate. (**d**) Inoculation hammers used. (**e**) Adapted procedure of inoculating standing trees (used in pulpwood stands)

respectively (Table 18.1). Although the sampling technique for the results in 2007 and 2008 included a higher proportion of bottom logs (where parasitism was higher), the results still showed a general improvement from the initial inoculations in 2004 and 2005, and showed an increase in parasitism over time. The majority of the increase could be attributed to the natural spread of the nematode, as inoculation success remained poor. This was evident from sites receiving their first inoculations where parasitism results were below 10% (authors, unpublished data). Inoculations continued in 2009, but at reduced levels due to the decrease in infestations of *S. noctilio*. In total, 8,300 trees were inoculated in the Eastern Cape and KwaZulu-Natal with 830 million nematodes. Results from these inoculations were not available at time of writing this review.

The reason for the decline in *S. noctilio* infestations is not known for certain. Increasing levels of parasitism by *D. siricidicola* could have contributed to the decline, but because parasitism has only recently increased and is still not near the over 90% infestation levels reported by Bedding and Iede (2005), especially from the middle and top sections of the trees, this may not be the main or only cause. Other possible factors include the decreasing supply of suitable hosts, with many suitable hosts previously attacked and killed by *S. noctilio* or removed by clear-felling of highly infested stands (see plantation management section).

18.4.2 Biological Control with Parasitic Wasps

Parasitic wasps are considered a key component for the control of *S. noctilio* (Taylor 1976, Chap. 8). Nine species were released in Australia and of these *Megarhyssa nortoni*, *Ibalia leucospoides* and *Rhyssa persuasoria* were considered the most effective (Taylor 1978; Neumann et al. 1987). *Ibalia leucospoides* and *M. nortoni* were introduced into South Africa, but only *I. leucospoides* has become established (Tribe and Cillié 2004) (Fig. 18.6d).

In 1998, *Ibalia leucospoides* and *M. nortoni* were imported, from Uruguay and Australia respectively (Tribe and Cillié 2004). From the 18 male and 19 female *I. leucospoides* imported, 456 progeny were released from late 1998–2001. These parasitoids were released from Cape Town to Riversdale (Fig. 18.5). The establishment of *I. leucospoides* was not confirmed until 2002, but the parasitiod has now been detected in numerous plantations of the Western Cape. From these emerging wasps, a further nine *I. leucospoides* were released in Knysna in 2002. From the 10 male and 44 female *M. nortoni* imported in 1998, only 79 progeny were released the following year in Van Rhynsdorp (Tribe and Cillié 2004). No further releases were made and *M. nortoni* has since then never been recovered from the field, thus leading to the assumption that this biological control agent has not become established in South Africa.

Ibalia leucospoides appears not to have moved with *S. noctilio* as it spread to the Eastern Cape and KwaZulu-Natal. *Sirex noctilio* infested billets collected from Knysna in 2001, the Eastern Cape and Weza (near Kokstad) in 2002, and from various sites in KwaZulu-Natal from 2003 to 2006 indicated the absence of *I. leucospoides* in these areas (Fig. 18.5). Plans were thus made to introduce *I. leucospoides* to these areas.

In January 2006, 138 *I. leucospoides* collected in the Western Cape were released at sites in KwaZulu-Natal and the Eastern Cape. A further 376 wasps were released from November 2006 to January 2007. *Ibalia leucospoides* were recovered from these release sites in late 2007 and released in the same area (a total of 201 wasps). In the subsequent *S. noctilio* emergence season, 74 *I. leucospoides* were recovered and these were released at sites in KwaZulu-Natal (Verleur and Kanzler 2008b).

Efforts are currently underway to develop a central rearing colony of *I. leucospoides*, from which the parasitoid can be sent to *S. noctilio* infested areas. This approach will provide a more sustainable and higher capacity to release the parasitoid than the capture and release method. A further aim is to increase the genetic diversity of the South African *I. leucospoides* population by introducing *I. leucospoides* from other countries into the rearing colony. The current genetic diversity of the South African population is expected to be very low due to the small number of wasps originally introduced. The influence of genetic diversity on the success of biological control agents is not clear, but could affect its ability to adapt to new environments and host types (Roderick and Navajas 2003).

18.4.3 Plantation Management

Stressed trees are known to be more susceptible to attack by *S. noctilio* (Madden 1968; Talbot 1977; Neumann and Minko 1981), and poor silviculture practices can be a major source of stress in plantations (Madden 1988). In South African pine forests, high stocking of pulpwood stands is likely one of the major reasons for the high levels of infestation of *S. noctilio* in regions where such stands are dominant. These stands are intentionally managed to have high stocking, with no thinning and no pruning or pruning only to allow access into the plantations. The goal here is to maximize revenue by maximizing fibre levels per hectare. The forest industry is currently investigating alternative management options for pulpwood stands, involving different thinning regimes in an effort to decrease stress in the stands while retaining economic feasibility. Saw timber stands are less prone to stress due to thinning, when these are conducted at the optimal time.

One of the management strategies of the forestry industry in response to *S. noctilio* infestations has been to clear-fell and harvest areas where infestations are above 15%. The harvested timber is moved to a mill where it was processed and the *S. noctilio* larvae destroyed. This strategy was used in KwaZulu-Natal, and likely contributed to the noticeable decline of *S. noctilio* in those areas in subsequent years.

18.4.4 Monitoring of **Sirex noctilio**

Effective management of *S. noctilio,* including the release of biological control agents, must rest on a clear knowledge of the distribution of the pest. In South Africa, monitoring of *S. noctilio* after its detection in 1994 was mainly by means of field surveys. These monitoring efforts resulted in the detection of *S. noctilio* in the Eastern Cape and KwaZulu-Natal (Fig. 18.5). Field surveys to detect *S. noctilio* in the northern Eastern Cape and KwaZulu-Natal started in 2002 after *S. noctilio* was unexpectedly detected in Umtata during a routine field excursion. The spread of *S. noctilio* to Umtata, far from the known range of *S. noctilio* the previous year, could have resulted from human-mediated transport of *S. noctilio*, but could also have resulted from the undetected movement of *S. noctilio* before 2002.

From 2004 onwards, various traps and lures based on volatile compounds from pine trees were tested for trapping *S. noctilio*. This work was conducted in collaboration with USDA-APHIS. These tests have resulted in the current black intercept (panel) trap, using a six-component blend lure containing general pine volatiles (Fig. 18.6e). Using this trap, it has been possible to detect new infestations of *S. noctilio* and this also led to the detection of *S. noctilio* in northern KwaZulu-Natal during 2007 and 2008 (see Fig. 18.5) and more recently in Mpumalanga (November 2009). Trials are underway to improve the current trap type and lures used to monitor the spread of *S. noctilio*. These new traps using generic or specific lures will be used to detect the presence of other wood borers.

In 2004, the use of trap trees was tested to monitor the spread of *S. noctilio* for the first time in South Africa. This is an approach described by Madden (1971), where trees are stressed to lure *S. noctilio* females. Trap trees can function as a means of detection as well as an inoculation source for *D. siricidicola*. The trap trees set in 2004 were established using previously described techniques (Anon 2002; Neumann et al. 1982), except that glyphosate was used as an alternative to dicamba, which is prohibited for use in South African plantations. Success with these trap trees was very poor. Because there were high numbers of naturally infested trees to inoculate and because of the current development of lure-based traps to detect new *S. noctilio* populations, efforts to use trap-trees were abandoned until 2007. A decision was then made to re-investigate trap trees as a method for monitoring, and to complement the lure-based traps. These trees could then also be used to provide an inoculation source in areas with new *S. noctilio* infestations, and where naturally infested trees were difficult to locate. Trap tree trials in 2007, showed that trap trees could be successfully used when applied a month before adult emergence using Dicamba/24-D/MCPA (Verleur and Kanzler 2009). Plans are underway to use traps and trap trees on a larger scale in areas ahead of the known limits of *S. noctilio* distribution.

18.5 Conclusions

It is clear that local adaptation of control strategies for *S. noctilio* is required where this pest has entered new environments. In South Africa, after the detection of *S. noctilio*, control strategies developed in Australia were rapidly adopted, specifically the introduction of biological control agents. Early failures with biological control underscored the importance of developing and modifying control strategies that are effective to local climate, environmental and management conditions, as these factors influence tree vigour and, therefore, susceptibility to *S. noctilio*.

Amongst the major adaptations to the *S. noctilio* control programme in South Africa, those pertaining to the application of the parasitic nematode *D. siricidicola* have been most significant. This nematode was highly successful as a biological control agent in other parts of the world, yet gave poor results in the first 2 years of inoculation in the summer rainfall area of South Africa. Investigations revealed that low moisture content in the wood at time of inoculation had influenced success. This situation could not be completely avoided, but inoculation techniques were adapted to increase the efficacy of inoculations by targeting the wetter bottom sections of standing trees in pulpwood stands.

Other possible factors influencing inoculation success include the influence of different strains of *A. areolatum* present in South Africa and Australia (Slippers et al. 2001, authors, unpublished data). Preliminary results have shown that the fungus associated with *S. noctilio* in South Africa may grow slower than the fungus on which *D. siricidicola* has been developed. The influence that this has on nematode survival and reproduction is being investigated. Furthermore, the competition between blue stain fungi commonly present in trees infested with *S. noctilio*, and *A. areolatum*, and the influence that this has on nematode survival is being considered. Both of these factors, namely *A. areolatum* strain and the presence and composition of blue stain fungi in infested trees can differ between and even within countries. In addition, plans to collect and screen other strains of *D. siricidicola* for their efficacy in South Africa's summer rainfall region are underway.

Insufficient effort has been made to introduce parasitic wasps for biological control of *S. noctilio* in South Africa. These wasps can play an important role in the control of the pest, and could be especially important given the low inoculation success with the nematodes to date. Increased effort is needed to establish a sustainable rearing colony of *I. leucospoides* from which large numbers of wasps can be released annually. Genetic diversity of the population of the parasitic wasps must also be considered. In addition, effort is needed to import and release other parasitic wasps, such as *Megarhyssa nortoni* and *Rhyssa persuasoria*. There is little question that the introduction of these parasitic wasps will contribute to a more robust and sustainable biological control effort.

Plantation management is a major factor determining infestations of *S. noctilio* and requires further attention by the South African forest industry. Clear-felling of highly infested sites provides an effective management approach, but the vast areas of highly stocked, stressed pulpwood stands will always provide an opportunity for *S. noctilio* and other stress-associated pests and diseases to establish. Management strategies are governed by economics, but the financial implications of pest outbreaks need to be considered in this equation. The current thinning trials aimed at finding a 'middle road' are an indication that the industry in South Africa is taking pest threats very seriously.

Effective monitoring tools were developed a decade after *S. noctilio* was first detected in South Africa. Yet the development of monitoring tools should be a priority for any new pest introduction. It is clear that if greater attention had been paid to monitoring after the detection of *S. noctilio* in South Africa, its movement into the Eastern Cape and KwaZulu-Natal would have been recognized earlier, which would have resulted in earlier management interventions and many thousands of trees might possibly have been saved. The monitoring of *S. noctilio* continues to be crucial as it moves northwards.

The extent to which *S. noctilio* will infest plantations in the Mpumalanga and Limpopo provinces is currently unknown. This region has a mixture of saw timber and pulpwood stands and other stress factors such as baboon damage (McNamara 2006) will also need to be taken into account. Past efforts to adapt control strategies for local conditions will certainly contribute to ensuring low populations of *S. noctilio*. *Deladenus siricidicola* is already present at the front of the *S. noctilio* distribution and with the assistance of monitoring tools it can be rapidly introduced, together with *I. leucospoides*, to new infestation sites. Thinning overstocked stands ahead of the front and clear-felling highly infested sites should also be considered. Continued efforts to improve inoculation success, including research on possible barriers to inoculation, the introduction of new nematode strains and the introduction of other parasitic wasps will also be important. Progress in these areas will be fundamental to the future management of *S. noctilio* in South Africa.

Acknowledgements We are most grateful to Dr. Ilaria Germishuizen (Institute for Commercial Forestry Research, South Africa) for providing the maps used in this review. We also appreciate the support of the South African Sirex Control Programme and particularly Prof. Colin Dyer, Chair of the South African Sirex Steering Committee, as well as many forest managers, field foresters and students who have contributed towards much of the research reported here. Nampak Limited is thanked for the provision of the picture used in Fig. 18.6a.

References

Anon (2002) National Sirex control strategy operation worksheets. National Sirex Co-ordination Committee, Australia

Bedding RA (1972) Biology of *Deladenus siricidicola* (Neotylenchidae) an entomophagous-mycetophagous nematode parasitic in siricid woodwasps. Nematol 18:482–493

Bedding RA, Iede ET (2005) Application of *Beddingia siricidicola* for Sirex woodwasp control. In: Grewel PS, Ehlers R-U, Shapiro-Ilan DI (eds) Nematodes as biocontrol agents. CAB International, London

Carnegie AJ, Matsuki M, Haugen DA, Hurley BP, Ahumada R, Klasmer P, Sun J, Iede ET (2006) Predicting the potential distribution of *Sirex noctilio* (Hymenoptera: Siricidae), a significant exotic pest of *Pinus* plantations. Ann For Sci 63:19–128

de Groot P, Nystrom K, Scarr T (2007) Discovery of *Sirex noctilio* (Hymenoptera: Siricidae) in Ontario, Canada. Great Lakes Entomol 39:49–53

Gilbert JM, Miller LW (1952) An outbreak of *Sirex noctilio* in Tasmania. Aust J Biol Sci 22:905–914

Godsmark R (2008) The South African forestry and forest products industry 2007. Forestry South Africa. http://www.forestry.co.za/website_industry_data. Accessed 20 Feb 2010

Haugen DA (1990) Control procedures for *Sirex noctilio* in the green triangle: review from detection to severe outbreak (1977–1987). Aust For 53:24–32

Hoebeke ER, Haugen DA, Haack RA (2005) *Sirex noctilio*: discovery of a palearctic siricid woodwasp in New York. Newsl Mich Entomol Soc 50:24–25
Hurley BP, Slippers B, Wingfield MJ (2007) A comparison of control results for the alien invasive woodwasp, *Sirex noctilio*, in the southern hemisphere. Agric For Entomol 9:59–171
Hurley BP, Slippers B, Croft PK, Hatting HJ, van der Linde M, Morris AR, Dyer C, Wingfield MJ (2008) Factors influencing parasitism of *Sirex noctilio* (Hymenoptera: Siricidae) by the nematode *Deladenus siricidcola* (Nematoda: Neotylenchidae) in summer rainfall areas of South Africa. Biol Control 45:450–459
Madden JL (1968) Physiological aspects of host tree favourability for the woodwasp, *Sirex noctilio* F. Proc Ecol Soc Aust 3:147–149
Madden JL (1971) Some treatments which render Monterey pine (*Pinus radiata*) attractive to the wood wasp *Sirex noctilio* F. Bull Entomol Res 60:467–472
Madden JL (1988) Sirex in Australasia. In: Berryman AA (ed) Dynamics of forest insect populations: patterns, causes, implications. Plenum Press, New York and London
McNamara LM (2006) Nutrient concentration of inner bark tissue in pine trees in Mpumalanga in relation to baboon damage. MSc dissertation, Faculty of Science, University of the Witwatersrand, South Africa
Miller D, Clarke AF (1935) *Sirex noctilio* (Hym.) and its parasites in New Zealand. Bull Entomol Res 26:149–154
Neumann FG, Minko G (1981) The Sirex wood wasp in Australian radiata pine plantations. Aust For 44:46–63
Neumann FG, Harris JA, Kassaby FY, Minko G (1982) An improved technique for early detection and control of the Sirex wood wasp in radiata pine plantations. Aust For 45:117–124
Neumann FG, Morey JL, McKimm RJ (1987) The Sirex wasp in Victoria. Lands and Forests Division, Department of Conservation, Forest and Lands, Melbourne
Owen DL, van der Zel DW (2000) Trees, forests and plantations in Southern Africa. In: Owen DL (ed) South African forestry handbook 2000. South African Institute of Forestry, Pretoria
Poynton RJ (1977) Tree planting in Southern Africa: the pines. Department of Forestry, Pretoria
Rawlings GB (1955) Epidemics in *Pinus radiata* forests in New Zealand. N Z J For 7:53–58
Roderick GK, Navajas M (2003) Genes in new environments: genetics and evolution in biological control. Nat Rev Genet 4:889–899
Schulze RE (2007) National Land Cover Database 2000. South African atlas of climatology and agrohydrology, Water Research Commission, WRC Report 1489/1/06, Pretoria
Slippers B, Wingfield MJ, Coutinho TA, Wingfield BD (2001) Population structure and possible origin of *Amylostereum areolatum* in South Africa. Plant Pathol 50:206–210
Talbot PHB (1977) The *Sirex-Amylostereum-Pinus* association. Annu Rev Phytopathol 15:41–54
Taylor JS (1962) *Sirex noctilio* F., a recent introduction in South Africa. Entomol Rec 74: 273–274
Taylor KL (1976) The introduction and establishment of insect parasitoids to control *Sirex noctilio* in Australia. Entomophaga 21:429–440
Taylor KL (1978) Evaluation of the insect parasitoids of *Sirex noctilio* (Hymenoptera: Siricidae) in Tasmania. Oecologia 32:1–10
Tribe GD (1995) The woodwasp *Sirex noctilio* Fabricius (Hymenoptera: Siricidae), a pest of *Pinus* species, now established in South Africa. Afr Entomol 3:215–217
Tribe GD, Cillié JJ (2004) The spread of *Sirex noctilio* Fabricius (Hymenoptera: Siricidae) in South African pine plantations and the introduction and establishment of its biological control agents. Afr Entomol 12:9–17
Verleur PM, Kanzler A (2006) Some observations on the emergence patterns of *Sirex noctilio* (Fabricius) and parasitism by *Beddingia siricidicola* on Sappi landholdings during 2005. Sappi Research Document 25/2006
Verleur PM, Kanzler A (2007) Assessment of the *Sirex noctilio* (Fabricius) infestations in the Western Cape and comparisons between the distribution of *Sirex noctilio* (Fabricius) larvae and levels of parasitism by *Beddingia siricidicola* in the Western Cape and Sappi landholdings in KwaZulu-Natal during 2006. Sappi Research Document 18/2007

Verleur PM, Kanzler A (2008a) Comparison of two inoculation methods with *Beddingia siricidicola* against *Sirex noctilio* conducted in Kwazulu-Natal pine plantations during 2006. Sappi Research Document 04/2008

Verleur PM, Kanzler A (2008b) Introduction and establishment of *Ibalia leucospoides* as an additional biological control agent against *Sirex noctilio* in KwaZulu-Natal. Sappi Research Document 03/2008

Verleur PM, Kanzler A (2009) Testing of six herbicides in order to establish trap tree plots for the control of *Sirex noctilio* in the summer rainfall regions of South Africa. Sappi Research Document 09/2009

Chapter 19
Sirex, Surveys and Management: Challenges of having *Sirex noctilio* in North America

Kevin J. Dodds and Peter de Groot

Abstract Since 2004, when the Sirex woodwasp, *Sirex noctilio*, was first discovered in North America, there have been intensive efforts to survey and determine the area infested and to assess management options. In this chapter we review the history of survey efforts in Canada and the USA from 2005 to 2009 and the challenges facing these surveys. Next we describe the significant differences between North America and the Southern Hemisphere (where this insect is a serious pest) in forest types, natural disturbance regimes, competing insects and disease, and forest management methods and how they affect surveys and management. We review the logistical issues of landscape and forest diversity, ownership, and access that affect the implementation of surveys, and biological issues of native siricids and other associated insects and diseases that complicate the use of trap trees and traps. We discuss the challenges of using silviculture and biological methods in North America to control *S. noctilio*. We conclude that management of *S. noctilio* will not be easy and must be multifaceted. The diversity, heterogeneity and complexity of North America's natural forests in terms of natural enemies and competing insects and diseases, may be a problem for survey efforts, but also a solution for management. Finally, the situation in North America is unique, allowing many interesting questions on invasion biology, community ecology, and management of an invasive species in native pine communities to be addressed.

K.J. Dodds (✉)
United States Department of Agriculture, U.S. Forest Service, 271 Mast Rd, Durham, NH 03824, USA
e-mail: kdodds@fs.fed.us

P. de Groot
Natural Resources Canada, Canadian Forest Service, Great Lakes Forestry Centre, 1219 Queen Street East, Sault Ste. Marie, ON, Canada P6A 2E5

B. Slippers et al. (eds.), *The Sirex Woodwasp and its Fungal Symbiont: Research and Management of a Worldwide Invasive Pest*, DOI 10.1007/978-94-007-1960-6_19,
© US Government 2012

19.1 Introduction

Sirex noctilio F. (Hymenoptera: Siricidae) was first detected in North America during a 2004 exotic species survey in New York state, USA (Hoebeke et al. 2005). This insect had been high on the exotic species target lists in Canada and the United States for a number of years because it has caused widespread economic losses in pine plantations in the Southern Hemisphere and also because the extensive areas of pine forests of North America are classified as high-risk (Haugen 1999; Borchert et al. 2007). It is of considerable concern because it is capable of establishing and flourishing in the prime industrial pine forest regions in the southern and western U.S. and over much of the boreal forest in Canada (Carnegie et al. 2006; Yemshanov et al. 2009a).

Despite the carefully considered risk assessments, the real impact of *S. noctilio* to pine in North America is still unknown and debatable. There are significant differences between the North American pine ecosystems and the Southern Hemisphere pine forests, where both pine and *S. noctilio* have been introduced. It is reasonable to expect that the impact will vary across the continent and that *S. noctilio* will cause different levels of tree damage and economic impact in areas where management, landscape heterogeneity, and tree species vary. Given the uncertainty, the rapid survey response to discovery of *S. noctilio* in North America has been prudent in regarding this as a serious risk requiring substantial effort to detect and delimit the area of infestation.

In this chapter, we review the general approaches and issues surrounding the survey for this insect in Canada and the USA. We then describe the types of forests where *S. noctilio* has already been found and the types of forests where it may be found, and could cause significant economic and ecological impact. We conclude by presenting many of the unique challenges facing survey and management efforts in North America.

19.2 Detection in North America

19.2.1 Agencies Involved in Detection and Delimitation Efforts in Canada and the United States

Both provincial/state and federal agencies are involved in detection efforts for *S. noctilio* in North America. The Canadian Food Inspection Agency (CFIA) is charged with protecting food, animals, and plants and has legislative authority to regulate invasive species in Canada. The companion agency in the USA is the United States Department of Agriculture (USDA), Animal and Plant Health Inspection Service (APHIS) and it is charged with safeguarding American agriculture and natural resources. The Canadian Forest Service (CFS) and USDA Forest Service (USFS) provide scientific support and expertise to the regulatory agencies when forest pests

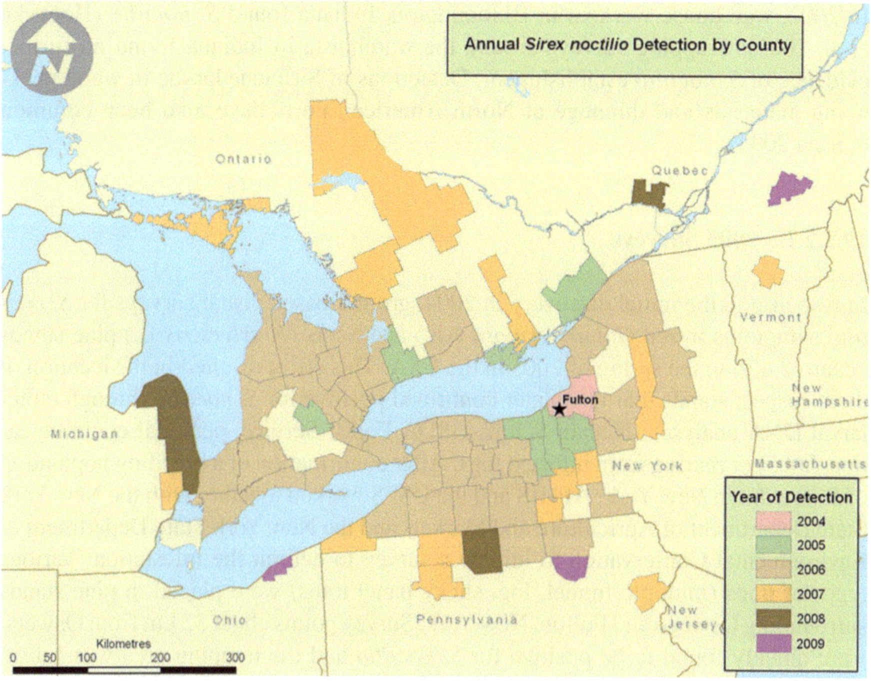

Fig. 19.1 Known distribution of *Sirex noctilio* in North America by county as of January, 2010

are involved. These federal agencies in Canada and the USA work closely with their provincial and state departments of agriculture and forestry to detect and respond to invasive species. International agreements under the North America Plant Protection Organization (NAPPO) facilitate cooperation between Canada, Mexico, and the USA when addressing plant health concerns, including invasive forest pests (NAPPO 2005).

19.2.2 Annual Surveys

Both Canada and the USA have annual exotic or invasive alien species surveys that use traps baited with bark beetle pheromones and/or host volatiles (CFIA 2006; USDA 2006). Traps are typically placed in "high-risk" areas (e.g., ports of entry, warehouse districts, forests adjacent to urban areas) to increase the likelihood of detecting an exotic species early during its establishment phase. On 7 September 2004, an individual *S. noctilio* female was captured in an exotic species survey trap placed in Fulton, New York (Hoebeke et al. 2005), an area that has limited industrial activity, and is 16 km inland from the port city of Oswego on Lake Ontario (Fig. 19.1). This was not the first time an adult *S. noctilio* female was found in North America.

In 2002, warehouse workers in Bloomington, Indiana found *S. noctilio* (Hoebeke et al. 2005). Trapping surveys around the warehouse in Indiana found no further evidence of *S. noctilio* establishment. Detections of Siricidae larvae in wood packaging materials and dunnage at North American ports have also been common (Ciesla 2003).

19.2.2.1 2005 Surveys

In response to the initial discovery in 2004, ground-based visual surveys for *S. noctilio* symptoms in red (*Pinus resinosa* Ait.) and Scots (*P. sylvestris* L.) pine stands occurred during the spring of 2005 in the USA. This effort resulted in the location of three suspect stands that were later confirmed positive for *S. noctilio* through either larval DNA analyses (Nathan Schiff, USDA Forest Service, personal communication, 2005) or rearing from infested logs. After confirmation of a breeding population of *S. noctilio* in New York, APHIS and the USFS worked together with the New York State Department of Agriculture and Markets and the New York State Department of Environmental Conservation to initiate a survey to delimit the infestation. Various types of traps (multiple-funnel, log, sticky panel traps) were placed in pine stands surrounding Oswego and Fulton, New York. Survey points about 32 km from Oswego were quickly found to be positive for *S. noctilio* and the trapping area was subsequently expanded with some traps out as far as 112 km. Other regional exotic bark beetle and woodborer survey traps were also screened for *S. noctilio*. Following confirmation of established populations in New York, a survey of 36 sites in Ontario across Lake Ontario from Oswego was conducted by CFS, CFIA and the Ontario Ministry of Natural Resources (de Groot et al. 2007a). During the 2005 survey, four Canadian and four U.S. counties were identified as having *S. noctilio* (Fig. 19.1). The combined results from Canadian and USA surveys suggested that a population of *S. noctilio* was established over a large geographic area covering portions of central New York and southern Ontario, and probably had been present for some time.

19.2.2.2 2006 Surveys

In the fall of 2005, APHIS organized a Sirex Science Panel that included members from APHIS, USFS, CFS, several universities, and international scientists (Bedding et al. 2006). The overall charge of the Sirex Science Panel was to provide science-based answers to questions posed by the U.S. State/Federal multi-agency Sirex Management Team, regarding such topics as pest biology, survey methods, and management options, and to recommend a *S. noctilio* response for the USA. A Sirex Management Team composed of scientists and regulatory officials from both state and federal U.S. agencies used information from the Sirex Science Panel report to create a national plan for *S. noctilio* population delimitation, and develop possible management options. A very similar structure of science and management teams with international participation was set up in Canada (de Groot et al. 2007b). One of the major initiatives to come from these meetings was the decision to conduct a more intense delimiting survey to define the *S. noctilio* population extent in both countries and to work in collaboration.

During the initial *S. noctilio* surveys in 2005, different lure and trap combinations were used in Canada and the USA. To produce comparable results between the two countries, the same traps and lures were used in 2006. While little data existed on trap/lure efficacy for *S. noctilio*, multiple-funnel and panel intercept traps baited with a 70:30 ratio of alpha-pinene:beta-pinene (Synergy Semiochemical 85 g ultra high release sleeve, alpha-pinene was 75%(+), beta-pinene was ~95% (–) or Pherotech 178 g ultra high release sleeve, alpha-pinene was 75% (+) and 25% (–), beta-pinene was nearly 100% (–)) were chosen as survey tools. This lure was based on findings by Simpson and McQuilkin (1976) that alpha- and beta-pinene elicited antennal responses from *S. noctilio* females. Propylene glycol was used with a wet-cup as a collection liquid for all surveys.

In the USA, a systematic survey grid (40 or 58 km^2) covered most of New York, parts of northern Pennsylvania, and western Vermont. Pine stands were prioritized for trap placement within each grid, with available host data used to assist in locating pine areas within a heterogeneous landscape. In some cases, traps were placed in non-host forest types. In addition to the systematic survey, high risk stands of stressed pine in other eastern U.S. states were surveyed using the same trapping methodology. Instead of using a systematic grid survey, the CFS and CFIA opted for a targeted approach in Ontario and Quebec, focusing on placing survey traps in high-risk Scots pine stands that showed symptoms of moderate decline. Survey traps were typically placed 10–20 m inside the stands, compared to the perimeter placement common in USA surveys.

Results from 2006 indicated that a widespread *S. noctilio* population was established in North America. Canadian surveys found *S. noctilio* in 17 counties in the province of Ontario and in the USA, 20 New York counties and two Pennsylvania counties were positive for the insect (Fig. 19.1).

19.2.2.3 2007 Surveys

The 2006 Canadian and American *S. noctilio* surveys provided a better picture of the distribution in North America. However, there were concerns that the western and northern population boundaries were not successfully delimited. This was in part because of the limited number of traps placed in these areas, but also because of doubts that the lure (a 70:30 ratio of alpha- and beta-pinene) was effective at capturing *S. noctilio* at endemic population levels. Ongoing studies by APHIS and their collaborators to provide a better attractant were inconclusive and consequently the same lure was used in 2007. The survey efforts shifted north in Canada with traps placed throughout most of northern Ontario, where forestry is important to the economy and where there are very extensive natural stands of jack pine, *P. banksiana* Lamb. Surveys also continued in Quebec and intensified in New Brunswick and Nova Scotia in eastern Canada. In the USA, *S. noctilio* delimitation efforts intensified in the west and south of New York, with a significant effort placed in Midwestern states (e.g., Michigan) where red pine is a more prominent commercial species and jack pine is present. Many states outside the delimitation survey area also conducted detection surveys of high-risk sites.

Traps in both countries were typically placed along the perimeter and in sun-exposed locations. Trap trees were also added as a survey tool in the USA. These trap trees were girdled in the spring, had a multiple-funnel trap placed on each tree to collect adults, and were examined for *S. noctilio* signs and symptoms over the course of the season.

Four additional counties or districts were found infested in Ontario in 2007 (Fig. 19.1). As in 2006, no *S. noctilio* were found east of Ontario. Several counties were added in New York and Pennsylvania, further expanding the known population south in the USA. One county in Vermont and two counties in Michigan expanded the known distribution east and west, respectively. Only one female *S. noctilio* was captured in Vermont and three in Michigan, so further delimitation efforts were planned for 2008.

19.2.2.4 2008 and 2009 Surveys

Sirex noctilio survey efforts in 2008 and 2009 continued to target high-risk sites for trapping in areas adjacent to positive finds from previous years. Canadian survey efforts focused on Quebec and northern Ontario. No new finds of *S. noctilio* were found in 2008 or 2009 in Ontario despite intensive efforts to survey the area north of the positive finds from 2005 to 2007 (Fig. 19.1). *Sirex noctilio* was first found in the province of Quebec in 2008 near La Chute about 75 km northwest of Montreal and about 100 km northeast of the closest known infestation in Ontario. In 2009, *S. noctilio* was found in one additional area about 95 km southeast of Montreal (Fig. 19.1). Survey efforts in the USA expanded slightly southward during 2008 with more effort along the northern border of Maryland and continued surveys in Midwest, Mid-Atlantic, and northeastern states. Vermont was intensively surveyed with traps placed throughout the state and trap trees deployed at locations adjacent to where *S. noctilio* was detected in 2007. The 2008 survey efforts in the USA resulted in one additional county in Pennsylvania and two in Michigan (Fig. 19.1). In 2009, survey efforts in the northeast were reduced while trapping in Maryland and the Midwest continued. Survey efforts in 2009 yielded an additional county in Pennsylvania and a new state record in Ohio where one *S. noctilio* specimen was captured. These findings only slightly altered the *S. noctilio* footprint determined from previous year survey efforts (Fig. 19.1).

19.3 North American Forests at Risk

Sirex noctilio has been problematic in countries where pine is grown intensively under pure even-aged silvicultural systems (Rawlings 1955; Haugen et al. 1990; Iede et al. 1998; Maderni 1998; Carnegie et al. 2005; Hurley et al. 2007). Southern

Hemisphere countries produce pine over large landscapes, where other than age variation, forests are generally homogeneous. The impact of *S. noctilio* on North American forests will vary with forest type, natural history of pine species present, natural disturbance regimes, competing insects and disease, and stand management regimes.

The forests of the northeastern USA and southern Canada, where *S. noctilio* is currently present, are much more diverse, both in species composition, landscape patterns, stand management plans, and uses (Foster and Aber 2004). The landscape of this area is heterogeneous with a mix of agricultural fields, urban areas, and forests forming a complex matrix of land types and uses. Much of the forest area in Ontario and the USA is managed according to a multiple-use ethic, with watershed and ecosystem protection, recreation areas, and timber production as common shared objectives. Forest management practices vary from excellent to nonexistent, with considerable areas receiving insufficient silvicultural attention since stand establishment.

Successful *S. noctilio* reproduction has occurred in eastern white (*Pinus strobus* L.), red, jack, and Scots pine in this area. Typical hard (yellow, or two and three needle pines including, red, jack and Scots) pine stands where *S. noctilio* has been found to date (2009) are dense, overstocked plantations with mortality due to overcrowded growing conditions. The poor growing conditions have been compounded by poor site quality, including a hard pan layer (a dense layer of soil, residing usually below the uppermost topsoil that is largely impervious to water), which has led to suppressed and dead trees. Winter ice storms and wind damage are common occurrences that further stress trees making them susceptible to bark beetles and other biotic organisms (Ryall and Smith 2005; Ryall et al. 2006). Interestingly, there have been several other invasive alien insects found throughout much of the same area, most notable is *Tomicus piniperda* (L.) (Coleoptera: Scolytidae) (Morgan et al. 2004; Haack 2006) creating a potential synergy between bark beetles debilitating trees (Czokajlo et al. 1997) and setting them up for mortality from *S. noctilio*. In addition to these factors, numerous native insects, especially wood and bark insects, and root and foliar diseases have taken advantage of the stress in these unthrifty stands (Whitney 1988; Klepzig et al. 1991, 1995; Erbilgin et al. 2002; Erbilgin and Raffa 2002). All these factors make it difficult to determine if *S. noctilio* is the primary cause of tree decline, or if it is amongst the suite of secondary insects and diseases that occur after a tree has been predisposed and weakened by other factors to attack.

Native pines, including eastern white and red, frequently occur in small patches dispersed across the landscape, while jack pine grows in much larger stands (often pure, fire-origin) in northern portions of Ontario. Scots pine, an introduced tree from Europe and the native host of *S. noctilio*, grows in small plantations dispersed throughout the landscape. This was once a popular Christmas tree, but there are now many abandoned plantations in eastern North America that are overstocked and serve as prime sites and hosts for *S. noctilio*. In Ontario, pine forests cover about 27 million ha (CFS 2001). In Michigan, New York, Pennsylvania, and Vermont approximately two million ha of forests are in pine or pine-mixed forests (Frieswyk and Widmann 2000; McWilliams et al. 2007; Hansen and Brand 2006; Alerich and Drake 1995).

Eastern white pine (a five-needle pine) is one of the most important economic species in eastern North America and is used for lumber, furniture making, veneer, log cabin timbers, and pulpwood. Natural eastern white pine stands are usually intermixed with other conifer and hardwood tree species, but pure stands do occur on the landscape. Commercially planted eastern white pine stands are managed either under even-aged or uneven-aged silvicultural regimes and have rotation ages between 90 and 120 years (Barrett 1995). While the potential impact of *S. noctilio* on eastern white pine is not currently understood, several biological factors could increase this tree's susceptibility to attack. Both white pine blister rust (caused by the fungus *Cronartium ribicola* J.C. Fischer: Rabenh.) and white pine weevil (*Pissodes strobi* Peck) attack eastern white pine in different parts of its range and can reduce vitality in trees, likely increasing their attractiveness and/or susceptibility to *S. noctilio*. Limited data suggest that eastern white pine is a suitable host for *S. noctilio* in North America (K. Zylstra, USDA APHIS, personal communication, 2008) and synergy with native and exotic organisms is a concern.

Red pine was widely planted for soil conservation, wind breaks, and reforestation efforts throughout northeastern USA and southern Ontario. As a commercial species, red pine is used for lumber, poles, log cabin timbers, railroad ties, and pulpwood. Often planted in areas to aid in soil stabilization after farm abandonment in the early twentieth century, red pine now suffers from being planted on poor quality sites, and overstocked growing conditions. Typically, red pine is found in pure or almost pure stands. Red pine rotation ages vary, but 60–90 years is common with some longer rotations of 100–150 years for large sawtimber stands (Ek et al. 2007). In the Midwestern USA, red pine is more intensively managed and is an important economic species with more than 758,000 ha planted in this forest type. Few biotic agents cause losses in red pine, but fungal diseases such as Armillaria root rot (Whitney 1988) and other diseases may stress stands and open up resources for *S. noctilio*. Red pine decline has been described in the Midwestern USA and is a function of interactions between insects and fungi (Klepzig et al. 1991).

Jack pine grows extensively in northern Ontario and throughout much of the vast boreal forest in Canada, where it is an important economic species, but in the northeastern USA it usually occurs in small isolated stands. In the Midwestern USA, jack pine is an important economic species and also provides habitat for the endangered Kirtland's warbler (Walkinshaw 1983). With a rotation of between 40 and 70 years, jack pine is primarily used for pulpwood, rough lumber, and pallet production. Jack pine is a fire-dependent species that is naturally regenerated through stand-replacing fires and often grows on poor sites (Benzie 1977; Rouse 1986). These stands are typically pure jack pine that develop into single-aged forests, thin themselves naturally, often have high basal areas, and could provide suitable habitat for *S. noctilio* in this region. In addition, biotic agents (e.g., *Choristoneura p. pinus* Freeman, *Neodiprion* spp., stem rusts) may also stress jack pine, opening resources for *S. noctilio* colonization (Cross et al. 1978; Wilson and Averill 1978; Gross 1992; Conway et al. 1999).

Surveys of infested stands in Ontario and New York during 2006–2009 indicated that *S. noctilio* behaved similarly to endemic populations documented in the Southern

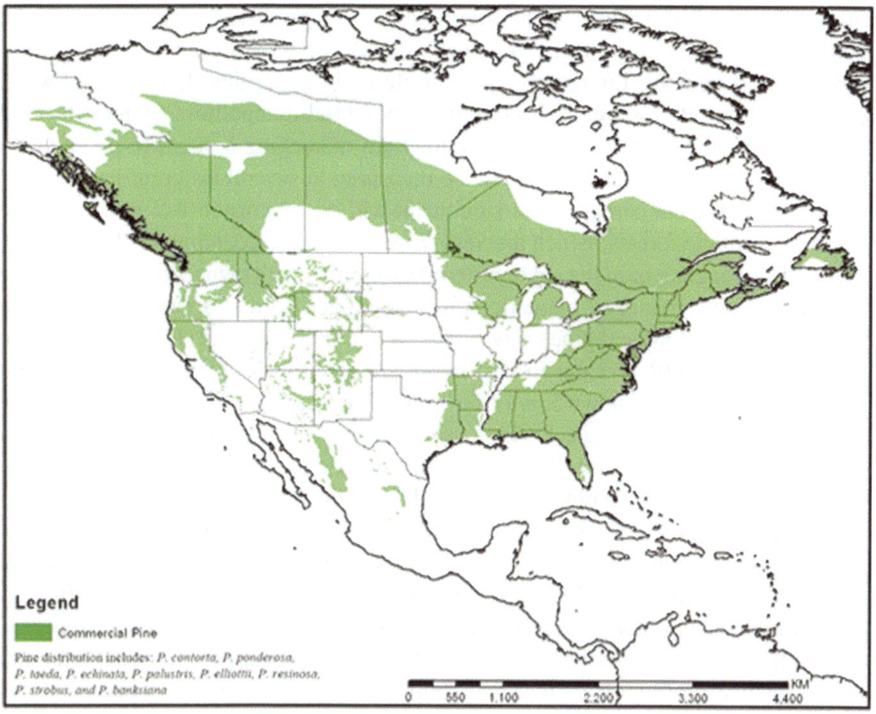

Fig. 19.2 Combined distribution of commercially important pine species in North America

Hemisphere (Jackson 1955; Neumann and Minko 1981; Neumann et al. 1987). Stands where *S. noctilio* has been located in North America are overstocked and generally in poor condition. At the individual tree scale, *S. noctilio* appears to be concentrating attacks on suppressed trees. Trees with smaller live crown ratios, smaller average diameters, and reduced incremental growth were generally more often attacked than more vigorous, larger trees with relatively healthy crowns (Dodds et al. 2010). Some larger apparently vigorous co-dominant trees in these stands were also attacked, but less frequently than smaller trees. *Sirex noctilio* had more of an impact in Scots pine stands compared to red pine stands in terms of percentage of trees attacked and basal area lost (Dodds et al. 2010). Interestingly, many overstocked red pine stands with larger over-story trees (> 30 cm dbh) and no overtopped trees are free of *S. noctilio*. These trees growing under overstocked conditions, sometimes with two times the recommended basal area and small live crown ratios, are not being exploited by *S. noctilio* at this point in the invasion.

The current area of infestation is within central North America, and represents only a small portion of pine forests on the continent. The other regions of concern are the southeastern USA, western North America, and eastern Canada, where pines are grown either under intensive silvicultural regimes or occur naturally in stands that may be more susceptible to *S. noctilio* attack (Fig. 19.2). Wildfire, bark beetle infestations, and insect-caused defoliation are common disturbances in many pine ecosystems in North America and likely would open resources for *S. noctilio* colonization causing interesting, yet problematic synergies. Conversely, well-established insect communities, containing potential competitors (both arthropod and fungal), predators, and parasitoids, may aid in keeping *S. noctilio* populations regulated or at low levels. Impact can be measured at several different levels using different criteria, but perhaps most significant will be the impact of *S. noctilio* in the southeastern pine forests, which are very important to the economy of this region.

Forest management in the southeastern USA varies with ownership type, ranging from intensively managed plantations to unmanaged natural areas. Loblolly pine (*Pinus taeda* L.), longleaf pine (*P. palustris* Mill.), shortleaf pine (*P. echinata* Mill.), and slash pine (*P. elliottii* Englem.) are important species throughout the southeastern region and several are known *S. noctilio* hosts (Haugen and Hoebeke 2005). Loblolly pine is the most common species in the southeastern USA and grows from southern New Jersey south to central Florida and west to eastern Texas (Baker and Langdon 1990). Loblolly pine is dominant on about 11.7 million ha and makes up over one-half of the standing pine volume in the southeastern U.S. (Baker and Langdon 1990). Many of these stands are overstocked and could provide excellent habitat for *S. noctilio*. Loblolly pine often grows in conjunction with other tree species, including hardwoods and other southeastern pines. Longleaf pine is an endangered ecosystem that occurs primarily in southeastern coastal areas. Because of the dramatic reduction in area of longleaf pine ecosystems, the USA Biological Survey listed this ecosystem as the third most endangered in the USA (Noss et al. 1995). Fire is a critical component of longleaf pine ecosystems and restoration efforts (Brockway et al. 2005) could result in trees stressed and attractive to *S. noctilio* infestation. Disturbance is common in southeastern forests, with abiotic factors such as fire and windstorms, and biotic factors such as insects and disease influencing forest structure, successional pathways, and possibly in the future the vulnerability to *S. noctilio*.

Like the southeastern USA, several known *S. noctilio* hosts grow over large areas in the western USA and Canada. Lodgepole (*Pinus contorta* Dougl.: Loud.) and ponderosa (*Pinus ponderosa* Dougl.: Laws.) pines are the primary economic pine species in the west and both are widely distributed across the landscape. Lodgepole pine forests cover an estimated 26 million ha in western North America (Lotan and Critchfield 1990). Lodgepole pine (like jack pine) often grows in almost pure stands that grow after stand-replacing fires. These stands either are left to self-thin or are mechanically thinned for timber production. Ponderosa pine has a large range that occurs throughout western North America from southern Canada south to Mexico. While often found in pure stands, ponderosa pine can also be found in mixed stands with other conifer species (Franklin and Dryness 1988). Fire is also important to ponderosa pine ecosystems and currently there is a movement to restore fire regimes into stands (Covington et al. 1997; Lynch et al. 2000; Allen et al. 2002). While most trees survive fire events, stand stress occurs, opening resources that are exploited by bark beetles (Breece et al. 2008) and might also be utilized by *S. noctilio*.

Bioeconomic analyses and simulation models of *S. noctilio* impacts on timber supply and harvesting in eastern Canada indicate that economic losses will vary significantly depending on the complex interactions among insect spread, tree mortality and adaptations of harvest schedules and approaches (Yemshanov et al. 2009a). In another simulation study, Yemshanov et al. (2009b) provided maps of the potential distribution of *S. noctilio* in North America over a 30 year time horizon based on the combination of three risk scenarios (high, medium and low) with two entry potential scenarios (entry from ports only and ports plus existing infestations). These and other modeling approaches are useful to provide advance warning of the potential impact of *S. noctilio* and no doubt will be improved as we continue to collect data from infested areas on the actual impact and ecological reasons why *S. noctilio* populations increased and decreased in areas.

In summary, there are vast areas of pine in North America that range from natural forests to intensively managed plantations. Within these forests, pest and forest management may range from nonexistent to the use of active and proactive measures. Furthermore, there exists a wide range of insect and disease communities present in these forests. While it is difficult to predict how *S. noctilio* will respond to these communities, it is likely that it will be significantly different from what has been recorded in the Southern Hemisphere. From a positive perspective, the native insects and fungi and natural enemies may help regulate *S. noctilio* populations. Conversely, the possible synergies and combined cumulative effects of native pests and *S. noctilio* damage are disconcerting. It is prudent to expect significant impact and to prepare strategies and tactics for the management of *S. noctilio* in these regions.

19.4 Survey Challenges in North America

Surveying for *S. noctilio* in North American forests has presented several unique challenges not encountered in other parts of the world where introductions have occurred. These challenges can be divided into two major components: (1) logistical issues of landscape and forest diversity, ownership, and access, and (2) biological issues of native siricids and other associated insects and diseases that complicate the use of trap trees and traps baited with attractants.

19.4.1 Logistical Issues

Pine forests in North America vary from small to large pine plantations, pure stands originating after mild to severe fire or wind disturbance events, to mixed conifer or mixed hardwood forests where the proportion of pines can vary. Coupled with this, is that ownership of the forests can reside with provincial, state, or federal governments, small to large forest industrial companies, or with private persons or family owners. Road access to pine forests can range from non-existent or poor (especially in much of the government-owned forests in the boreal region in Canada), to excellent, particularly in the rural areas of southern Canada and much of the eastern USA. Land ownership may also prohibit or restrict access to the site and the establishment of trap trees or even placement of removable traps. For example, on some publicly owned land, the use of herbicides is prohibited, thus effectively precluding the use of trap trees. Determining ownership of private land requires time to search land registry records and then tracking the owners, some who may live considerable distances away. Overlaid on this mosaic is the information on tree and stand attributes (e.g., age, species composition, basal area, vitality), which can vary depending on ownership and location. This presents a further challenge of determining high-risk sites where survey efforts should be prioritized.

Some of these logistical challenges can be mitigated by working closely with the various agencies responsible for forest management and forest health. In Canada and the USA, there are several organizations that already have co-operative arrangements in place to assess forest health conditions. By working with people who have local knowledge, suitable sites for *S. noctilio* surveys can be tentatively identified. Once potential sites have been identified, a field inspection is required to further assess the health of the stand. If the site is deemed suitable, then the owner has to be located, and consulted, and either verbal or written permission obtained. This hierarchal process is time-consuming, costly, and is further complicated by the biological issues inherent to North America.

19.4.2 Biological Issues

In comparison to the Southern Hemisphere countries where *S. noctilio* has become invasive, the pine forests of North America are complex ecosystems consisting of mostly native species with well-developed arthropod, plant and disease communities that influence stand structure, tree growth, vitality, and mortality. On the other hand, the exotic pure pine plantations in the Southern Hemisphere typically have few pine-killing insects and diseases present (Neumann 1979; Neumann and Marks 1976). Consequently, identifying and attributing tree mortality to *S. noctilio* is much easier and definitive in the Southern Hemisphere than it is in North America. Visual inspection of trees for round exit holes (signs) of adult *S. noctilio*, is not only difficult to do in the first place, but is confounded in North America because several native species of Siricidae are also present (Smith and Schiff 2002), and several species of Cerambycidae (e.g., *Monochamus notatus*, *M. scutellatus*) make similar sized round exit holes (Wilson 1975; Drooz 1985). Resin beads, typically formed after the bark and wood have been penetrated by the ovipositor of female *S. noctilio*, are often natural occurrences on eastern white pine or occur after infection by white pine blister rust or other agents. To date, in North America, we have not observed the classic pine needle wilting and drooping symptoms that are seen in the Southern Hemisphere. What is seen is a general chlorosis and browning of the needles, but there are several diseases (e.g., *Diplodia pinea* blight), insects, and abiotic stressors that can also show the same symptoms. Not only do these factors that produce similar signs and symptoms complicate any ground inspections, they essentially rule out any chance of rapid aerial or remote sensing techniques being successful in unequivocally identifying the presence of *S. noctilio* on the landscape. Surveying for visual symptoms is thus limited to ground based methods, but this requires good training and experience to distinguish damage by *S. noctilio* from the many other possibilities and is time-consuming (Iede et al. 1998). It also requires numerous trained staff that can be mobilized to cover large areas, a capacity that is not present in Canada or the USA. For these reasons, and other practical considerations, surveys in Canada and the USA have relied on traps baited with host volatiles, and to a very limited degree on trap trees.

Trap trees have been the primary survey method for *S. noctilio* management in most regions of the Southern Hemisphere. Essentially, a tree is injected with an herbicide so that the tree becomes stressed and thus attractive to *S. noctilio* (Neumann et al. 1982). A significant challenge to the use of herbicides in North America is the complex of wood and bark boring insects already present in forests that will also utilize these trees. Many of these insects and diseases are active before or during the same time as *S. noctilio*. Thus, unlike in the Southern Hemisphere where trees can be injected with herbicide at least 3 months in advance of the *S. noctilio* attack period (Neumann et al. 1982), timing of an herbicide application is much more critical and has a much smaller window of opportunity in North America. A three or more month window simply will not be an option in North America, but creating a stressed tree more quickly over a narrower time frame is possible (Zylstra et al. 2010). This narrower window also constrains the application of the trap tree method where there are limits to the number of field crews that can establish trap trees. It is possible that trap trees may simply not be as effective in North America as in the Southern Hemisphere because of these biological and logistical constraints.

Forest insect surveys in North America rely widely on the use of traps that emit insect attractants. Most often, these attractants are insect pheromones, which in the majority of cases have proved to be very efficient in detecting incipient populations of a specific insect species (e.g., gypsy moth). To a much smaller degree, attractants based on host-associated odors are used for insect surveys, and usually where a survey for a general group of insects is required, or where the pheromone has not been identified. The discovery of *S. noctilio* in North America was in a trap that contained both pheromones and host volatiles, including alpha-pinene, ipsdienol, cis-verbenol, and 2-methyl-3-buten-2-ol. Although portable insect traps baited with attractants can have their own limitations, they can offer some distinct advantages in cost, ease of use, and earlier detection of populations. Mainly for this reason, traps baited with attractants have been used extensively in North America.

A formidable challenge to the use of traps is that lures and trapping methods for *S. noctilio* are still in the early stages of development. While a contact sex pheromone has been identified for *S. noctilio* (Böröczky et al. 2009), a long-range sex pheromone more useful for detection purposes has not yet been identified, and may not exist. Consequently, lure development has focused on the use of host attractants, which may be used by *S. noctilio* during host finding and selection. The current lure combination of alpha- and beta-pinene are compounds that are common to other trees and thus also attract numerous other wood-boring species. While these collections provide opportunities to survey for other insects along with *S. noctilio*, sorting through these samples is laborious and expensive.

Research to develop lures is time consuming and expensive. There are also practical and statistical constraints in the number of treatments (chemicals, blends and release rates), layout of traps, and replications possible in field experiments where low numbers of insects are often captured, especially in forests. Furthermore, little is known about effective methods to trap siricids, which can present difficulties in evaluating attractants. In general, an important component of the development of a trap-based detection program for insects is the design and placement of the trap itself. Attractive chemicals placed in an ineffective trap may result in poor trap catches, and conversely, a well-designed trap, baited with unattractive chemicals or incorrect release rates may also result in poor trap catches. Several trap types were tested in North America in an attempt to determine the optimal trap for *S. noctilio* surveys. Several variations of sticky traps (panel, drainpipe, log) and flight intercept (funnel, panel intercept, aerial malaise) traps were tested in research trials (Dodds and de Groot, unpublished). At the same time, Canada and the USA split their operational survey traps between panel intercept and 12-unit multiple-funnel traps. Data from these research trials were inconclusive, but similar numbers of *S. noctilio* were captured in funnel and panel intercept traps in the operational survey. This same pattern was seen in both the Canadian and USA surveys where almost equal numbers of *S. noctilio* were captured in the two trap types. Consequently, surveys rely on either multiple-funnel or panel intercept traps.

Competition between volatiles emanating from stressed trees in a stand adds another variable in a survey plan. Severely stressed stands that are typical of the habitat for *S. noctilio* in North America will release a full and complex blend of host volatiles, which most likely out-compete simple traps baited with a few compounds. Because of this, surveyors might be more inclined to deploy traps in stands that are not in overly poor condition where volatile-baited traps would be more apparent. However, we know little about *S. noctilio* behavior in stands with a range of susceptibility and the presence of this insect in "healthy" stands. Whatever approach is taken it is important to understand that a host volatile baited trap is unlikely to attract insects over trees that emanate stress volatiles. In some stands in Canada and the USA, traps baited with alpha- and beta-pinene failed to capture *S. noctilio* even though they were readily seen in the heavily infested stands (de Groot and Dodds, personal observations, 2006), resulting in false negatives.

The current approach taken in Canada and the USA has been to target stressed or moderately stressed pine stands for trap placement. While this is subjective, placing traps near a potential source of insects or potential habitat increases the likelihood of detection. On a landscape level, traps have been placed in or adjacent to potential habitat patches (i.e., stressed pine stands). At the stand level, placement has varied from interior placement to stand edges. Systematic surveys have resulted in traps being placed first in a hard pine stand, second another pine, and third conifer stands.

The presence of native Siricidae further complicates survey and detection efforts. In Canada and the USA, there are 23 species/subspecies of Siricidae (Schiff et al. 2006). In eastern North America, three other *Sirex* species have been collected (*S. edwardsii* Brullé, *S. juvencus* (L.), *S. nigricornis* F.) and all have been captured in *S. noctilio* detection traps. Although some native siricids can be distinguished with the naked eye, one species, *S. juvencus* is very similar to *S. noctilio* and requires examination under the microscope to identify it. Although there have been two recent keys to the Siricidae in North America (Smith and Schiff 2002; Schiff et al. 2006), this group is under revision again, which will result in some synonymy of species and recognition of new species (H. Goulet, Canadian National Collection of Insects, Agriculture and Agri-Food Canada). Therefore, all specimens from the survey must be kept for future examination. While adult *Sirex* species can be differentiated by morphological characteristics, molecular techniques must be implemented to distinguish larval specimens and would help with closely related and similar appearing adults. Molecular barcoding techniques based on sequence data from the mitochondrial COI gene has recently been developed to distinguish individual siricid species (Wilson and Schiff 2010). The technique has also been shown to be useful for larval identification and successfully used to confirm the presence of *S. noctilio* in suspect trees found in New York (N. Schiff, personal communication, 2005).

In summary, there are several interesting and formidable logistical and biological challenges that face the implementation of a survey for *S. noctilio* in North America. Notwithstanding these constraints, the surveys for *S. noctilio* since detection in 2004 have been remarkable in that this insect has been detected over a wide area with limited tools and knowledge. Many improvements are needed, perhaps no more so than in the development of better attractants, knowledge of where traps should be placed, and understanding the limits (efficiency) of detection with these methods.

19.5 Management Challenges in North America

The management of *S. noctilio* in the Southern Hemisphere is based on three pillars: (1) survey to monitor population levels, (2) silvicultural control by maintaining tree vitality and stand health, and (3) biological control through the introduction of a parasitic nematode, *Deladenus (=Beddingia) siricidicola* (Bedding 1974; Bedding and Akhurst 1974; Neumann 1979; Haugen 1990; Neumann et al. 1987), and to a lesser degree insect parasitoid populations (Hurley et al. 2007). Attempts at eradication have failed where it has been tried on a large scale (Neumann et al. 1987; Haugen et al. 1990), and it certainly is not an option in North America given the very large area of infestation (Fig. 19.1). The application of pesticides, either contact or systemic, also presents very significant economic, practical, and environmental impediments so their use can also effectively be ruled out. In North America, then, the apparent options are silviculture and biological control. Before these two control methods can be implemented, an assessment on the natural level of control (real or expected) by native mortality agents, economic damage attributed to *S. noctilio*, environmental impact, and cost/benefit analysis of pest management techniques must be conducted.

19.5.1 Silviculture

To maintain overall forest health, intermediate silvicultural treatments are commonly recommended as a stand management tool in North America. Similar stand management practices would also be beneficial for reducing the effects of *S. noctilio* on North American pine stands (Dodds et al. 2007). Treatments focused on increasing residual tree vitality to increase host tree defenses and simultaneously remove potential habitat (i.e., suppressed trees) could reduce the impact *S. noctilio* may have in a stand. Stand treatments, such as thinning from below as a part of a timber stand improvement treatment could be helpful for increasing residual tree vitality (Dodds et al. 2007). Because many of the trees growing in these overstocked stands have small live crown ratios and have been growing very slowly for many years (Dodds et al. 2010), it is unknown how well they will respond to silvicultural treatments. However, preliminary results from a stand thinning study in New York suggest that thinning stands in relatively poor condition does reduce the impact of *S. noctilio* in these forests (K. Dodds, unpublished data). If trees are beyond a point where thinning will help stand health, other options such as salvage and cutting followed by replanting, or converting off-site planted pine stands back to hardwoods should be considered.

Several native North American forest insects can reach epidemic populations where economic and ecological damages can be widespread and severe. Insects likely to interact with *S. noctilio* as populations spread include problematic species like defoliators (e.g., *Choristoneura p. pinus*, etc.), as well as several bark beetle species (e.g., *Dendroctonus ponderosae* Hopkins, *Dendroctonus frontalis* Zimmermann). Active stand management is currently suggested for these species and management recommendations are in line with what would be suggested for *S. noctilio*. Forest stand management focused on improving the overall vitality of trees could reduce stand susceptibility to numerous pests at once, increasing the attractiveness of such measures for landowners.

Forest stand management objectives in North America, however, are often driven by considerations other than tree health. Intermediate stand treatments are often not economically viable and are therefore either delayed or ignored. Market values and conditions or costs of treatments may preclude proactive actions, even though they are desired. Unfortunately, many stands in the currently known infested area lack a management plan, are owned by absentee landowners who know little about silviculture or the conditions in their forests, or landowners that have no desire or means to manage their forest. Consequently, many stagnant pine stands remain on the landscape with no hopes of increasing the overall health in the stand. Perhaps concern over *S. noctilio* damage in a stand will move concerned landowners to consider thinning to maintain forest health and reduce damage not only from *S. noctilio*, but native species as well.

19.5.2 Biological Control

In addition to *S. noctilio* living amongst and competing with other wood-boring insects, all its major parasitoids are known to be naturally present in North America (Cameron 1962; Kirk 1974, 1975). Field-based observations and laboratory rearing suggest that indigenous parasitoids are successfully locating and parasitizing *S. noctilio* eggs and/or larvae (Long et al. 2009, P. de Groot and co-workers, unpublished). Competition with woodborers and parasitism by indigenous parasitoids needs to be evaluated and done so across different pine ecosystems to determine the level of natural control already present. Against this background, we can then determine if further augmentation of parasites or the introduction of the parasitic nematode is warranted and where.

In New Zealand, the nematode *D. siricidicola* was found along with *S. noctilio* (Zondag 1969) and probably came with the original or possible subsequent introductions of *S. noctilio*. Fortunately, this introduction was of a strain that parasitized the eggs. Not all strains of *D. siricidicola* are parasitic or result in 100% parasitism (Bedding and Iede 2005), and the current commercial strain is a result of extensive screening and proper culturing to ensure high levels of control. When the original screening work took place, many localities were examined in Europe to find the best strain.

In 2006, work was initiated in Canada and the USA to determine if the nematode was present, and if so, what levels of parasitism existed and how pervasive it was in the populations. In Canada, the nematode has been found and identified as *D. siricidicola* (Yu et al. 2009). This task was complicated by the fact that a morphologically similar species, *D. wilsoni* Bedding, is present in North America (Bedding and Akhurst 1978). Considerable taxonomic expertise is required to distinguish species. Further work is ongoing in Canada and the USA to develop molecular tools to help identify the juvenile form. Early and very preliminary data in Canada suggests that the strain found to date may have low virulence (de Groot et al., unpublished). If so, then at least in Canada, a considerable effort may be required to displace the low virulent strain with the commercial strain that has higher efficacy. This "swamping out" of the natural strain is similar in many respects to the situation in Australia where a "defective strain" developed in the culturing process and had to be replaced by a more virulent strain (Bedding and Iede 2005).

While the nematode has been very effective in Australia, there have been some early problems in South America, and more recently in South Africa (Hurley et al. 2007; Hurley et al. 2008) in establishing populations and getting the desired level of control necessary to minimize *S. noctilio* damage. It is reasonable to expect that there will be some early "start-up" problems with the application of nematodes in North America as well.

One issue that can be easily overlooked, yet is critical to the success of the nematode controlling an ascending population of *S. noctilio*, is that early detection is imperative. In Australia, the recommended objective is to detect *S. noctilio* in a locality before any plantation reaches 0.1% annual Sirex-associated mortality, which is about 1–2 trees per ha in an un-thinned stand in Australia (Haugen et al. 1990) and about 2–3 trees in North America. In the Southern Hemisphere, early surveillance and establishment of trap-trees are used in conjunction with a biocontrol program. Field observations from surveys in Canada and the USA suggest that the current trapping systems do not detect populations at these low levels.

19.6 Conclusions

Sirex noctilio is clearly well-established in North America. Surveys to continue to detect and then delimit the infested area are ongoing and despite the numerous obstacles and challenges, new areas of infestation have been discovered every year. Continued research on development of new survey tools is urgently needed to be effective within the various forest types in North America.

Management of *S. noctilio* will not be easy and must be multifaceted. The diversity, heterogeneity and complexity of North America's natural forests in terms of natural enemies and competing insects and diseases, may be a problem for survey efforts, but also a solution for management, at least in some areas. Diverse pre-existing communities of natural enemies, as well as the potential use of *D. siricidicola* as a biological control agent, stand management to improve individual stand vitality, and the implementation of quarantines to slow the spread of *S. noctilio* may reduce or limit the impact of this insect in North America. Because of the complexities of managing an insect over large areas and diverse habitats, pragmatic and adaptable plans will have to be developed to fit the needs of a given forest type or ecosystem.

Much remains to be observed and discovered about *S. noctilio* in North America. The situation in North America is unique, allowing for many interesting questions to be addressed on invasion biology, community ecology, and management of an invasive species in native pine communities that have not been previously addressed elsewhere. Only through continued national and international collaboration at the research and operational levels, will solutions and further understanding of *S. noctilio* biology and management in North America be achieved.

Acknowledgements This chapter summarizes the efforts of many individuals and agencies involved with *S. noctilio* detection and survey efforts in North America. In the USA, federal and state departments of agriculture and natural resources participated in various aspects of delimitation and detection efforts. Response to the *S. noctilio* detection relied upon the efforts of many people, including Ethan Angell and Bob Mungari (New York State Department of Agriculture & Markets); Jerry Carlson (New York Department of Environmental Conservation); Dennis Haugen, Noel Schneeberger, and Robert Rabaglia (U.S. Forest Service); and Leon Bunce, Vic Mastro, Yvonne DeMarino, and Lynn Evans-Goldner (APHIS PPQ). In Canada, Rob Favrin and Troy Kimoto (CFIA) and Pierre Therrien (Quebec Ministry of Natural Resources and Wildlife) and Taylor Scarr (Ontario Ministry of Natural Resources).

References

Alerich CL, Drake DA (1995) Forest statistics for New York: 1980 and 1993 Resource Bulletin NE-132. USDA Forest Service Northeastern Experiment Station, Radnor, PA

Allen CD, Savage M, Falk DA, Suckling KF, Swetnam TW, Schulke T, Stacey PB, Morgan P, Hoffman M, Klingel JT (2002) Ecological restoration of Southwestern Ponderosa pine ecosystems: a broad perspective. Ecol Appl 12:1418–1433

Baker JB, Langdon OG (1990) *Pinus taeda* L. loblolly pine. In: Burns RM, Honkalatech BH (eds) Silvics of North America: 1. Conifers. Agriculture handbook 654. U.S. Department of Agriculture, Forest Service, Washington, DC

Barrett JW (1995) Regional silviculture of the United States. Wiley, New York

Bedding RA (1974) Five new species of *Deladenus* (Neotylenchidae), entomophagous-mycetophagous nematodes parasitic in siricid woodwasps. Nematology 20:204–225

Bedding RA, Akhurst RJ (1974) Use of the nematode *Deladenus siricidicola* in the biological control of *Sirex noctilio* in Australia. J Aust Entomol Soc 13:129–135

Bedding RA, Akhurst RJ (1978) Geographical distribution and host preferences of *Deladenus* species (Nematoda: Neotylenchidae) parasitic in siricid woodwasps and associated hymenopterous parasitoids. Nematology 24:286–294

Bedding RA, Iede E (2005) Application of *Beddingia siricidicola* for Sirex woodwasp control. In: Grewal PS, Ehlers RU, Shapiro-Ilan DI (eds) Nematodes as biocontrol agents. CABI Publ, Cambridge, pp 385–399

Bedding R, Carnegie A, de Groot P, Dodds K, Germain R, Hain F, Haugen D, Iede E, Lance D, Mastro V, Reardon R, Rogers D, Schiff N, Tumlinson J, Williams D (2006) Sirex science advisory panel. Report, Annapolis, MD, 9–10 Jan 2006

Benzie JW (1977) Manager's handbook for jack pine in the north central states. General Technical Report NC-32. U.S. Department of Agriculture, Forest Service, North Central Forest Experiment Station, St. Paul, MN

Borchert D, Fowler G, Jackson L (2007) Organism pest risk analysis: risk to the conterminous United States associated with the woodwasp, *Sirex noctilio* Fabricius, and the symbiotic fungus, *Amylostereum areolatum* (Fries: Fries) Boidin. Pest Risk Analysis (USDA-APHIS-PPQ-EDP), p 41

Böröczky K, Crook DJ, Jones TH, Kenny JC, Zylstra KE, Mastro VC, Tumlinson JH (2009) Monoalkenes as contact sex pheromone components of the woodwasp *Sirex noctilio*. J Chem Ecol 35:1202–1211

Breece CR, Kolb TE, Dickson BG, McMillin JD, Clancy KM (2008) Prescribed fire effects on bark beetle activity and tree mortality in Southwestern Ponderosa pine forests. For Ecol Manage 255:119–128

Brockway DG, Outcalt KW, Tomczak DJ, Johnson EE (2005) Restoration of longleaf pine ecosystems. General Technical Report SRS-83. U.S. Department of Agriculture, Forest Service, Southern Research Station, Asheville, NC

Cameron EA (1962) North American survey for natural enemies of the Siricidae. Commonwealth Institute of Biological Control Mimeo, May–Oct 1962, 21 pp

Carnegie AJ, Eldridge RH, Waterson DG (2005) The history and management of sirex wood wasp, *Sirex noctilio* (Hymenoptera: Siricidae), in pine plantations in New South Wales, Australia. N Z J For Sci 35:3–24

Carnegie AJ, Matsuki M, Haugen DA, Hurley BP, Ahumada R, Klasmer P, Sun J, Iede ET (2006) Predicting the potential distribution of *Sirex noctilio* (Hymenoptera: Siricidae), a significant exotic pest of *Pinus* plantations. Ann For Sci 63:119–128

CFIA (2006) 2006 Plant protection report. Available online. http://www.inspection.gc.ca/english/plaveg/pestrava/surv/sit2006e.shtml#exo. Accessed 29 Jan 2010

CFS (2001) Canada's forest inventory. Available online. https://nfi.nfis.org/home.php?lang=en. Accessed 4 Mar 2010

Ciesla WM (2003) European woodwasp: a potential threat to North America's conifer forests. J For 101:18–23
Conway BE, McCullough DG, Leefers LA (1999) Long-term effects of jack pine budworm outbreaks on the growth of jack pine trees in Michigan. Can J For Res 29:1510–1517
Covington WW, Fule PZ, Moore MM, Hart SC, Kolb TE, Mast JN, Sackett SS, Wagner MR (1997) Restoring ecosystem health in ponderosa pine forests of the southwest. J For 95:23–29
Cross HL, Patton RF, Ek AR (1978) Reduced growth, cull and mortality of jack pine associated with sweet fern rust cankers. Can J For Res 8:47–53
Czokajlo D, Wink RA, Warren JC, Teale SA (1997) Growth reduction of Scots pine, *Pinus sylvestris*, caused by the larger pine shoot beetle, *Tomicus piniperda* (Coleoptera, Scolytidae), in New York State. Can J For Res 27:1394–1397
de Groot P, Nystrom K, Scarr T (2007a) Discovery of *Sirex noctilio* (Hymenoptera: Siricidae) in Ontario, Canada. Great Lakes Entomol 39:49–53
de Groot P, Dodds KJ, Haugen DA, Humble L, Mastro V, Smith SM, Turgeon J (2007b) Sirex science panel. Report to the Canadian food inspection agency. Sault Ste Marie, Ontario, Canada, 2 Apr 2007
Dodds KJ, Cooke RR, Gilmore DW (2007) Silvicultural options to reduce pine susceptibility to attack by a newly detected invasive species, *Sirex noctilio*. North J Appl For 24:165–167
Dodds KJ, de Groot P, Orwig DA (2010) The impact of *Sirex noctilio* in *Pinus resinosa* and *Pinus sylvestris* stands in New York and Ontario. Can J For Res 40:212–223
Drooz AT (1985) Insects of eastern forests. U.S. Department of Agriculture, Forest Service, Washington, DC, Miscellaneous Publication 1426
Ek AR, Palik BJ, Katovich SA, Kilgore MA, Domke GM, Finley SR, Benzie JW, Blinn CR, Burk TE, David AJ, Harder AL, Hoganson HM, Jacqmain B, Jones B, Lundgren AL, Mack TJ, Ostry ME, Zenner EK (2007) North central region red pine web-based forest management guide. U.S. Department of Agriculture, Forest Service. Available online. http://www.ncrs.fs.fed.us/fmg/nfmg/rp. Accessed 29 Jan 2010
Erbilgin N, Raffa KF (2002) Association of declining red pine stands with reduced populations of bark beetle predators, seasonal increases in root colonizing insects, and incidence of root pathogens. For Ecol Manage 164:221–236
Erbilgin N, Nordheim EV, Aukema BH, Raffa KF (2002) Population dynamics of *Ips pini* and *Ips grandicollis* in red pine plantations in Wisconsin: within- and between-year associations with predators, competitors, and habitat quality. Environ Ent 31:1043–1051
Foster DR, Aber JD (2004) Forests in time: the environmental consequences of 1000 years of change in New England. Yale University Press, New Haven
Franklin JF, Dryness CT (1988) Natural vegetation of Oregon and Washington. Oregon State University Press, Corvallis, OR
Frieswyk T, Widmann R (2000) Forest statistics for Vermont: 1983 and 1997. Northeastern research station resource Bulletin NE–145. U.S. Department of Agriculture, Forest Service, Newtown Square, PA
Gross HL (1992) Impact analysis of a jack pine budworm infestation in Ontario. Can J For Res 22:818–831
Haack RA (2006) Exotic bark- and wood-boring Coleoptera in the United States: recent establishments and interceptions. Can J For Res 36:269–288
Hansen MH, Brand GJ (2006) Michigan's forest resources in 2004. Resource Bulletin NC-255. U.S. Department of Agriculture, Forest Service, North Central Research Station, St. Paul, MN
Haugen DA (1990) Control procedures for *Sirex noctilio* in the Green Triangle: review from detection to severe outbreak (1977–1987). Aust For 53:24–32
Haugen DA (1999) *Sirex noctilio*. Record in the exotic forest pest information system. Published by North American Forest Commission. www.spfnic.fs.fed.us/exfor/. Accessed 9 Mar 2008
Haugen DA, Hoebeke ER (2005) Pest alert: Sirex woodwasp – *Sirex noctilio* F. (Hymenoptera: Siricidae). U.S. Department of Agriculture, Forest Service, Newtown Square, PA NA-PR-07-05
Haugen DA, Bedding RA, Underdown MG, Neumann FG (1990) National strategy for control of *Sirex noctilio* in Australia. Aust For Grower 13:1–8, Special liftout section

Hoebeke ER, Haugen DA, Haack RA (2005) *Sirex noctilio*: discovery of a palearctic siricid woodwasp in New York. Newsl Mich Entomol Soc 50:24–25

Hurley BP, Slippers B, Wingfield MJ (2007) A comparison of control results for the alien invasive woodwasp, *Sirex noctilio*, in the southern hemisphere. Agric For Entomol 9:159–171

Hurley BP, Slippers B, Croft PK, Hatting HJ, van der Linde M, Morris AR, Dyer C, Wingfield MJ (2008) Factors influencing parasitism of *Sirex noctilio* (Hymenoptera: Siricidae) by the nematode *Deladenus siricidicola* (Nematoda: Neotylenchidae) in summer rainfall areas of South Africa. Biol Control 45:450–459

Iede ET, Penteado SRC, Schaitza EG (1998) *Sirex noctilio* problem in Brazil – detection, evaluation and control. In: Iede E, Shaitza E, Penteado S, Reardon R and Murphy T (eds) Proceedings of a conference: training in the control of *Sirex noctilio* by use of natural enemies, FHTET 98-13, USDA Forest Service, Morgantown, WV, pp 45–52

Jackson DS (1955) The *Pinus radiata/Sirex noctilio* relationship at Rotoehu forest. N Z J For 7:26–41

Kirk AA (1974) Siricid woodwasps and their associated parasitoids in the southeastern United States (Hymenoptera: Siricidae). J Georgia Entomol Soc 9:139–144

Kirk AA (1975) Siricid woodwasps and their associated parasitoids in the southwestern United States (Hymenoptera: Siricidae). Pan-Pacific Entomol 51:57–61

Klepzig KD, Raffa KF, Smalley EB (1991) Association of an insect-fungal complex with red pine decline in Wisconsin. For Sci 37:1119–1139

Klepzig KD, Smalley EB, Raffa KF (1995) *Dendroctonus valens* and *Hylastes porculus* (Coleoptera: Scolytidae): vectors of pathogenic fungi (Ophiostomatales) associated with red pine decline disease. Great Lakes Entomol 28:81–87

Long SJ, Williams DW, Hajek AE (2009) *Sirex* species (Hymenoptera: Siricidae) and their parasitoids in *Pinus sylvestris* in eastern North America. Can Entomol 141:153–157

Lotan JE, Critchfield WB (1990) *Pinus contorta* Dougl. ex. Loud. Lodgepole pine. In: Burns RM, Honkalatech BH (eds) Silvics of North America: 1. Conifers. (Agriculture handbook 654). U.S. Department of Agriculture, Forest Service, Washington, DC

Lynch DL, Romme WH, Floyd ML (2000) Forest restoration in Southwestern Ponderosa pine. J For 98:17–24

Maderni JFP (1998) *Sirex noctilio* F. present status in Uruguay. In: Iede E, Shaitza EP, Penteado S, Reardon R, Murphy T (eds) Proceedings of a conference: training in the control of *Sirex noctilio* by use of natural enemies, USDA Forest Service, Morgantown, WV

McWilliams WH, Cassell SP, Alerich CL, Butler BJ, Hoppus ML, Horsley SB, Lister AJ, Lister TW, Morin RS, Perry CH, Westfall JA, Wharton EH, Woodall CW (2007) Pennsylvania's forest 2004. Resource Bulletin NRS-20. U.S. Department of Agriculture, Forest Service, Northern Research Station, Newtown Square, PA

Morgan RE, de Groot P, Smith SM (2004) Susceptibility of pine plantations to attack by the pine shoot beetle (*Tomicus piniperda*) in southern Ontario. Can J For Res 34:2528–2540

NAPPO (2005) NAPPO constitution and bylaws. Available online. http://www.nappo.org/OfficialDoc/constitution-e.htm. Accessed 29 Jan 2010

Neumann FG (1979) Insect pest management in Australian radiata pine plantations. Aust For 42:30–38

Neumann FG, Marks GC (1976) A synopsis of important pests and diseases in Australian forests and forest nurseries. Aust For 39:83–102

Neumann FG, Minko G (1981) The sirex wood wasp in Australian radiata pine plantations. Aust For 44:46–63

Neumann FG, Harris JA, Kassaby FY, Minko G (1982) An improved technique for early detection and control of the sirex wood wasp in radiata pine plantations. Aust For 45:117–124

Neumann FG, Morey JL, McKimm RJ (1987) The sirex wasp in Victoria. Department of Conservation, Forests and Lands, Victoria

Noss RF, LaRoe ET III, Scott JM (1995) Endangered ecosystems of the United States: a preliminary assessment of loss and degradation. Biological Report No. 28. USDA National Biological Service. Washington, DC

Rawlings GB (1955) Epidemics in *Pinus radiata* forests in New Zealand. New Zealand J For 7:3–58

Rouse C (1986) Fire effects in northeastern forests: jack pine. General Technical Report NC-106. U.S. Department of Agriculture, Forest Service, North Central Forest Experiment Station, St. Paul, MN

Ryall KL, Smith SM (2005) Patterns of damage and mortality in red pine plantations following a major ice storm. Can J For Res 35:487–493

Ryall KL, de Groot P, Smith SM (2006) Sequential patterns of colonization of coarse woody debris by *Ips pini* (Say) (Coleoptera: Scolytidae) following a major ice storm in Ontario. Agric For Entomol 8:89–95

Schiff NM, Valley SA, LaBonte JR, Smith DR (2006) Guide to the siricid woodwasps of North America. U.S. Department of Agriculture, Forest Service, Morgantown, WV FHTET-2006-15

Simpson RF, McQuilkin RM (1976) Identification of volatiles from felled *Pinus radiata* and the electroantennograms they elicit from *Sirex noctilio*. Entomol Exp Appl 19:205–213

Smith DR, Schiff NM (2002) A review of the Siricid woodwasps and their Ibalid parasitoids (Hymenoptera: Siricidae, Ibaliidae) in the Eastern United States, with emphasis on the mid-Atlantic region. Proc Entomol Soc Wash 104:174–194

USDA (2006) Exotic woodborer bark beetle national survey field manual. Available at: http://www.docstoc.com/docs/1087960/Exotic-Wood-Borer-Bark-Beetle-National-Survey-Field-Manual. Accessed 29 Jan 2010

Walkinshaw LH (1983) Kirtland's warbler: the natural history of an endangered species. Cranbrook Institute of Science Bulletin No. 58, Bloomfield Hills, MI

Whitney RD (1988) Armillaria root rot damage in softwood plantations in Ontario. For Chron 64:345–351

Wilson LF (1975) White-spotted sawyer. U.S. Department of Agriculture, Forest Service, St. Paul, MN Forest Pest Leaflet 74

Wilson LF, Averill RD (1978) Redheaded pine sawfly. U.S. Department of Agriculture, Forest Service, Broomall, PA, Forest Insects and Disease Leaflet 14

Wilson AD, Schiff NM (2010) Identification of *Sirex noctilio* and native North American woodwasp larvae using DNA barcodes. J Entomol 7:60–79

Yemshanov D, McKenney DW, de Groot P, Haugen D, Sidders D, Joss B (2009a) A bioeconomic approach to assess the impact of an alien invasive insect on timber supply and harvesting: a case study with *Sirex noctilio* in eastern Canada. Can J For Res 39:154–168

Yemshanov D, Koch FH, McKenney DW, Downing MC, Sapio F (2009b) Mapping invasive species risks with stochastic models: a cross-border United States-Canada application for *Sirex noctilio* Fabricius. Risk Anal 29:868–884

Yu Q, de Groot P, Leal I, Davis C, Ye W, Foord B (2009) Characterization of *Deladenus siricidicola* (Tylenchida: Neotylenchidae) associated with *Sirex noctilio* (Hymenoptera: Siricidae) in Canada. Int J Nematol 19:23–32

Zondag R (1969) A nematode infection of *Sirex noctilio* (F.) in New Zealand. N Z J Sci 12: 732–747

Zylstra KE, Dodds KJ, Francese JA, Mastro VC (2010) The effect of chemical girdle timing on the attractiveness and suitability of *Sirex noctilio* (Hymenoptera: Siricidae) trap trees in the northeastern United States. Agric For Entomol 12:243–250

Chapter 20
Sirex Research and Management: Future Prospects

Bernard Slippers and Michael J. Wingfield

Abstract The Sirex woodwasp, *Sirex noctilio* with its mutualistic fungal symbiont *Amylostereum areolatum* is the most damaging invasive pest in Southern Hemisphere *Pinus* plantations. Despite a century of work, many unanswered questions and numerous challenges remain to be addressed. These are especially linked to changes associated with the continuing spread of the pest. The *S. noctilio* complex also presents unique research opportunities to study the dynamics and genetics of invasive pest populations, and symbioses (both mutualism and parasitism), perhaps more effectively than in any other system known. These opportunities and challenges, enhanced by an explosion in the availability of new molecular and chemical technologies, will shape the next phase of research on this pest, its symbiotic fungus and their parasites.

20.1 Introduction

After its first invasion into new pine environments in the Southern Hemisphere, a considerable amount of research was undertaken on *Sirex noctilio*, its *Amylostereum areolatum* symbiont and other associated organisms. This work was predominantly conducted in Australia and New Zealand where the first invasions occurred and it was strongly focused on reducing the impact of the pest. A point was, however, reached where *S. noctilio* appeared to be relatively well controlled and research efforts slowed. As the pest appeared in new areas, existing knowledge and biological control agents were transferred without much adaptation or research input.

B. Slippers (✉) • M.J. Wingfield
Department of Genetics, Forestry and Agricultural Biotechnology Institute,
University of Pretoria, Pretoria, South Africa
e-mail: bernard.slippers@fabi.up.ac.za

B. Slippers et al. (eds.), *The Sirex Woodwasp and its Fungal Symbiont: Research and Management of a Worldwide Invasive Pest*, DOI 10.1007/978-94-007-1960-6_20,
© Springer Science+Business Media B.V. 2012

An undue reliance on existing knowledge persisted. At least amongst many forest managers, there was also a somewhat naïve belief that *S. noctilio* invasions could be relatively easily controlled.

Prior to the meeting that gave rise to this book, there was a distinct lack of focus on the Sirex woodwasp, its biology and its management. The increasing areas of invasion of the wasp and growing examples of biological control failures provided substantial motivation for a review of the current *S. noctilio* situation. More importantly perhaps, it became clear that there was an urgent need for a revival in research efforts to better understand this pest.

Chapters of this book have reconsidered virtually every aspect of the biology and management of the Sirex woodwasp and its associated organisms. They treat past experiences reaching back to the groundbreaking research conducted in New Zealand and Australia, which has provided the foundation for Sirex woodwasp management for 50 years. They have also considered more recent research and contemporary issues relating to the expanding non-native range of *S. noctilio*. This chapter attempts to highlight some of the knowledge gaps and research opportunities that lie ahead. It also considers the power of new technologies, not available at the time when the first *S. noctilio* invasions occurred and that will deeply influence a growing knowledge of this remarkable pest.

20.2 Questions Relating to the Natural History of Siricids and Their Symbionts

The systematics of the Siricidae remains one of the main challenges facing research on these organisms. Recent work treating the North American *Sirex* species (Chap. 1), has revealed the need for revisions in a number of taxa. However, beyond the North American species, there are many others that remain to be re-evaluated, and discovered. Northern Asia for example is expected to harbor a great diversity of Siricids. While enlightening studies on the biology of Siricids and the fungal symbionts have emerged from this region (e.g., see review in Chap. 7), very little research has been done on the diversity of the Siricids and their fungi in this region. It is unfortunate that there appears to be little access to material for current molecular and taxonomic studies from vast areas including Russia, China and elsewhere. Similarly, isolated regions of Central and South America might be expected to yield new species (Chap. 1), some of which might threaten trees in the future.

Accompanying a re-evaluation of the Siricid diversity world-wide, a re-evaluation of the diversity and specificity of the association of these insects with *Amylostereum* spp. should be undertaken. Recent studies have shown that *A. laevigatum* is associated with Siricids in Japan, a fact that was not expected based on patterns observed in Europe (Chaps. 5 and 7). It has also been known for some time that at least one undescribed species of *Amylostereum* is present in North America (Slippers et al. 2000). Furthermore, recent work has shown that some native North American Siricid species can carry *A. areolatum* (Nielsen et al. 2009). Could this imply that

A. areolatum is also native to North America, or did these wasps acquire *A. areolatum* horizontally after invasion of the region by species such as *S. noctilio* and *S. juvencus*? These and other questions relating to the diversity of the fungal symbiont must be at the forefront of research to understand the potential impact and control of the *S. noctilio* invasion in North America.

Despite years of detailed study, some of the basic aspects of the ecology and biology of *S. noctilio* remain unclear (Chaps. 2 and 4). For example, key factors influencing the rate of spread and population dynamics of *S. noctilio*, including flight distance, nematode infection, chemical queues, distance between plantations, outbreak dynamics and others remain to be studied quantitatively in a regions (see an example of such studies summarized for Argentina in Chap. 4).

Some key life history traits for *S. noctilio* also remain poorly studied. For example, it is known that the insect, like other Hymenoptera, has a haplodiploid sex determination system (Chap. 2). Consequently, it is expected that there will be strong male bias in the populations in early phases of invasion due to difficulty in mate finding. But in some cases, extreme variation in sex ratios persists across regions and over time, varying from 1 female: 1 male to 1 female: >10 males (Chaps. 2, 18). These sex ratios can also vary sharply from the bottom to the top of infested trees. This ratio has a significant influence on the reproductive potential of populations, yet we lack an understanding of the factors that determine variation in patterns of sex ratio.

An intriguing problem that has raised speculation, but that has never conclusively been answered, is how the adult female wasp acquires the fungus in the internal mycangia (Chap. 2). It is known that the larvae maintain spores of the fungus in external hypopleural sacs, in a waxy matrix. The developing pupae are, however, free from the fungus. After emerging from the pupal case, the fungus thus needs to be acquired by the mycangium, probably via the ovipositor. It has been suggested that movement of the abdomen and ovipositor can accomplish this, but this remains speculation. A related question can then be raised regarding the purity of the fungus inside the mycangia. If the fungus is acquired from an external source, why are other fungi and bacteria not also acquired into this same space at the same time? Detailed studies of the chemistry, biological composition and behavior in the mycangium just before emergence will be needed to address this intriguing question.

The transmission of *Amylostereum* genotypes and species at both an ecological and evolutionary timescale might not be as clear as it appeared a decade ago. It was accepted that transmission is mostly (if not exclusively) vertical, through asexual spores, and that the association is species specific and stable for long periods of time (see reviews of previous work in Chaps. 2 and 5). The fact that there is no strict pattern of co-evolution between genera of Siricids and *Amylostereum* spp., however, suggests that horizontal exchange of *Amylostereum* spp. has occurred at an evolutionary timescale (Slippers et al. 2003, Chap. 6). Some contemporary studies also provide a hint of a more frequent acquisition of sexual propagules, or genotypes associated with other wasps (Nielsen et al. 2009, Chap. 6). The co-existence of apparently free-living, sexual populations of the fungus in some areas (e.g., in central Europe; B. Slippers and R. Vasaitis, personal observation; Chap. 6) alongside those occurring in close mutualism with the wasp is unusual when compared with other similar mutualistic systems. There is clearly much more to be learned regarding the basic biology and history of co-evolution of Siricids and *Amylostereum* spp. Both detailed population genetic studies and broad phylogenetic investigations, in both cases including not only the wasps, but also their associated fungi, are needed to resolve these questions. The tools are now available to do this.

It has been known for some time that there are many microbes associated with *S. noctilio* larvae (Madden 1975). Yet, there has been no effort to understand their role in the system, partly because no clear pattern of association has been obvious. A recent study that makes use of "next generation" DNA sequencing technologies has shown that a multitude of bacteria are present in galleries and the guts of *S. noctilio* larvae in North Amarica (Adams et al. 2011). Some of these bacteria are excellent cellulose degraders, some even more so than *A. areolatum*. Cellulose degradation has always been thought to be a key service provided to *S. noctilio* by the fungus. The discovery of these bacteria raises a number of questions regarding the nature and evolution of the mutualism between *A. areolatum*, *S. noctilio* and these other microbes. In particular, the patterns of association between *S. noctilio* and other microbes remain to be understood. For example, important questions include how widespread these bacteria are, how consistently and at what stages of the life-cycle they are associated with *S. noctilio* and how they are transmitted.

An area of research that has received insufficient attention in recent decades is the host relationships of *S. noctilio* and *A. areolatum*. Superb work in this regard was conducted in the 1960s and 1970s (Chap. 3). However, currently available chemical and molecular techniques now provide powerful tools to answer questions regarding the interaction of the fungus and the tree, and the mucus and the tree than was true four decades ago.

Throughout their ranges as an invasive aliens, *S. noctilio* and *A. areolatum* have encountered and have also overcome the defenses of *Pinus* species that do not occur where the pest is native e.g., *P. elliottii*, *P. patula*, *P. radiata*, *P. taeda* and others (Chap. 5). In most cases, these invaded hosts occur as vast tracks of monocultures, with little room for choice by the wasp. In North America, host preference is, however, of significant importance to management programs. This is because the wasp invades areas of native forest including host and non-host trees, as well as areas planted to native and non-native conifers. Consequently, a better understanding of host choice and host resistance is urgently needed.

The chemical ecology of the interaction between the Sirex woodwasp and its host trees has received attention in the past, but this complex process is only now gradually beginning to be unraveled to a point where it can be utilized in management schemes (Chaps. 11 and 12). But this information has yet to lead to a selective trapping method, which is the ultimate goal, alongside a deeper understanding of the biology of the organisms involved. Information on a previously uncharacterized aspect of the biology of Siricids, namely the role of pheromones in mating, is now emerging and might well contribute to the discovery of more specific trapping lures (Chap. 11). An aspect of the chemical ecology of Siricids that has, however, not received attention is the possible role of fungal volatiles in the attraction of Siricids carrying the fungus. *Xeris* spp. (a Siricid that does not carry the fungus itself) and parasitoids of Siricids are known to respond to volatiles of *A. areolatum* and it seems probable that Siricid females might also be attracted to previously infested wood.

The cornerstone of the biological control of *S. noctilio* has been the use of the nematode parasite, *Deladenus siricidicola*. It has also been a source of joy for biologists studying this system, who marvel at the level of adaptation of the nematode, having distinct morphological forms that feed on either the fungus or the larvae of the wasp; followed by juvenile development in the wasp that coincides with wasp egg development and in which the nematodes are then spread. Given its importance for control, but also its biological novelty, it seems unfortunate that the diversity, biology and systematics of this fascinating nematode genus was last studied in its native environment three decades ago. It is known for example that there are at least seven species of *Deladenus* associated with Siricids and *Amylostereum* (Bedding 1974), but today only *D. siricidicola* is available for molecular genetic studies from the Southern Hemisphere biological control programs, as well as from recent and surprising collections in North America (Yu et al. 2009, Chap. 10). Recent studies on the population diversity of *D. siricidicola* in the Southern Hemisphere have revealed that the Kamona strain, which is so widely used in the area, is a highly inbred line that is devoid of any diversity (Mlonyeni et al. 2011, Chap. 9). The failure of the biological control program in some areas of the Southern Hemisphere (Chap. 9) and the questions surrounding biological control of *S. noctilio* in North America (Chaps. 10 and 19) make this an area of research that must be addressed urgently.

20.3 Continued Risk and Challenges Facing Management

Anthropogenic movement has resulted in *S. noctilio* invading new regions for over a century. In the process, this has revealed the weaknesses of current quarantine systems. Importantly, the wasp would not have been on any quarantine list for Southern Hemisphere countries with *Pinus* plantations if its distribution and damage in its native areas were considered. Even after it was recognized as a serious threat following the invasion of New Zealand plantations, its appearance on quarantine lists did not stop its invasion into new regions. This is simply a consequence of the fact that not all woody material, in which Sirex larvae can be transported, can be effectively treated or inspected at the volumes that are moving around the world. This suggests that completely new tools and technologies will be needed to reveal hidden pests in products of trade. There will clearly also need to be a deeper focus on pathways of movement rather than on lists of potentially threatening organisms, as has frequently been suggested in recent years (e.g., Evans 2010).

There has been an overall increase in numbers of introductions and establishments of forest pests in many parts of the world. It is consequently not surprising that the rate of spread of *S. noctilio* has also increased over time, taking 50 years after its first invasion of New Zealand to reach Australia, 30 years more after that to South America and 10 years later to Africa (Hurley et al. 2007). Only a decade later, it had also appeared in North America (Hoebeke et al. 2005, Chap. 19). As with other insects, this increasing rate is most likely a consequence both of increasing general world trade, but also an ever-growing pool of source populations from which *S. noctilio* can enter trade pathways.

The Sirex woodwasp continues to spread into previously unaffected plantation areas on the continents where it now occurs (Chaps.13–18). It is clear that *S. noctilio* is adapting well to all the climatic regions it has encountered. This includes Mediterranean climates similar to its origin in Europe and Africa, but also covers cool to warm temperate to semi-tropical areas. As it has moved into these new areas, its phenology has shifted, it has infested previously un-encountered *Pinus* species, and it has overcome changing communities of interacting fungi and insects. Unfortunately, the biological control agents, which have at times been viewed as 'silver bullets', have not adapted equally well in all cases. This has led to a level of enemy release and consequent serious outbreaks in some areas, such as has occurred in some plantations in South Africa (Hurley et al. 2007; Chaps. 8 and 9). Such shifting phenologies, host associations and parasite interactions will be even more complex as the Sirex woodwasp invades North America, an area having native *Pinus* spp., Siricids and parasitoids. Not all authors contributing to this book agree on the extent of damage that might be expected from *S. noctilio* in North America. However, what is clear is that this new invasion will add additional pressure to an already stressed system in many areas. It will also lead to vastly more complex management issues than those that have been considered in the Southern Hemisphere (Chaps. 10 and 19). Importantly, effective management of the Sirex woodwasp in this environment will require a much deeper level of understanding of this pest than has been required elsewhere.

The history of the introductions of *S. noctilio* has mostly been studied through the population genetic diversity of its associated fungus. Early work in this regard indicated that all introductions in the Southern Hemisphere share a common origin (Slippers et al. 2002). Subsequent research has revealed minor variations in molecular (AFLP) markers between isolates from Australia and New Zealand, vs South America and Africa (authors unpublished). But even this supports a close, and even single, origin of the introduction across the Southern Hemisphere. Recent work in North America also indicates limited diversity in *A. areolatum* populations in that region, but that there are at least two distinct genotypes of the fungus associated with invasive *S. noctilio* populations (Nielsen et al. 2009; Bergeron et al. 2011). Given the extremely close relationship of these fungi, a deeper analysis of molecular diversity appears to be needed to resolve this question. What is missing, however, is a study of the diversity of the wasp populations themselves. The possibility of horizontal symbiont exchange (see below, and Chap. 6) means that the pattern of fungal and wasp population diversity might not be parallel.

The pattern of spread of *S. noctilio* discussed above should provide a warning that this pest will continue to spread. Apart from invasion of new areas, secondary introductions into already affected regions should also be expected. Indeed they may already have occurred. Such secondary introductions will lead to changes in diversity and structure of the invasive populations, and there will most likely be consequences for control measures. In addition, global climate change is bound to add more stress to plantations in some areas, increasing their susceptibility to *S. noctilio* attack. These changes will also affect the interaction patterns of Sirex, its parasitoids and other organisms in the environment. Areas such as Europe and New Zealand that are minimally affected, could become more vulnerable in the future.

The above assessment might appear somewhat bleak, painting a picture of a continuing, and in some cases increasing threats of *S. noctilio* to *Pinus* plantations. What this book, however, illustrates is that there is a large, growing and active global research community that has placed a renewed focus on this problem. It also illustrates how current and new technologies can be used to address many unanswered questions. If these efforts can be optimally combined with government and forestry processes, then many of the chapters in this book show that the threat from the Sirex woodwasp and other forest pests can be reduced.

20.4 New Technologies and Unexplored Opportunities

Molecular tools will continue to contribute substantially to a better understanding of the taxonomy, biology and distribution of *S. noctilio* together with its associated symbionts and parasites. For example, DNA barcoding is already impacting on the understanding of Siricid diversity in North America (Wilson and Schiff 2010) and it will help to resolve many of the remaining questions facing the systematics of this group. Given the difficulty of collecting fresh material to represent all taxa, this work could focus also on analyzing preserved museum specimens from many parts of the world, including currently under-represented areas such as eastern Eurasia. Such data, analysed in a phylogenetic context, will also yield valuable insights into the evolution of the group, time since separation and spread across the world, areas of origin and diversity hotspots.

Genomics has revolutionized the biological sciences over the past decade. As yet it has not impacted on research on the Sirex system in a major way, but this is about to change very rapidly. Already research is ongoing to analyze the recently sequenced full genome of *D. siricidicola* (authors unpublished). The phylogenetic position of this nematode at the base of the Tylenchida, a group of mainly plant pathogenic nematodes and including a number of important plant pathogens, as well as its unique bi-cyclic (mycetophagous and entomophagous) life cycle, provides an opportunity to understand some of the basic changes that occurred during the evolution of plant pathogenic systems.

Genome sequencing of a number of strains of *A. areolatum* has begun (authors unpublished). A well-annotated full genome of the fungus will first be used in comparative studies with completed genomes of closely related taxa, such as various *Heterobasidion* spp., and other symbiotic basidiomycetes. The aim will be to understand changes that might be associated with the symbiotic life-style of *Amylostereum*. A second project will produce genome sequences for a number of strains from South

Hemisphere clones. This will make it possible to more deeply study the effect of the symbiosis on the genetics of the fungus.

Only parts of the genome of *S. noctilio* have thus far been characterized, but plans are already advanced to also obtain a full genome sequence for the wasp. This is likely to be amongst the first members of the Symphyta to be sequenced, and the project promises to yield insights into the evolution of the Hymenoptera. These will emerge for example from comparisons with the various ant, bee and parasitic wasp genomes that have already been sequenced. It is likely that within the next few years, not only these genomes will be available, but most likely also those of the other parasitoids. It is, however, especially the resequencing of strains from different populations of all these organisms that might hold some of the most interesting insights into the effect of invasion and symbiosis on these organisms at the genome level.

20.5 Conclusions

For many years, it has been considered that the final word has been spoken on the biological control programs for *S. noctilio*. Since its development in the 1960s and 1970s, and after some slight adaptation in the late 1980s in Australia, very little has been done to adapt or improve these systems. Biological control species that were introduced up until the 1970s have simply been distributed amongst countries in the Southern Hemisphere. Emerging research is showing that at least in some cases, this has led to severely inbred and genetically depauperate populations. The question as to whether this is influencing the ability of these parasites to adapt to the new environments that *S. noctilio* is entering must thus be asked. Furthermore, it is necessary to understand what effect the increasing diversity in *S. noctilio* via secondary introductions might have in the future. A re-evaluation of these systems is long overdue. Extension of existing biological control programmes, for example by increasing diversity in terms of populations, or of species of parasites will be essential in the future.

The existence of populations of an introduced mutualistic system such as that of *S. noctilio* and its associates across the Southern Hemisphere, isolated for different periods of time in distinct regions, provides a model system for study, the potential of which has yet to be tapped. For example, understanding the evolution of virulence or resistance in biological control systems, the consequences of long term clonality in a mutualistic fungus, the influence of diversity on adaptability of parasites, and many more basic ecological and evolutionary questions are waiting to be addressed through studies on this unique and intriguing system. In order to fully utilize this opportunity, further ties should be sought between researchers focusing on theoretical questions in invasion biology and biological control. In addition, reinforcing relationships between researchers and practitioners that have access to and understand the Sirex system, will lead to many new and unexpected opportunities in the future.

References

Adams AS, Jordan MS, Adams SM, Suen G, Goodwin LA, Davenport KW, Currie CR, Raffa KF (2011) Cellulose-degrading bacteria associated with the invasive woodwasp *Sirex noctilio*. ISME J 5:1323–1331

Bedding RA (1974) Five new species of *Deladenus* (Neotylenchidae), entomophagous-mycetophagous nematodes parasitic in siricid woodwasps. Nematologica 20:204–225

Bergeron M-J, Leal I, Foord B, Ross G, Davis C, Slippers B, de Groot P (2011) Putative origin of clonal lineages of *Amylostereum areolatum*, the fungal symbiont associated with *Sirex noctilio*, in eastern Canada. Fungal Biol 115:750–758

Evans HF (2010) Pest risk analysis – organisms or pathways? NZ J For Sci 40:S35–S44

Hoebeke ER, Haugen D, Haack RA (2005) *Sirex noctilio*: discovery of a Palearctic siricid woodwasp in New York. Newsl Mich Entomol Soc 50:24–25

Hurley BP, Slippers B, Wingfield MJ (2007) A critical review of control of the alien invasive woodwasp, *Sirex noctilio*, in the southern hemisphere, with a special emphasis on biological control. Agr For Entomol 9:159–171

Madden JL (1975) Bacteria and yeasts associated with *Sirex noctilio*. J Invert Pathol 25:283–287

Mlonyeni XO, Wingfield BD, Wingfield MJ, Ahumada R, Klasmer P, Leal I, de Groot P, Slippers B (2011) Extreme homozygosity in Southern Hemisphere populations of *Deladenus siricidicola*, a biological control agent of *Sirex noctilio*. Biol Control (in press)

Nielsen C, Williams DW, Hajek AE (2009) Putative source of the invasive *Sirex noctilio* fungal symbiont, *Amylostereum areolatum*, in the eastern United States and its association with native siricid woodwasps. Mycol Res 113:1242–1253

Slippers B, Wingfield MJ, Wingfield BD, Coutinho TA (2000) Relationships among *Amylostereum* species associated with siricid woodwasps inferred from mitochondrial ribosomal DNA sequences. Mycologia 92:955–963

Slippers B, Wingfield BD, Coutinho TA, Wingfield MJ (2002) DNA sequence and RFLP data reflect relationships between *Amylostereum* species and their associated wood wasp vectors. Mol Ecol 11:1845–1854

Slippers B, Coutinho TA, Wingfield BD, Wingfield MJ (2003) A review of the genus *Amylostereum* and its association with woodwasps. SA J Sci 99:70–74

Wilson AD, Schiff NM (2010) Identification of *Sirex noctilio* and native North American woodwasp larvae using DNA barcode. J Entomol 7:60–79

Yu Q, de Groot P, Leal I, Davis C, Ye W, Foord B (2009) Characterization of *Deladenus siricidicola* (Tylenchida: Neotylenchidae) associated with *Sirex noctilio* (Hymenoptera: Siricidae) in Canada. Int J Nematol 19:23–32

References

Index

B. Slippers et al. (eds.), *The Sirex Woodwasp and its Fungal Symbiont: Research and Management of a Worldwide Invasive Pest*, DOI 10.1007/978-94-007-1960-6,
© Springer Science+Business Media B.V. 2012

MIX
Papier aus verantwortungsvollen Quellen
Paper from responsible sources
FSC® C105338

If you have any concerns about our products,
you can contact us on
ProductSafety@springernature.com

In case Publisher is established outside the EU,
the EU authorized representative is:
Springer Nature Customer Service Center GmbH
Europaplatz 3, 69115 Heidelberg, Germany

Printed by Libri Plureos GmbH
in Hamburg, Germany